Lecture Notes in Mobility

Series Editor

Gereon Meyer

For further volumes:
http://www.springer.com/series/11573

Michael Hülsmann · Dirk Fornahl
Editors

Evolutionary Paths Towards the Mobility Patterns of the Future

Springer

Editors
Michael Hülsmann
Systems Management
  International Logistics
Jacobs University Bremen
Bremen
Germany

Dirk Fornahl
Business Studies and Economics
University of Bremen
Bremen
Germany

ISSN 2196-5544 ISSN 2196-5552 (electronic)
ISBN 978-3-642-37557-6 ISBN 978-3-642-37558-3 (eBook)
DOI 10.1007/978-3-642-37558-3
Springer Heidelberg New York Dordrecht London

Library of Congress Control Number: 2013948619

© Springer-Verlag Berlin Heidelberg 2014
This work is subject to copyright. All rights are reserved by the Publisher, whether the whole or part of the material is concerned, specifically the rights of translation, reprinting, reuse of illustrations, recitation, broadcasting, reproduction on microfilms or in any other physical way, and transmission or information storage and retrieval, electronic adaptation, computer software, or by similar or dissimilar methodology now known or hereafter developed. Exempted from this legal reservation are brief excerpts in connection with reviews or scholarly analysis or material supplied specifically for the purpose of being entered and executed on a computer system, for exclusive use by the purchaser of the work. Duplication of this publication or parts thereof is permitted only under the provisions of the Copyright Law of the Publisher's location, in its current version, and permission for use must always be obtained from Springer. Permissions for use may be obtained through RightsLink at the Copyright Clearance Center. Violations are liable to prosecution under the respective Copyright Law.
The use of general descriptive names, registered names, trademarks, service marks, etc. in this publication does not imply, even in the absence of a specific statement, that such names are exempt from the relevant protective laws and regulations and therefore free for general use.
While the advice and information in this book are believed to be true and accurate at the date of publication, neither the authors nor the editors nor the publisher can accept any legal responsibility for any errors or omissions that may be made. The publisher makes no warranty, express or implied, with respect to the material contained herein.

Printed on acid-free paper

Springer is part of Springer Science+Business Media (www.springer.com)

Acknowledgments

This book is based on the conference "Evolutionary Paths Towards the Mobility Patterns of the Future", which was held at Jacobs University in Bremen/Germany in August 2011. The conference and the associated project "Module 4: Traffic concepts and business models" was funded by the Federal Ministry of Transport, Building and Urban Development in the context of the 'Model Regions Electric Mobility'. Responsible for program coordination was the NOW GmbH National Organization Hydrogen and Fuel Cell Technology.

Therefore, we are indebted to the Ministry of Transport, Building and Urban Development that provided us with the opportunity to organize the conference and brought scientists working on electric mobility together to discuss ideas and to set up this book. Furthermore, we want to thank all presenters and participants of the conference for their interesting presentations and fruitful discussions.

Contents

Introduction 1
Christian Hanke, Michael Hülsmann and Dirk Fornahl

Part I Mobility Needs and Mobility Concepts

Socio-Economic Aspects of Electric Vehicles: A Literature Review ... 13
Christian Hanke, Michael Hülsmann and Dirk Fornahl

Mobility Scenarios for the Year 2030: Implications
for Individual Electric Mobility 37
Tobias Kuhnimhof, Irene Feige and Peter Phleps

How Does the Actual Usage of Electric Vehicles Influence
Consumer Acceptance? 49
Uta Schneider, Elisabeth Dütschke and Anja Peters

Identifying Consumer Groups with Satisfactory Characteristics
for Electric Mobility Usage 67
Dominik Santner and Dirk Fornahl

What Do Potential Users Think About Electric Mobility? 85
Christian Hoffmann, Daniel Hinkeldein,
Andreas Graff and Steffi Kramer

Electric Car Sharing as an Integrated Part of Public Transport:
Customers' Needs and Experience 101
Steffi Kramer, Christian Hoffmann, Tobias Kuttler
and Manuel Hendzlik

Transforming Mobility into Sustainable E-Mobility: The Example
of Rhein-Main Region 113
Birgit Blättel-Mink, Monika Buchsbaum, Dirk Dalichau,
Merle Hattenhauer and Jens Weber

**Validation of Innovative Extended Product
Concepts for E-Mobility** 131
Jens Eschenbaecher, Stefan Wiesner and Klaus-Dieter Thoben

Part II Challenges for Companies and Politics

**Strategic Perspectives for Electric Mobility: Some Considerations
About the Automotive Industry** 155
Richard Colmorn and Michael Hülsmann

**How Knowledge-Based Dynamic Capabilities Help to Avoid
and Cope with Path Dependencies in the Electric Mobility Sector** 169
Philip Cordes and Michael Hülsmann

**Safety Aspects of Electric Vehicles: Acoustic Measures,
Experimental Analysis, and Group Discussions** 187
Kathrin Dudenhöffer and Leonie Hause

**Transition Management Towards Urban Electro Mobility
in the Stuttgart Region** 203
Alanus von Radecki

**Assessment of CO_2-Emissions from Electric Vehicles:
State of the Scientific Debate** 225
Jürgen Gabriel, Philipp Wellbrock and Marius Buchmann

How to Integrate Electric Vehicles in the Future Energy System? 243
Patrick Jochem, Thomas Kaschub and Wolf Fichtner

**An Approach Towards Service Infrastructure Optimization
for Electromobility** 265
Tim Hoerstebrock and Axel Hahn

Part III Practical Insights

**Coping with a Growing Mobility Demand in a Growing City:
Hamburg as Pilot Region for Electric Mobility** 283
Sören Christian Trümper

The City of Bremen and Its Approach Towards E-Mobility 307
Michael Glotz-Richter

Acceptance of Electric Vehicles and New Mobility Behavior:
The Example of Rhine-Main Region 319
Petra K. Schaefer, Kathrin Schmidt and Dennis Knese

Introduction

Christian Hanke, Michael Hülsmann and Dirk Fornahl

Abstract For decades electric mobility is discussed as a solution for future mobility. But over the years the technology could only be enforced by rail. On the road combustion engines were more powerful, cheaper and offered a higher flexibility. But in terms of an increasing world population, further worldwide urbanization and economic growth as well as climate change on the whole it is common sense that in a few decades the oil era comes to an end. Combustion engines, based on the use of fossil fuels can no longer satisfy all the future mobility needs. Electric mobility as a solution approach becomes a new recovery. This chapter introduces electric mobility issues and provides an overview of the contributions of this book.

1 Introduction

For several years electric mobility has been recognized more and more as a solution for individual transportation that might meet environmental and resource requirements as well as economic and social aspects of modern societies. But the

---

C. Hanke (✉)
IPMI—Institut für Projektmanagement und Innovation, Universität Bremen,
Wilhelm-Herbst-Str. 12, 28359 Bremen, Germany
e-mail: christian.hanke@innovation.uni-bremen.de

M. Hülsmann
School of Engineering and Science, Systems Management, International Logistics,
Jacobs University Bremen, Campus Ring 1, 28759 Bremen, Germany
e-mail: m.huelsmann@jacobs-university.de

D. Fornahl
Centre for Regional and Innovation Economics, Universität Bremen, Wilhelm-Herbst-Str.
12, 28359 Bremen, Germany
e-mail: dirk.fornahl@uni-bremen.de

high potential of electric mobility is only visible in connection with the use of renewable energies (Pehnt et al. 2007; Schindler et al. 2009; Whitmarsh and Koehler 2010). Electric mobility is also a driver for innovation and the establishment of new technologies, which in turn help to create and maintain jobs and would support prosperity (Pehnt et al. 2007; Schindler et al. 2009; Whitmarsh and Koehler 2010).

The changed conditions lead the automobile manufacturers, customers and legislature to realize that in future there are more options as only conventional drives. Therefore, a need of change of the fossil fuels oriented drivetrian exists. The emerging path is still at the beginning, that means there are lots of options. One of the potential solutions, full or hybrid electric vehicles, have to take into account to customer needs and by technological and regulatory frameworks. Requirements exist between the particular customer's needs, the manufacturer's technological environment and the regulatory framework (Bakker et al. 2012; Wallentowitz et al. 2010).

A special thrust to the issue has been given in Germany with the establishment of the National Development Plan of Electric Mobility in 2009. This plan promotes to establish the German economy as the lead market for technologies of electric mobility and is aimed to bring one million electric vehicles on German roads till 2020 (Federal Goverment of Germany 2009). For this purpose, in the context of the promotional program of the Federal Ministry of Transport, Building and Urban Development "Electric Mobility in Model Regions" eight model regions are established to found prototypical markets and impulse zones for gaining scientific progress and economic success (Federal Ministry of Transport Building and Urban Development 2012). The scientific studies within these model regions and lots of other scientific or industrial projects reach from technological to ecological issues and aren't exploited with the contemplation of socio-economic aspects.

In addition to the exploration and development of usability properties like safety characteristics, speed, ranges, charging times or charging options of electric vehicles (Pearre et al. 2011; Schraven et al. 2011) and questions of standardization regarding for example to batteries or charging cables (Brown et al. 2010) there are also numerous socio-economic, ecological, political and legal issues. From the perspective of strategic management the evaluation of new technologies in terms of cost-benefit calculations or the investigation of required conditions for the use of new technologies analyzes technology-driven value networks and their contribution to enterprise value, positioning and differentiation of companies (Beria et al. 2012; Knoppe 2012). From an economic point of view analysis based of value chains, industries or clusters can show the development and innovation potential of the economy and science in the field of electric vehicles. This may be given recommendations for the creation of efficient structures for the transfer of innovation and regional policies to introduce and establish new technologies (Arnold et al. 2010; Brauner 2011).

Furthermore, questions on ecology, energy economics and policy like the reduction of local noise and emissions during the vehicle operation, the development of renewable energy, progression on energy efficiency, guarantee of grid stability or building of a charging infrastructure are of increasing strategic importance not only for regions which are characterized by larger wind parks like the Northwestern Germany (Brady and O'Mahony 2011; Held and Baumann 2011; Kley 2011; Pehnt et al. 2011, 2007). Moreover, social conditions are of significance. Here is studied for examples who could be potential users under the given conditions of limited range, incomplete infrastructure or comparatively expensive electric cars, who will be the potential users of electric mobility in 5, 10 or 20 years and what mobility needs the customers will have (Franke et al. 2012; Pierre et al. 2011; Sammer et al. 2008).

2 Background and Structure

This book is essentially the result of the conference 'Evolutionary Paths towards the Mobility Patterns of the Future' in August 2011 which was held at the Jacobs University in cooperation with the Centre of Regional and Innovation Economics of the University of Bremen. The Focus is to provide new insights and highlighting trends and challenges for companies or policy in the field of electric mobility in relation to mobility needs, new mobility concepts, needed infrastructures or the interplay with renewable energies. Despite a strong socio-economic view of the topic, the technical aspects of electro mobility playing also a role in the view of future business models or the combination with renewable energies. The book reflects the "State of the Art" in socio-economic research and provides valuable impulses for new; respectively further fields of research in the field of electric mobility. It's also offering theoretically well-substantiated, modern approaches and knowledge for the business world and gives impetus for new ideas and fields of positioning in a dynamic future industry. The book is aimed to the sub-disciplines of economics, social and technical science. The target group of the intended book is to be seen in all individuals with a scientific and/or practical background which deal with the future of mobility.

The book is subdivided into three parts. A big part of the socio-economic research occupied around mobility needs and mobility concepts. For this, part 1 discusses the acceptance of electric vehicles and possible changes in the coming years, changing in customer needs, new forms of vehicle operation like eCar-Sharing or effects on spatial and settlement structures. The chapter also gives a short overview about the activities in the program of the electric mobility model regions in Germany and reflects the state of the art of the socio-economic research in this context to lay the foundation for the subsequent articles.

Part 2 deals with the challenges for companies and politics. In this context, the discussion leads to new business models in the transport sector and approaches to the financing of new mobility services. Additionally safety aspects, the role of different stakeholders or the construction and needs of refueling infrastructure will be parts of this chapter.

Finally part 3 gives practical insights into the topic. Here the viewpoint of a municipality in terms of the change of propulsion technologies is shown, the potential of the use of electric cars is illustrate to the behavior of traffic and the developments and activities in a pilot region are illustrate.

3 Mobility Needs and Mobility Concepts

Part 1 of the book deals with mobility needs and mobility concepts. Several studies show that people have a strong need to move (Ahrens et al. 2011; Follmer et al. 2010). Whether the way to work or shopping, everyone wants to reach the destination as quickly as possible. Therefore, different modes of transport are used or combined. Rising energy costs, daily traffic jams in urbanized areas or new labor market requirements could lead to new individual mobility needs in the near future. In the following causes or influences are analyzed. These include the investigation and assessment of mobility concepts, the study of acceptance and degrees of mobility technology-based services as well as assessing the effectiveness of traffic-related measures. In particular, the following aspects are considered:

- **Hanke et al.** give an overview of important topics and issues related to the introduction of electric vehicles and focus primarily to socio-economic issues. For this the article shows current discussions of usability (range of electric vehicles, charging time and charging options etc.), standardization (battery, charging stations etc.) and economic, social, political, ecological and legal conditions. Last but not least the developments are considered from the perspective of management, e.g., the development of value-added networks or the management of networks and supply chains.
- **Kuhnimhof and Feige** pursue the goal of a long-term perspective of future mobility. Here the authors use the technique of scenario planning and show results of three produced scenarios of an extensive process for the future of mobility in Germany. In all three scenarios, the German population aging, is concentrated in urban areas and is increasingly aware of environmental issues with implications for the consumer choice. The scenarios differ especially in terms of immigration to Germany, the global and national economies and of the political decision-making. It is shown that this leads to various projections of future mobility behavior.
- **Schneider et al.** focusses their attention on the acceptance of electric vehicles. The article deals with the question, what the users or those who intend to use an electric vehicle in the near future, really think about electric vehicles. It is shown which changes the perception has when the electric vehicles are used in everyday life over a certain period. Schneider et al. use a series of longitudinal survey data from field trials of the eight model regions for electric vehicles in

Germany. All in all almost 1,000 German drivers were interviewed. It is found that the actual experience has a positive influence on the acceptance.
- **Santner and Fornahl** identify those groups which are most likely to use or buy an electric car in the near future. This question is addressed by identifying those societal groups which show socioeconomic characteristics, mobility patterns and attitudes that are compatible with a future usage of electric cars. To answer this question a survey in the model region Bremen/Oldenburg is conducted to gather the necessary information. They group the more than 700 respondents according to selected socioeconomic characteristics and then analyze whether their needs, resources and attitudes are in favor of electric mobility. They come to the conclusion that urban singles, urban seniors and rural families are three user groups with the highest likelihood to employ electric mobility.
- **Hoffmann et al**. direct their attention to the views of the potential electrical transport users. However, analysis of integrated mobility products are in focus, e.g., what the products should afford or what target groups can be addressed. The article shows a closer look by the introduction of integrated mobility products for potential users. As a result, five different product concepts have been developed. There were also discussions with members of predefined focus groups on issues such as pricing models, business models, compatibility with the daily use or reliance of the potential operators.
- **Kramer et al**. investigate the benefits of electric vehicles as part of the public transport system. The survey of users of the public car-sharing system in Berlin shows that unique user groups can be identified. This group characterizes their travel behavior with the domination of the use of multi-modal public transport system. These users have positive expectations and experiences with electric vehicles. The costs of the use are seen critical, but overall, however, the integration of electric car-sharing in public transport in Berlin is seen positive.
- **Blaettel-Mink et al**. deal with issues of acceptance of electric mobility. The goal is to give suggestions for strategies, which supports the applicability of electric mobility concepts and detection of improvement potentials. For these purpose exploratory in-depth interviews with selected people in the region, focus groups with users of electric vehicles and creative workshops with lead users of the electric mobility were done. The results show that the central motive for the choice of transport is convenience. Although electric vehicles brings fun, but the sales and operation are still too expensive.
- **Eschenbaecher and Wiesner** focus on the analysis of various Extended Product (EP) concepts for electric mobility. They identify and analyze the most valuable and feasible ones for implementation. The authors show the reflecting changes in the dynamics and integration in this new sector and that the establishment of Extended Product is necessary. For this EP modeling concepts were promised. The example car-sharing shows the general applicability.

4 Challenges for Companies and Politics

Part 2 shows challenges for companies and politics in connection with the electrification of the powertrain. On the one hand, there are exists risks with the introduction of electric cars, but on the other hand, electrical mobility opens opportunities facing the future competitiveness of regions. Especially the automotive and energy industry will be assigned a great potential. But the full potential can be possible developed only in cooperation with the information- and communication industry. While the industry drifts technological innovations and business models forward, there is no easy task for public authorities to develop appropriate frameworks and infrastructure (Arnold et al. 2010). In the following the contents of the articles are illuminating:

- **Colmorn and Huelsmann** discuss strategic options in terms of technological capabilities, cost realizations and innovative concepts in the context of electric mobility. They focus of the contribution is whether new technologies to penetrate the market and what are determinants and enabler for electric vehicles and the resulting strategic implications. The authors respond the positioning of competing forces and determine what are the structures and processes of global value creation networks. As a result, it is assumed that increased complexity will only lead to limited changes in the strategic behavior of the players.
- **Cordes and Huelsmann** deal with the risks of technological and institutional path dependencies and the resulting lock-in situations for players of electric mobility. The authors use the approach of knowledge-based dynamic capabilities to show that the companies can avoid or manage lock-ins. The contribution poses the exemplary effect which obtains knowledge of such activities to the abilities or increases their technological and strategic flexibility. As a result, it is shown that companies in the field of electric mobility have the option to reduce risks of path-dependent development by sharing their knowledge internally and with other actors in the electric mobility network.
- **Dudenhoeffer and Hause** discuss security aspects of electric vehicles. It is shown that the nearly silent propulsion at low speeds is beneficial for residents in the immediate neighborhoods of roads, but blind and partially sighted people have problems to locate the vehicles. The article describes the effective noise perceptions and feelings of security of people and presents risks of electric cars. The investigation includes five pairs of vehicles, each consisting of a battery electric vehicle (BEV) and one or two identical vehicles with internal combustion engine (ICE). As a result, it is shown that there are problems of perception for quiet electric vehicles for higher speeds of 30 km per hour.
- **Von Radecki** analyzed the introduction of electric mobility in the Stuttgart region. This article emphasizes who are the relevant stakeholders which create management structure or assist the implementation process. The contribution aimed on the existing structure and represents if it is a good basis for the change of the urban mobility system in direction of electric vehicles. The major

challenges with the introduction of electric vehicles are the transparency of information and the development of a holistic vision of all stakeholders.
- **Gabriel et al**. focus on the ongoing debate about the environmental impact of electric vehicles. In recent years the ecological comparison of conventional combustion engine vehicle with a battery-electric has been made in different studies. However, it appears that the studies use different parameters and it is quite difficult to compare the results. The authors want to contribute to a better understanding on the impacts of CO2 emissions associated with the electrification of the transport sector. Therefore, they give an introduction to these parameters and discussed, how (almost) zero-emission can be reached in connection with the use of renewable energy.
- **Jochem et al**. give an overview of the future electricity market with a focus on electric vehicles. In this case challenges are described, which required different approaches like measures in the electricity supply side or in line extensions, which are expensive and time consuming due to the local acceptance. It is shown that an automatic deferred loading of electric vehicles can contribute to peaks in the household load curves and increase the low power requirements during the night. This helps to an easier integration of volatile renewable energy sources.
- **Hoerstebrock and Hahn** point a methodology for the development and evaluation of a charging infrastructure layout, without requiring a large number of electric vehicles. The basis for this approach is a planning model. This model uses concepts and parameters such as user mobility patterns, regional road layout, car parking infrastructure and the electric car and its technical characteristics such as range or battery capacity. The subsequent simulation allows the analysis and presentation of sufficient regional charging infrastructure.

5 Practical Insights

Part 3 allows some practical insights into three model regions in Germany. Here challenges and initial success with the introduction of electric cars are demonstrated. The contributions focus on:

- **Truemper** first gives an overview about the German funding system and then moves on to the model region Hamburg. Here he presents special features and goals of the model region and gives insights into the used pilot-vehicles in Hamburg. A special feature in Hamburg are two different systems of charging infrastructure, a private with limited access and a public charging infrastructure which everyone can use. The author also shows economic benefits which deals with the establishment of knowledge and competences.
- **Glotz-Richter** discusses that the transport sector is a politically highly sensitive area in all cities and therefore high expectations from the electrical mobility are also linked. However, it can capture that the change of drive technology not

solve all the traffic problems per se. The potential is only visibly in connection with comprehensive strategies which include traffic reduction or traffic shift to more environmentally friendly modes of transport. For this Bremen is a good visual example. Bremen itself presents as a cycling city and has a broader concept of sustainable mobility—including the benefits from further electrification.

- **Schaefer et al.** present the project "Elektrolöwe 2010", which shows the potential for electric cars in Hesse. The data is evaluated on the mobility behavior in traffic. The focus thereby is the contribution to a single-centeric (Cassel) and a polycentric city (Frankfurt) and a rural region (Lauterbach). It turns out that the majority of respondents could well cope the daily necessity with an electric car. The operating range is thus less of a problem especially in connection with the generally positive attitude towards electric mobility of the users, but few would consider buying an electric car because of the high cost. However, the authors certify that electric vehicles have a great potential for daily travel to work.

Acknowledgments This article was supported by the federal program "Elektromobilität in Modelregionen". The Federal Ministry of Transport, Building and Urban Development (BMVBS) is providing a total of 130 million euros from the Second Stimulus Package. The program is coordinated by the NOW GmbH Nationale Organisation Wasserstoff- und Brennstoffzellentechnologie.

References

Ahrens G-A, Bäker B, Fricke H, Schlag B, Stephan A, Stopka U, Wieland B (2011) Zukunft von Mobilität und Verkehr. Technische Universität Dresden, Dresden
Arnold H, Kuhnert F, Kurtz R, Bauer W (2010) Elektromobilität: Herausforderungen für Industrie und öffentliche Hand. PricewaterhouseCoopers AG, Stuttgart
Bakker S, van Lente H, Engels R (2012) Competition in a technological niche: the cars of the future. Technol Anal Strateg Manag 24(5):421–434
Beria P, Maltese I, Mariotti I (2012) Multicriteria versus cost benefit analysis: a comparative perspective in the assessment of sustainable mobility. Eur Transp Res Rev 4:1–16
Brady J, O'Mahony M (2011) Travel to work in Dublin: the potential impacts of electric vehicles on climate change and urban air quality. Transp Res D: Transp Environ 16(2):188–193
Brauner G (2011) Nachhaltige Mobilitätsstrategien. e & i Elektrotech Informationstechnik 128(1):36–39
Brown S, Pyke D, Steenhof P (2010) Electric vehicles: the role and importance of standards in an emerging market. Energ Policy 38(7):3797–3806
Federal Goverment of Germany (2009) Nationaler Entwicklungsplan Elektromobilität. Berlin
Federal Ministry of Transport Building and Urban Development (2012) Modellregionen Elektromobilität. http://www.bmvbs.de/SharedDocs/DE/Artikel/UI/modellregionen-elektromobilitaet.html

Follmer R, Gruschwitz D, Jesske B, Quandt S, Lenz B, Nobis C,. Mehlin M (2010) Mobilität in Deutschland 2008: Ergebnisbericht. Struktur—Aufkommen—Emissionen—Trends. Ergebnisbericht. Struktur—Aufkommen—Emissionen—Trends. Bonn

Franke T, Neumann I, Buehler F, Cocron P, Krems JF (2012) Experiencing range in an electric vehicle: understanding psychological barriers. Appl Psychol Int Rev: Psychol Appl Rev Int 61(3):368–391

Gnann T, Plötz P (2011) Status Quo und Perspektiven der Elektromobilität in Deutschland Sustainability and Innovation, vol S14. Fraunhofer Institute for Systems and Innovation Research (ISI), Stuttgart

Held M, Baumann M (2011) Assessment of the environmental impacts of electric vehicle concepts. In: Finkbeiner M (ed) Towards life cycle sustainability management. Springer, Netherlands, pp 535–546

Kley F (2011) Neue Geschäftsmodelle zur Ladeinfrastruktur Sustainability and Innovation, vol S5. Fraunhofer Institute for Systems and Innovation Research (ISI), Stuttgart

Knoppe M (2012) E-mobility generates new services and business models, increasing sustainability. In: Wellnitz J, Leary M, Koopmans L, Subic A (eds) Sustainable automotive technologies. Springer, Berlin, pp 275–281

Kroos K (2012) Mobilität als Wachstums- und Werttreiber. In: Proff H, Schönharting J, Schramm D, Ziegler J (eds) Zukünftige Entwicklungen in der Mobilität. Gabler Verlag, Wiesbaden, pp 41–59

Pearre NS, Kempton W, Guensler RL, Elango VV (2011) Electric vehicles: how much range is required for a day's driving? Transp Res C: Emerg Technol 19(6):1171–1184

Pehnt M, Höpfner U, Merten F (2007) Elektromobilität und erneuerbare Energien. Arbeitspapier Nr. 5. Retrieved from http://www.bmu.de/files/pdfs/allgemein/application/pdf/elektromobilitaet_ee_arbeitspapier.pdf

Pehnt M, Helms H, Lambrecht U, Dallinger D, Wietschel M, Heinrichs H, Behrens P (2011) Elektroautos in einer von erneuerbaren Energien geprägten Energiewirtschaft. Z Energiewirtschaft 35(3):221–234

Pierre M, Jemelin C, Louvet N (2011) Driving an electric vehicle: a sociological analysis on pioneer users. Energ Effi 4(4):511–522

Sammer G, Meth D, Gruber CJ (2008) Elektromobilität—Die Sicht der Nutzer. e & i Elektrotech Informationstechnik 125(11):393–400

Schindler J, Held M, Wuerdemann G (2009) Postfossile Mobilität—Wegweiser für die Zeit nach dem Peak Oil. VAS—Verlag für Sozialwissenschaften, Bad Homburg

Schraven S, Kley F, Wietschel M (2011) Induktives Laden von Elektromobilen—Eine techno-ökonomische Bewertung. Z Energiewirtschaft 35(3):209–219

Wallentowitz H, Freialdenhoven A, Olschewski I (2010) Zunehmende Elektrifizierung des Antriebsstranges. In: Wallentowitz H, Freialdenhoven A, Olschewski I (eds) Strategien zur Elektrifizierung des Antriebsstranges. Vieweg + Teubner, Wiesbaden, pp 35–70

Whitmarsh L, Koehler J (2010) Climate change and cars in the EU: the roles of auto firms, consumers, and policy in responding to global environmental change. Cambridge J Reg Econ Soc 3(3):427–441

Author Biographies

Christian Hanke is member of the research staff at the Institute for Project Management and Innovation (IPMI) at the University of Bremen and was also research associate at the workgroup "System Management" at the School of Engineering and Science at Jacobs University Bremen. His research interests are in the field of innovation, strategic management, future studies and urban and regional development in the framework of mobility and energy systems.

Michael Hülsmann holds the chair of "System Management" at the School of Engineering and Science at Jacobs University Bremen. He focuses on Strategic Management of Logistics Systems. Additionally he leads the sub-projects "Business model concept/product idea" and Sustainable Innovation and technology strategies" in the framework of the electric mobility model region Bremen/Oldenburg.

Dirk Fornahl leads the Centre for Regional and Innovation Economics (CRIE) at the University of Bremen. His work is settled on cluster analysis, analysis of regional supply chains and potential, innovation processes and networks. He also coordinates the socio-economic research in the electric mobility model region Bremen/Oldenburg.

Part I
Mobility Needs and Mobility Concepts

Socio-Economic Aspects of Electric Vehicles: A Literature Review

Christian Hanke, Michael Hülsmann and Dirk Fornahl

Abstract The dependence on fossil fuels and the climate change led not only in Germany to a debate on how the future of mobility might be designed. In this context the importance of electric mobility has especially grown in the public perception since the National Development Plan for Electric Mobility was published by the German federal government in 2009 and the eight model regions for electric mobility were established. The goal of the federal government is to bring a total of 1 million electric and hybrid vehicles on German roads by 2020 and establish Germany as lead market and lead provider of mobile electric mobility solutions. To achieve these objectives many questions on technical feasibility, ecological impacts or on the acceptance of the products by customers need to be answered for a successful market launch. Therefore this contribution gives an overview of important topics and issues related to the introduction of electric vehicles with a focus primarily on socio-economic topics. The literature review addresses existing studies and classifies main insights in seven thematic areas.

C. Hanke (✉)
IPMI—Institut für Projektmanagement und Innovation, Universität Bremen,
Wilhelm-Herbst-Street 12, 28359 Bremen, Germany
e-mail: christian.hanke@innovation.uni-bremen.de

M. Hülsmann
School of Engineering and Science, Systems Management, International Logistics,
Jacobs University Bremen, Campus Ring 1, 28759 Bremen, Germany
e-mail: m.huelsmann@jacobs-university.de

D. Fornahl
Centre for Regional and Innovation Economics, Universität Bremen,
Wilhelm-Herbst-Street 12, 28359 Bremen, Germany
e-mail: dirk.fornahl@uni-bremen.de

1 Introduction

In Germany, being able to use a car is one element of individual freedom and is a cornerstone of individual mobility. 82 % (2008) of all German households possess a car and in just over a third of these households two or more cars are available. Despite the already high degree of car availability in the past, the motorization has even further increased in recent years (Follmer et al. 2010). In the new millennium the high dependence of individual mobility on private cars led to a discussion of alternatives to the internal combustion engine, initiated particular by significantly increasing oil prices. Worldwide, many governments and interest groups moved the issue of sustainable mobility in the foreground (Greenpeace 2010, Kendall 2008). Hybrid and electric vehicles (EV) are seen as an important part of a technology portfolio targeted at reducing greenhouse gas emissions as well as the dependence on oil by funding the expansion of renewable energies, developing new mobility concept, e.g. for public transport, or supporting technological developments (Greenpeace 2010, Kendall 2008). The electrification of the drivetrian could lead to a sustainable technology path which benefits the consumer by lower and less volatile (electricity) prices for mobility (Schill 2010a, c). This discussion is encouraged by the fact that electric mobility is benefitting from high oil prices (Conrady 2012, Dijk et al. 2012).

The rise of interest in the topic electric mobility may be particularly well traced by looking to the internet. This can be done by considering first the number of search inquiries of internet users over a time span and second by analyzing the number of hits displayed by search engines over a period for a specific topic. We have first investigated different frequently used keywords with Google Insight for Search (Fig. 1). There are three search inquiries related to electric mobility that were most often searched in Google: electric mobility, electric car and electric vehicle. After that we have investigated the number of searches in the period from the beginning of 2004 till the end of June 2012. It turns out that especially the term 'electric car' was asked with peaks in 2006 and 2008.

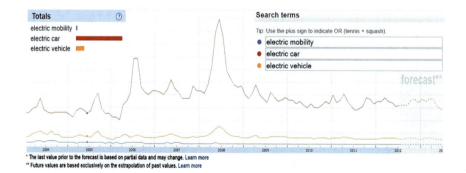

Fig. 1 Search request "electric car" (Google insight 2012)

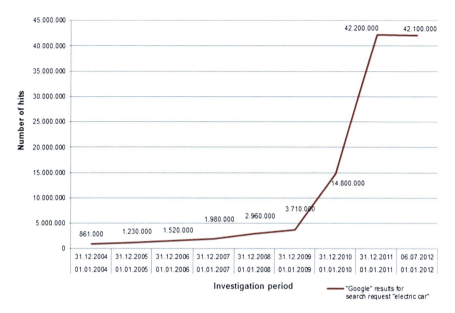

Fig. 2 Number of hits "electric car" (Google search 2012)

After that we analyzed how many entries the search engine Google could find for the most searched term 'electric car'. In Fig. 1 it is shown that the entries only slowly grow up to the year 2009 and then sharply increase. The figure also indicates that although for the year 2012 only the first half could be examined, currently the hits continue to increase. However, it is suspected that the increase levels off. This suggests that the topic could reach its climax in the public debate in the near future. Nevertheless, this excursion suggests that the public has a great interest in the topic of electro mobility. Such an interest will also have an impact on many companies, e.g. managers consider how their company can benefit from the latest developments. Especially for such managers but as well for policy makers it is important to understand the actual (scientific) debate in the field of electric mobility.

Consequently the following search questions arise: What are important socio-economic fields in the area of electric mobility? Which state-of-the-art results can be derived from research in these areas? In order to answer these questions the contribution aims at summarizing key topics and presenting the main findings.

In the following, we turn to the scientific examination of the topic in Sect. 2 to Sect. 8 by reviewing current literature on electric vehicles. The Sect. 2 focusses on the management perspective (e.g. innovation management, technology strategies), the Sect. 3 addresses usability topics (e.g. range or safety) and in the Sect. 4 issues of standardization are presented (e.g. charging cable and battery). Afterwards economic conditions (e.g. business models) (Sect. 5), social conditions (e.g. customer needs) (Sect. 6), ecological conditions (e.g. emissions) (Sect. 7) and politic conditions (e.g. subsidies or regulations) in Sect. 8 are discussed. We searched by key words in online databases like sciencedirect.com, springerlink.com and webofknowledge.com to receive an overview on the current debate. As a supplement grey literature was also partly used by employing for example scholar.google.de. Section 9 finally summarizes the developments and presents implications for the management of companies within the field of electric mobility as well as for further research requirements.

2 Management Perspectives

From a management perspective, electric mobility especially relates to the development, planning and implementation of content, objectives and guidelines in the company based on such new technological trends (Johnson et al. 2011). It is necessary to examine how the developments of electric mobility affect strategies (e.g. orientation of a company), structures (e.g. design of a company) or systems (e.g. infrastructures of a company).

Strategic opportunities for companies and public organizations arise especially from an understanding of the functioning and production of components of electric cars, which can e.g. be demonstrated by technology roadmaps (Bai et al. 2012). This helps to get deeper insights into specific aspects of the developments and also to establish support schemes according to the forecasts that could lead to a breakthrough of that technology. To achieve this goal it is important to recognize key factors (such as the oil price) which are affecting the mobility of the future and to compare potentials and limitations of different propulsion concepts (Hanselka and Jöckel 2010, Li 2007). As major driving forces Hanselka and Jöckel mention the growth of world population and urbanization, i.e., mobility in large cities is becoming increasingly important and is linked with the desire for a livable urban environment without traffic noise or fine dust (Dudenhöffer 2010, Hanselka and Jöckel 2010). However, it is still not identified which future technological solution ensures such future mobility.

Electric mobility is still in a niche and competes for example with the hydrogen car (Bakker et al. 2012). Bakker et al. show that the niche technology is shielded from normal market forces by government subsidies on the one hand and compete with alternative drive technologies in terms of R&D funding, regulation and the construction of supporting infrastructure on the other hand. It turns out that big hopes follow later disappointments of various component technologies. Therefore, it is important to monitor sales and product evolution. With a growing market companies are motivated to develop new strategies based on market preferences (Orbach and Fruchter 2011). In terms of innovation management it is important to understand the process of technological change. This includes questions such as how the involved actors manage their early moves and how the knowledge base could rise to achieve competitive advantages. Also spill-over and their consequences for business sectors, e.g. sectors like energy or IT, should be analyzed (Pohl 2012, Pohl and Elmquist 2010).

Furthermore, it should be noted that certain developments require different materials and resources. The raw material of lithium which is widely used for most batteries and accumulators (Peters 2010), is still available for decades, but otherwise it is located only in a few global regions which usually are known as (politically) less stable and reliable regions. In order to secure the raw materials, considerations are to be made how such batteries can be effectively recycled or which alternative battery systems exists. In addition, the entire system of electric mobility should be examined in terms of shortages of raw materials (e.g. cobalt, copper, indium and neodymium) and also answers have to be developed how actors could strategically respond to these shortages (Angerer et al. 2009).

With a look on the structures of companies, the challenge for the automotive industry is the emergence of new value chains. Automobile manufacturers must prepare themselves technologically and organizationally for an increasing demand for green technologies, but otherwise it is also important that they focus on their competencies which are conventional drives (Bechmann and Scherk 2010). Thus, there are opportunities and risks for the automobile industry which could lead to positional changes in the value chain. The impacts for innovation and marketing strategies are controversial discussed. Kasperk and Drauz (2012) e.g. assume that the automotive value chain will change by shifting the share of value added upstream (Kasperk and Drauz 2012). Therefore, the new car designs will have a completely new architecture around the power train and will have a huge impact on manufacturers and supply industry. Especially in the area of battery development and production European companies face major challenges, because appropriate technologies have been neglected for years and especially Asian manufacturers gained great experience and thus have strategic advantages (Deutsche Bank Research 2011). Changes are also expected in the energy sector, because an increase in the number of electric cars raises many questions regarding the network integration of these vehicles (Leitinger and Litzlbauer 2011).

Huelsmann and Colmorn suppose that competition in the automotive sector will increase and the forces will shift in favor of customers and suppliers. The authors assume that in the short-term the strategic and in the long-term the logistical complexity of the value-added networks increase, but they also come to the conclusion that only limited changes are taking place in the strategic behaviour of the firms and therefore it is expected that there are only perceptible changes in the value chain (Huelsmann and Colmorn 2011).

Also an important challenge for companies is to be prepared for volatile future automobile markets in an ongoing globalization and in terms of economic ups and downs of local economies as in the economic and financial crisis. This should not lead to an ignorance of future trends to develop sustainable products and alternative mobility concepts by using green technologies (Bechmann and Scherk 2010).

In conclusion it can be said that the firms' managements are responsible to develop strategies for a **payable sustainable mobility**. But electric mobility is not the only option to achieve this goal. Management strategies must also consider other **alternatives like biogas or fuel cell** (Avadikyan and Llerena 2010,

Contestabile et al. 2011, Shinnar 2003). This can be achieved on the one hand in connection with the **development of mobility technologies**, e.g. power drive, battery, chassis, driving safety, or infrastructures like charging infrastructure. On the other hand it can be done by focusing on the development of **sustainable business concepts**, e.g. new mobility offers like car sharing (Canzler 2010, Canzler and Knie 2010). Another challenge is the potential change of the value chain. Companies must therefore derive consequences from the current developments for their business and examine which **capabilities** from other companies or sectors could be integrated (Knoppe 2012b, Thoma and O'Sullivan 2011). For all these issues, it is important to answer the question **which target groups** should be addressed and to analyze what customer classification can be employed (Knoppe 2012a). It is also interesting how the **technical understanding** of the (final) customer is addressed (Franke et al. 2012, Zhang and Shi 2011), because this determines how the products will be accepted which is one of the most crucial factors for the successful market launch (Dütschke et al. 2012, Franke et al. 2012, Sammer et al. 2008, Zhang et al. 2011).

3 Usability

In the following aspects of usability are considered. Usability means how a product or a service system can be used by certain users to achieve goals like effectiveness, efficiency and satisfaction. The usability of electric cars is closely related to the acceptance of the user. Thus, a positive attitude towards the development and diffusion of new technologies and consumer products as well as the consequent behavior and actions based upon these attitudes is central for market penetration. Acceptance also relates to the advantages of the consumers' use of these electric automobiles in comparison to other mobility means (e.g. trains, or even other drive systems in the automotive sector). The level of acceptance is affected by hard factors such as safety or range and soft factors such as design or image (Sammer et al. 2008, 2011).

Since electric cars still have disadvantages with regard to the hard factors, it is possible that individual factors such as comfort or a better personal image could convince the customer of the new products (Knoppe 2012a). Here also arises the question, how the general mobility orientation of the society looks like which is strongly based on the desired or frequently cited role models in the media. Sustainable aspects of public and personal perceptions play an increasingly important role, but only a few people use electric cars currently. Since it turns out that the experience of the new technology by driving electric vehicles triggers learning processes maybe mobility routines can change in the future (Zimmer and Rammler 2011). Hence, perhaps in the future the range will no longer be quite so critical for the customer.

But today one of the main characteristics of usability is still the range which is significantly affected by driving style and battery capacities (Bingham et al. 2012,

Chlond et al. 2012). Others are charging time and the amount and distribution of charging possibilities or safety issues which in addition influence the price of electric cars and have a big influence on public awareness and acceptance (Franke et al. 2012, Lunz et al. 2012, Zhang et al. 2011).

The charging times of electric car batteries are significantly different from those related to refueling internal combustion engines. Usually a distinction is made in the literature between the loading scenarios based on conventional household sockets with 230 volts or those at fast-charge stations with up to 400 volts (Fraunhofer-IAO 2010a). The charging takes between 6 and 8 hours (230 volt socket) and is re-duced to 30 minutes with a capacity of approximately 80% by fast charging infrastructure (Fraunhofer-IAO 2010b). The downside of such a fast charging process is the reduction of the life time of the batteries. Another concept relies on the replacement of the whole battery at battery-changing stations. The battery replacement just takes a few minutes (Betterplace 2012, Johnson and Suskewicz 2009).

Based on observed mobility patterns it is expected that electric vehicles will be available especially for driving short distance for individual transport due to the limited range of the batteries and drive systems (Raykin et al. 2012). Therefore the vehicles are likely to operate in cities and are used as a second car (acatech 2010a). There are also studies that deal with the optimal range of electric vehicles in relation to the development of the oil price. It turns out that the oil price level and the driving habits have a strong impact on the cost-performance and the ideal electric driving range (Özdemir and Hartmann 2012).

In connection with the discussion of the necessary range there are also questions of the optimal state of charge. Studies have examined e.g. the optimizing of charging cycles, how the increased demand of renewable energy can be satisfied (Finn et al. 2012), how charging operations can be controlled intelligently (Schill 2010a), how a design system of charging infrastructures should be layout in terms of numbers of charging station in an area (Brauner 2008, Hiwatari et al. 2012) or how fast loading should take place (Schroeder and Traber 2012).

Another aspect of usability is security, which can be viewed from two perspectives. First there are questions on the failure of components of the electric drive train and how safe the electrical components are in accidents. In contrast to the voltage employed in household sockets (230 V) electric cars are powered with 400 V. The potential additional threat can be countered with a special risk switch, switching off the current electric flows in the car completely in case of an accident (Kulkarni et al. 2012, Peters 2010).

Second there exist safety aspects focusing on the safety of other traffic users like pedestrians and cyclists. Current models of electric cars are already almost silently and can be a risk for blind and visually impaired persons (ANEC 2010).

In summary it can be stated that the user must gain **experience** with the electric vehicles so that in addition to price or range some positive experiences of everyday usability are perceived. However, crucial questions remain unanswered: When starts the nationwide introduction of electric vehicles (Driscoll et al. 2012, Hugosson and Algers 2012)? How will the user charge their car (Axsen and Kurani

2012, Richter et al. 2012)? Which **charging standards** (charging infrastructure) will prevail (Brown et al. 2010)? Could the car also be recharged during long distance travels to other areas or who provides a comprehensive **infrastructure** (Reichert et al. 2012, Wang et al. 2011)? Which **business models** (owned vehicles vs. car sharing) will prevail (Baur et al. 2012, Dosch 2012)?

4 Standardization

Standardization, which designated the unification of products and product components, as well as adherence to rules in the manufacturing and administrative processes in enterprises, is critical to the successful introduction of electric mobility (Arnold et al. 2010, Reichert et al. 2012). Uniform standards will facilitate market participants to use their resource development efficiently and give customers transparent and interoperable offerings. To find the optimal time to establish standards is very difficult, because an early setting may exclude new and relevant (technical) knowledge. On the other hand, it is also problematic to set standards later, because if various solutions are on the market there are numerous resistance which leads to increased costs and coordination time until standards are agreed upon (acatech 2010a).

Standardization is an established process and in Germany these processes are promoted by the German Institute for Standardization (DIN 2012). The private club offers business, science, consumers, test institutes and authorities the possibility to work in an organized way on the development of consensus standards. The special challenge in the field of electric mobility issues is that many players from different business sectors need to work effective and purposeful together. In Germany a committee that developed a coordinated standardization roadmap was established under the National Development Plan for Electric Mobility (Federal Goverment of Germany 2009). The first version of the roadmap was presented in November 2010 as a result of the "Working Group 4 standardization" of the National Platform for Electric Mobility (AG 4 „Normung Standardisierung und Zertifizierung" 2010), in January 2012 an update and supplement was published (AG 4 „Normung Standardisierung und Zertifizierung" 2012). This roadmap is the central element of standardization for the coming years in Germany. The roadmap also provides an overview of the standardization landscape and identifies blank spots and demonstrates recommendations and actions to fill them. Accordingly, the standardization activities focus on topics that result in standard interfaces (AG 4 „Normung Standardisierung und Zertifizierung" 2012) in order to provide an unrestricted worldwide secure and free use of vehicles using alternative fuels.

It is discussed that standardization can create synergies and avoid duplication of efforts (Brown et al. 2010). Furthermore, requirements for cross-border payment systems (e.g. for data protection) or auditing standards for comparability of product characteristics (e.g. regarding the performance and life of batteries) are defined. Standardization also seeks to limit the variety of alternatives. Consumers will benefit from such an approach because a comparison of products is much easier. Organizations benefit because they do not have to offer multiple options of their products and thus can often reduce costs. This also contributes to the reduction of investment risks and to reduce trade barriers. The key issues (AG 4 „Normung Standardisierung und Zertifizierung" 2012 see pp. 23–38) of standardization are:

- **Electric vehicle and smart grid** with sub-themes such as charging (locations, functions or vehicle functions during operation on stationary power), billing or electrical safety (for developments in this area see also (Brown et al. 2010, Gallardo-Lozano et al. 2012, He et al. 2012b, Ramezani et al. 2011, Wang et al. 2012),
- **Interfaces, energy flows and communication** with sub-themes such as vehicle versus charging infrastructure, driver versus customer, vehicle versus energy trading (pricing), charging infrastructure versus network (Brauner 2008, Bunting 2012, Wang et al. 2012),
- **Electric vehicles** with sub-themes such as system approaches to power drive, system approaches for loading, safety (electrical safety, crash), fuel-cells, battery, capacitors (Axsen and Kurani 2012, Werther and Hoch 2012),
- **Charging stations** with the sub-themes provision of the energy flow or control (Dong and Lin 2012, Hiwatari et al. 2012, Leitinger and Litzlbauer 2011, Lunz et al. 2012, Schuster and Leitinger 2011).

Currently the utilization of used batteries ("second life") is discussed as a stationary back-up (e.g. for wind and solar energy). In addition energy recovery or communication and inductive charging are current issues of standardization.

Other relevant sources of information are:

- **DIN-study** on "standardization needs for alternative power and electric mobility", which identified relevant standards and gives an overview of standards and recommendations for the development of a standardization roadmap (Bremer 2009).
- **VDE study** "E-Mobility 2020: Technology—Infrastructure—Markets", which shows estimates of the current technology as well as to Germany's position and also discusses opportunities and challenges of electric vehicles (VDE 2010a).
- **VDE study** "electric vehicles", which represented the potentials of the electric vehicles in connection with batteries (VDE 2010b).
- **PwC study** on socioeconomic components of electric mobility. This study examines mainly socioeconomic issues such as business, law, policy and user in connection with standardization (PwC AG 2012).

In summary it can be said that standards need to be developed which support the introduction of electric vehicles, e.g. those standards which are comparable to those of internal combustion engines with regard to security issues. As a key challenge the establishment of **international standards** can be identified. For this the current situation in the respective countries must be analyzed. The **energy systems** in the various countries strongly differ from one another, which has consequences for the etablishment of an optimal **charging infrastructure**. Further, the most important issues of standardization are the charging stations in terms of structure and protection, the **charging plug**, the **communication infrastructure** between the vehicles, power stations, infrastructure and network operators (acatech 2010a).

5 Economic Conditions

Economic conditions of electric mobility, which means the planned masures and actions covering the human needs, are examined in this section. This includes issues like integration of the automotive manufacturing in the economy or the development of fuel prices. Other aspects are costs for electric cars or challenges of the future such as the finiteness of resources.

As one of the major economic sectors in Germany and Europe the automotive industry faces enormous challenges. Besides economic globalization trends (Bechmann and Scherk 2010) this also includes the politically supported issues like the reduction of waste emissions (Bernhart and Zollenkop 2011). Many OEMs have reacted and introduced new car models with reduced emissions and more fuel-efficient technology in their portfolio (Baur et al. 2012). This development also occurs against the background of rising oil prices (Bunting 2012, Dijk et al. 2012), so that for the companies the question arises, which are the automobile technologies of the future and in which field should they invest (Avadikyan and Llerena 2010). It is quite controversial which trends in the energy sector like biofuels and biomass to liquids have a major influence on current developments in the mobility market. However, it should be noted that the energy landscape changed in terms of energy prices or local energy use preferences (Bunting 2012).

Companies must also be aware that changing mobility needs and new value creation potential in the entire mobility system lead to changes in the customer's decision criteria (Canzler 2010, Dütschke et al. 2012). New business models allow not only to define some elements of the new mobility, it could also come to an altered interaction which creates business models such as car sharing, mobility shops or e-mobility provider services (Canzler and Knie 2010, Knoppe 2012a). New business models also aim at the combining mobility with the mobile internet (Heinrichs et al. 2012) or with charging infrastructure (Kley 2011). Further, companies also need to initiate R&D in new business fields without begin able to anticipate a clear market for each innovation and at the same time they must continue to grow their daily business, e.g. in terms of increased market turnover or by increasing the efficiency of existing vehicles (Deutsche Bank Research 2011).

It also raises the question of availability of the necessary raw materials. For example current electric cars or those close to market introduction mostly use lithium-ion batteries. However, it should be noted that in the strict sense the rechargeable energy storage are not batteries, which cannot be re-charged, but accumulators (Peters 2010). The preferred use of the lithium technology could be explained by the experience in other areas such as the computer industry, which already resulted in a learning curve (Peters 2010). It is forcasted that in the next four decades there are sufficient reserves of lithium. However, these reserves are concentrated in a few countries and some of them are located in geopolitically unstable regions (Angerer et al. 2009, Kushnir and Sanden 2012). So it is of strategic importance for the economic success to ensure the necessary resource base early.

As a key issue for the future market success, the direct and indirect costs of these vehicles can be identified (Biere et al. 2009). This includes purchasing costs and variable costs such insurance rates, energy prices or costs for maintenance and repairs. Others are costs for the construction of the necessary charging and intelligent navigation infrastructure and last but not least costs for training of skilled workers. Basically the question arises how much consumers would be willing to pay for vehicles or services. Some surveys suggest that first mover would pay up to 3,000 € as a premium in comparison with conventional vehicles (Schlick et al. 2011). But how the loss in value of the vehicle will develop largely dependents on the attractiveness of the overall system and the individual brands (Hanselka and Jöckel 2010). The purchasing costs at the market launch phase of electric vehicles are much higher due to high battery costs and high expenses for R&D in comparison to conventional vehicles. That will have a major impact on customer's willingness to buy an eletric vehicle (Hidrue et al. 2011). The price of the battery and its expected life time have significant influence on the overall purchasing and probably maintenance costs (DeLuchi et al. 1989, Hidrue et al. 2011). It is forecasted that these costs will be reduced significantly in the coming years (Baker, Chon and Keisler 2010) which in turn should increase the attractiveness of buying an electric car. This is of course closely connected to the development of the energy price, which is influenced by factors such as the development of the world economy, or local preferences (Bunting 2012, Özdemir and Hartmann 2012).

Economic conditions are also affected by political and economic institutions. For example there are mobility restrictions such as city tolls for cars in many cities (Sammer 2012). An increase in user costs leads to a reduction in car use and selection of alternative modes of transport. Furthermore it is important how the value chain of electric mobility is established in the region. From a regional perspective the question rises if there exist only sales and service institutions or if there develops a value chain which includes manufacturer of raw materials and auxiliary materials, component manufacturers and OEMs (Cooke 2011, Knoppe 2012a, b).

It can summarize that electric mobility enable the reduction of dependence on fossil fuels. The **change of the value chain architecture** is a challenge and also a chance for the automotive industry. On the one hand many manufacturers and suppliers in Germany and also in Europe own **key technologies** for future mobility developments, have **market leadership** and invest heavily in this approach, on the other hand threats arise mainly by the massive **government support** of new actors, especially in China and the USA (acatech 2010, Yang 2010). The challenge for companies is to find a sustainable and economically competitive solution.

6 Social Conditions

Under social conditions the requirements and necessities of the users as well as mobility behavior is summarized. The success of electric mobility products depends on the fulfillment of users' expectations (Pierre et al. 2011, Sammer et al. 2008). But in the past, electric vehicles were not able to meet all the needs of users. Hence, in Germany the population of electrically powered vehicles accounts for less than 1 % of all cars (Sammer et al. 2008).

Based on the current mobility patterns and behavior projections for future developments are made, e.g. in terms of willingness to pay or ecological settings. From this the market potential of electric vehicles can be deduced (Lieven et al. 2011, Link et al. 2012). Not only in Germany there exists social movements which are dedicated to increase environmental awareness by educating consumers and thereby influencing development processes of companies which provide green technologies (Ustaoğlu and Yıldız 2012).

Based upon acceptance and attractiveness studies the potential of electric vehicles, especially among private users, are forecasted (He et al. 2012a). For example in a simulation based on interviews of new car buyers it could be shown that in 2020 two-thirds of all new car buyers in Germany would opt for an electric vehicle, battery electric vehicle or a plug-in hybrid car (Götz et al. 2012). Depending on the vehicle class the likelihood to choose a battery electric vehicle is between 12 and 25 %. Especially for smaller cars, the acceptance is already high in 2020. For larger vehicles an increasing acceptance is forecasted until 2030. As one reason for this increase it is stated that the technological progress leads to falling prices. Besides the price, enviromental aspects also play an increasing role for purchasing decisions. The analysis demonstrates that there is already a large acceptance potential for electric vehicle concepts and that this may even further increase in the future. It is also shown that the acceptance of electric vehicles is influenced by different individual factors such as academic degree, annual income and number of previous vehicles or government policies. The price sensitivity for electric vehicles additionally depends on the number of family members, the opinion of peers, maintenance costs and degree of safety (Zhang et al. 2011). Besides the price, buyers also consider subjective aspects such as prestige or style (Sammer et al. 2011) and it is also crucial how people get information on electric vehicles, e.g. by interpersonal communications or by the media, and how preferences, e.g. on environmental issues, change over time (Gould and Golob 1998).

This is consistent with studies on the mobility needs. Electric vehicles already meet the daily requirements in terms of average required ranges and flexibility, especially if the electric vehicles can be charged at home (Arnold et al. 2012, Pearre et al. 2011). However, a main obstacle is still the limited range of

150–300 km (Hawkins et al. 2012). Users have become accustomed to the fact that conventional vehicles have a range of up to 1,000 km (Chlond et al. 2012). That means that for some long trips, e.g. for vacations, electric vehicles do not seem suitable. To increase the acceptance for electric mobility, the development of integral mobility systems is required to offer alternatives for long-distance travel, e.g. combinations of electric vehicles with the use of public transportation or car sharing (Canzler 2010, Canzler and Knie 2010).

Overall it can be said that electric vehicles have great potential from a user perspective. But for a successful market penetration numerous technological and economic challenges must still be managed. Particularly critical are the attitudes towards the **range** as well as towards the **high cost of the vehicles**, mainly caused by the **high cost of the battery**. For the future it is still to be analyzed how the customer will accept **new mobility concepts**, which combine electric vehicles with other **services** to compensate the weaknesses of electric mobility. Furthermore, it seems necessary to investigate how the previous mobility habits which lead to **physiological barriers** could break up so that the people rethink their mobility decisions.

7 Ecological Conditions

The full potential of electric mobility is only visible in connection with innovative business plans which include the use of renewable energies (Barkenbus 2009, Brady and O'Mahony 2011, Pehnt et al. 2007, 2011, Schill 2010a). Electric vehicles have a high efficiency and by charging renewable electricity they almost create no emissions. This of course assumes that the required sustainable electricity can be provided in the right amount at the right time.

Nevertheless, it can hardly be estimated how big the ecological potential of electric mobility really is. In fact there is only few data on environmental effects based on empirical cases and experiments, e.g. the CO_2 emissions of electric cars, so that many studies deal with model calculations (Kudoh et al. 2001, Ramezani et al. 2011, Schill 2010a, Yabe et al. 2012). It is also little known about the impacts of electric vehicles and their components over their life cycles (Held and Baumann 2011).

Furthermore, the question arises, what the market potential of electric mobility is to make realistic assumptions of future electricity demand (Lieven et al. 2011, Link et al. 2012). Although the real consumption of electric vehicles in practice is higher than according to manufacturer's instructions, the potential for reducing emissions is very high. This applies even when the energy would originate from fossil fuels. At the same time it is pointed out that the increasing use of electric cars would not radically increase electricity demand (Nischler et al. 2011).

More central questions relate to the geographic distribution of charging stations as well as to the impact of charging strategies on the utilization and stability of the electricity network (Hartmann and Özdemir 2011). Without a smart control of the charging processes demand peaks might be generated at points in time when there already exists a high demand for energy originating from other applications. A solution could be to use intelligent charging to move the load peaks to different time windows, e.g. those with a high feed-in of wind energy (Pehnt et al. 2007, Schill 2010a).

Another issue is the analysis of the efficiency of the drive train. Here the focus is on the development of efficient vehicle systems, which allow a good driving performance with low energy and resource usage (Estima and Marques Cardoso 2012). In this context the energy consumption under different environmental conditions is also studied (Hwang and Chang 2012). The energy efficiency in terms of driving patterns should be measured from the provision of primary energy up to the wheel so called 'well to wheel' contemplation (Raykin et al. 2012, van Wee et al. 2012). In addition to the vehicle, the entire infrastructure for fueling and maintenance of vehicles could also be analyzed in terms of resource consumptions. It can be shown that the infrastructure for electric vehicles is more carbon and energy intensive than for diesel and petrol vehicles (Lucas et al. 2012).

Some open issues are the investigation of mobility patterns and climate effects in terms of using electric vehicles. That means how and to which extent the use of electric cars really substitutes other motorized individual mobility or whether it rather leads to additional traffic. It should also be considered how the **resource requirements** develop with regard to particular **(rare) commodities** and which environmental impacts result from a rising demand. Another often neglected field deals with the recycling of components, vehicles or infrastructures. Resource-efficiency or the prevention of supply constraints make it necessary to develop **strategies for recycling systems**.

8 Politic Conditions

The promotion of electric mobility by policy makers is not without controversy (Arnold et al. 2010, Dudenhöffer et al. 2012, Indra 2012, Krutilla and Graham 2012, Schill 2010a, b, Yang 2010). To support the development and the launch of electric cars, there are different approaches. Current projects e.g. in Germany aimed at the development and demonstration of technology, infrastructure and business models (Brauner et al. 2012, Knoppe 2012a, b). This overview also includes other measures such as financial support, the provision of information and the granting of privileges.

There is a lot of controversy whether and how financial support of electric vehicles should take place. Policy might subsidize the purchase price, support market penetration, support the market launch and increase production to achieve economies of scale and by public procurement (Dosch 2012, Lieven et al. 2011). The discussion focuses primarily on the promotion of sustainable mobility systems, which includes motorized vehicles, electric bikes and cars and also heavy commercial vehicles (Buller et al. 2009). The importance of electric vehicles for climate protection is often highlighted (Ahrens et al. 2011, Gnann and Plötz 2011, Indra 2012). It is also emphasized repeatedly that the automotive industry in many countries is an important sector of the economy and the public assistance shall be seen as a stimulus program to help an injured industry to recover from the financial and economic crisis (Baur et al. 2012, Canzler and Knie 2010).

The financial assistance can take different forms. First, purchase subsidies are granted to reduce the purchase price the customer pays for an electric car in order to narrow the price gap between conventional and electric vehicles and hence to make electric cars more attractive for customers (Kley et al. 2010). While in Germany no such subsidies are granted at the moment, they already exist in many other countries. But the approaches differ considerably between the countries. While for example in the U.S. direct grants are awarded, the Japanese government supports car buyers by paying parts of the price gap between electric and conventional cars. China mainly supports electric mobility while other alternatives are limited (Yang 2010).

In addition to direct funding, there are also indirect incentives. In the discussion are measures such as a waiver of taxes, which incurred on the purchase of an electric car in the respective States (e.g. sales tax). Furthermore the users of electric vehicles could be supported by a reduction of the motor vehicle tax. Taxes could also be designed for a differential promotion of electricity and gasoline or diesel to achieve savings by driving electric vehicles compared to internal combustion engines (Gallagher and Muehlegger 2011). Of course this can lead to a decline in the revenue from oil in the future (Indra 2012). Depending on the budget situation states maybe faced to develop new sources of income. From a financial view other measures are the promotion by low-interest loans or improved possibilities of tax depreciation.

Because it is important for the market introduction that customers are willing to buy the products, they must be supplied with information in advance. Measures which are affecting the attitude and the willingness to buy are the coverage in everyday media (e.g. by the provision of information on television or online portals) or the own driving experience of electric cars (e.g. during open house presentations or as a test driver within government-funded research projects in the model regions). It is central that the new technology is "experienced" and thus directly affects the perception of the consumer (Dütschke et al. 2012). The public sector has also the possibility to directly affect market demand by public procurement and equip public carpools with electro mobiles, for example busses for public transport or vehicles for waste management (Rudolph 2012). The increase in consumer acceptance can further be achieved by exhibition performances, where visitors can collect information on electric vehicles and the necessary infrastructure.

In addition to the financial aspects of the promotion there are also non-monetary options of support. These include the joint use of bus or taxi lanes or parking facilities with charging infrastructure. Furthermore, the toll exemption or the driving in green zones are considered as incentives to use or buy electric cars (Sammer 2012). Specifically in city centers the possibility to access pedestrian zones with electric delivery vehicles at night could represent an important sales stimulus for logistic firms (Clausen and Schaumann 2012). For all this a marker for the distinction of electric vehicles would be necessary.

Political measures can also address the transparency for customers, for example the different energy consumption of drive types are made comparable. Hence, in electric battery-powered vehicles the indication of consumption can be made in "kWh/km" or in kilowatt-hours per mile. However, it is important that the customer can compare different parameters between battery-powered, hybrid and conventional vehicles (Chlond et al. 2012, Contestabile et al. 2011).

In summary it can be said that a variety of monetary and non-monetary conveying opportunities exist or are discussed to support the introduction of electric vehicles. However, it is still unresolved how the concrete funding scenery shall look like, especially in Germany, so that other alternative drives are discriminated or also claim special rights. Kley, Wietschel and Dallinger analyze also the effectiveness, efficiency, flexibility and practical political acceptance of electric mobility funding in Europe and come to the conclusion that the **efficiency of the funding must be increased**. In addition, **non-monetary incentives** such as free parking help to overcome existing technical and economic barriers (Kley et al. 2010). In the future it has to be clarified whether and how the sale of electric vehicles should be supported, where the money to support the market launch can come from, which **reasons for or against** a government support exist and how the efficiency of the subsidy programs can be measured. Last but not least it remains to be investigated how **government objectives** can be achieved and whether they result in **ecological and macroeconomic benefits** (Raich et al. 2012).

9 Conclusions

Due to the dynamic development in recent years, it is difficult to get an overall perspective on the development of electric mobility. This dynamic can be illustrated by the google search terms and search results. To receive a common understanding, the aim of this article was to get some insights in the influential socio-economic fields in the area of electric mobility and identify those fields which are currently under investigation. The research questions which were addressed by this contribution were the following: What are important socio-economic fields in the area of electric mobility? Which state-of-the-art findings can be identified? The short answer to the last question are as follows:

Management perspectives: Managements should consider different future mobility options such as electric mobility, biogas or fuel cell and their complementarities. Companies are challenged with a magnitude of consequences such as changes of value chains or radical innovations changing the market drastically. Strategies should also aim at the identification of new capabilities and their timely integration.

Usability: It shows that the user must gain experience with the electric vehicles so that perceived negative aspects like price or range can be opposed by some positive experiences of everyday usability. The user friendliness can also be evaluated as positive especially with regard to new business models and services like car sharing.

Standardization: For a successful market launch international standards, e.g. in terms of structure and protection, the charging plug, the communication infrastructure between the vehicles, power stations, infrastructures and network operators, are essential. By standardization electric vehicles must be put into a position comparable to internal combustion engines.

Economic conditions: The automotive industry is facing a radical turn. The value chain architecture changes due to the development of sustainable mobility systems. Indeed some incumbent companies possess important technologies, but they are challenged especially by new competitors from other countries and sectors. From the perspective of the user the purchasing and utilization costs must be transparent and comparable to other mobility systems or conventional drives.

Social conditions: Electric vehicles have great potential from a user perspective. Particularly users who drove electric vehicles are already excited about the new technology. But due to existing weaknesses in comparison to competitive technologies, e.g. with regard to the range, it should be analyzed how the customers will accept new mobility concepts, which combine electric vehicles with other services. Furthermore, psychological barriers must break up, so that people reconsider their mobility behavior.

Ecology conditions: The environmental considerations of electric mobility target primarily the eco-balance and the question where the electricity comes from or should come from. Challenges exist in terms of resource requirements and by the environmental impacts generated by a rising demand for electric vehicles.

Political conditions: There exists a variety of monetary and non-monetary opportunities to support electric mobility. The efficiency of the funding programs is currently not always guaranteed. Moreover, it is clearly important which causes are for or against a government funding and where the money might come from.

Consequently, further research requirements result mainly from the comprehensive issue of the research. It is first necessary to analyze separate aspects in more detail to get deeper insights in single challenges. Further it may be advantageous to **monitor** some **topics** like business models which are subject to a constant change because of **persistent technology changes**. For such a monitoring the initially introduced techniques for searching catchwords in the internet could be a good approach. Therefore, the research landscape could benefit from further research in all presented topics. This is also affected by ongoing dynamic technological developments. Many **research questions** and the gained insights **need to be adjusted to new progresses** e.g. in battery technology or the introduction of new service or business models. On the other hand it is also shown that it is probably still a long way for the electric mobility to have a **significant market share**. On the economic side companies may have to pay attention that the foundation of their success of tomorrow has to be laid today. That means that the developments must be followed precisely to compete with old and new rivals and with focus of **core competencies** or core products it is very important that necessary skills are developed in time if electric mobility leaves the niche.

Acknowledgments This article was supported by the federal program „Elektromobilität in Modelregionen". The Federal Ministry of Transport, Building and Urban Development (BMVBS) is providing a total of 130 million euros from the Second Stimulus Package. The program is coordinated by the NOW GmbH Nationale Organisation Wasserstoff- und Brennstoffzellentechnologie.

References

acatech (2010a) Grundzüge zukünftiger Mobilität Wie Deutschland zum Leitanbieter für Elektromobilität werden kann. Springer, Berlin, Heidelberg, pp 15–17
acatech (2010b) Standardisierung und Normung Wie Deutschland zum Leitanbieter für Elektromobilität werden kann. Springer, Berlin, Heidelberg, pp 27–29
acatech (2010c) Wertschöpfung Wie Deutschland zum Leitanbieter für Elektromobilität werden kann. Springer, Berlin, Heidelberg, pp 30–32
AG 4 Normung Standardisierung und Zertifizierung (2010) Die deutsche Normungs-Roadmap Elektromobilität. Version 1 from GGEMO http://www.elektromobilitaet.din.de/cmd?level=tpl-home&contextid=emobilitaet
AG 4 Normung Standardisierung und Zertifizierung (2012) Die deutsche Normungs-Roadmap Elektromobilität. Version 2 from GGEMO http://www.elektromobilitaet.din.de/cmd?level=tpl-home&contextid=emobilitaet
Ahrens G-A, Bäker B, Fricke H, Schlag B, Stephan A, Stopka U, Wieland B (2011) Zukunft von Mobilität und Verkehr. Technische Universität Dresden, Dresden
ANEC (2010) Silent but dangerous: when absence of noise of cars is a factor of risk for pedestrians. http://www.anec.eu/attachments/ANEC-DFA-2010-G-043final.pdf
Angerer G, Marscheider-Weidemann F, Wendl M, Wietschel M (2009) Lithium für Zukunftstechnologien: Nachfrage und Angebot unter besonderer Berücksichtigung der Elektromobilität. Fraunhofer ISI, Karlsruhe
Arnold H, Kuhnert F, Kurtz R, Bauer W (2010) Elektromobilität-Herausforderungen für Industrie und öffentliche Hand. Pricewaterhouse-Coopers AG, Fraunhofer-Institut für Arbeitswirtschaft und Organisation IAO, Stuttgart
Arnold H, Schäfer PK, Höhne K, Bier M (2012) Elektromobilität-Normen bringen die Zukunft in Fahrt PricewaterhouseCoopers AG
Avadikyan A, Llerena P (2010) A real options reasoning approach to hybrid vehicle investments. Technol Forecast Soc Change 77(4):649–661
Axsen J, Kurani KS (2012) Who can recharge a plug-in electric vehicle at home? Transp Res Part D: Transp Environ 17(5):349–353
Bai Q, Zhao S, Xu P (2012) Technology roadmap of electric vehicle industrialization. In: Jin D, Lin S (eds) Advances in computer science and information engineering, vol 169. Springer, Berlin, Heidelberg, pp 473–478
Baker E, Chon H, Keisler J (2010) Battery technology for electric and hybrid vehicles: expert views about prospects for advancement. Technol Forecast Soc Change 77(7):1139–1146
Bakker S, van Lente H, Engels R (2012) Competition in a technological niche: the cars of the future. Technol Anal Strateg Manage 24(5):421–434
Barkenbus J (2009) Our electric automotive future: CO_2 savings through a disruptive technology. Policy Soc 27(4):399–410

Baur F, Koch G, Prügl R, Malorny C (2012) Identifying future strategic options for the automotive industry. In: Proff H, Schönharting J, Schramm D, Ziegler J (eds) Zukünftige Entwicklungen in der Mobilität. Gabler Verlag, Wiesbaden, pp 273–286

Bechmann R, Scherk M (2010) Globalization in the Automotive Industry–Impact and Trends. In: Ijioui R, Emmerich H, Ceyp M, Hagen J (eds) Globalization 2.0. Springer, Berlin Heidelberg, pp 177–192

Bernhart W, Zollenkop M (2011) Geschäftsmodellwandel in der Automobilindustrie-Determinanten, zukünftige Optionen, Implikationen. In: Bieger T, zu Knyphausen-Aufseß D, Krys C (eds) Innovative Geschäftsmodelle. Springer, Berlin, Heidelberg, pp. 277–298

Betterplace (2012) Batteriewechselstationen. Retrieved 11 June 2012 http://deutschland.betterplace.com/the-solution-switch-stations

Biere D, Dallinger D, Wietschel M (2009) Ökonomische Analyse der Erstnutzer von Elektrofahrzeugen. Zeitschrift für Energiewirtschaft 33(2):173–181

Bingham C, Walsh C, Carroll S (2012) Impact of driving characteristics on electric vehicle energy consumption and range. IET Intell Transp Syst 6(1):29–35

Brady J, O'Mahony M (2011) Travel to work in Dublin. The potential impacts of electric vehicles on climate change and urban air quality. Transp Res Part D: Transp Environ 16(2):188–193

Brauner G (2008) Infrastrukturen der Elektromobilität. e i Elektrotechnik Informationstechnik 125(11):382–386

Brauner G, Geringer B, Schrödl M (2012) Elektromobilität III. e i Elektrotechnik Informationstechnik 129(3):107

Bremer W (2009) Normungsbedarf für alternative Antriebe und Elektrofahrzeuge. Deutsches Institut für Normung e.V, Berlin

Brown S, Pyke D, Steenhof P (2010) Electric vehicles: the role and importance of standards in an emerging market. Energy Policy 38(7):3797–3806

Buller U, Hanselka H, Dudenhöffer F, John EM, Weissenberger-Eibl MA (2009) Zukunftstechnologien: Förderung von Elektroautos-wie sinnvoll ist die Unterstützung einzelner Technologien? Ifo Schnelldienst 62(22):03–10

Bunting BG (2012) Recent trends in emerging transportation fuels and energy consumption. In: Subic A, Wellnitz J, Leary M, Koopmans L (eds) Sustainable automotive technologies 2012. Springer, Berlin, Heidelberg, pp 119–125

Canzler W (2010) Mobilitätskonzepte der Zukunft und Elektromobilität. In: Hüttl RF, Pischetsrieder B, Spath D (eds) Elektromobilität. Springer, Berlin, Heidelberg, pp 39–61

Canzler W, Knie A (2010) Grüne Wege aus der Autokrise. Vom Autobauer zum Mobilitätsdienstleister. Ökologie, Band 4. Retrieved 11 June 2012 http://www.boell.de/downloads/Autokrise_Endf%281%29.pdf

Chlond B, Kagerbauer M, Vortisch P (2012) Welche Anforderungen sollen Elektrofahrzeuge erfüllen? In: Proff H, Schönharting J, Schramm D, Ziegler J (eds) Zukünftige Entwicklungen in der Mobilität. Gabler Verlag, Wiesbaden, pp 445–454

Clausen U, Schaumann H (2012) Entwicklung eines Konzepts zur Innenstadtbelieferung mittels Elektromobilität. In: Proff H, Schönharting J, Schramm D, Ziegler J (eds) Zukünftige Entwicklungen in der Mobilität. Gabler Verlag, Wiesbaden, pp 467–478

Conrady R (2012) Status Quo and future prospects of sustainable mobility. In: Conrady R, Buck M (eds) Trends and issues in global tourism 2012. Springer, Berlin, Heidelberg, pp 237–260

Contestabile M, Offer GJ, Slade R, Jaeger F, Thoennes M (2011) Battery electric vehicles, hydrogen fuel cells and biofuels. Which will be the winner? Energy Environ Sci 4(10):3754–3772

Cooke P (2011) Transition regions: regional-national eco-innovation systems and strategies. Prog Plann 76:105–146

DeLuchi M, Wang Q, Sperling D (1989) Electric vehicles: performance, life-cycle costs, emissions, and recharging requirements. Transp Res Part A: Gen 23(3):255–278

Deutsche Bank Research (2011) Elektromobilität. Sinkende Kosten sind conditio sine qua non. Frankfurt a.M

Dijk M, Orsato RJ, Kemp R (2012) The emergence of an electric mobility trajectory. Energy Policy
DIN (2012) Wir über uns, from Retrieved 11 June 2012 http://www.din.de/cmd?level=tpl-artikel&cmsdintextid=impressum_de&bcrumblevel=1&languageid=de
Dong J, Lin Z (2012) Within-day recharge of plug-in hybrid electric vehicles: energy impact of public charging infrastructure. Transp Res Part D: Transp Environ 17(5):405–412
Dosch B (2012) Eco-mobility-Will new choices of electric drive vehicles change the way we travel? In: Conrady R, Buck M (eds) Trends and Issues in global tourism 2012. Springer, Berlin, Heidelberg, pp 261–270
Driscoll PA, Theodorsdottir AH, Richardson T, Mguni P (2012) Is the future of mobility electric? Learning from contested storylines of sustainable mobility in Iceland. Eur Plann Stud 20(4):627–639
Dudenhöffer F (2010) Batteriespitzentechnologie für automobile Anwendungen und ihr Wertschöpfungspotential für Europa. Ifo Schnelldienst 63(11):19–27
Dudenhöffer F, Bussmann L, Dudenhöffer K (2012) Elektromobilität braucht intelligente Förderung. Wirtschaftsdienst 92(4):274–279
Dütschke E, Schneider U, Sauer A, Wietschel M, Hoffmann J, Domke S (2012) Roadmap zur Kundenakzeptanz: Zentrale Ergebnisse der sozialwissenschaftlichen Begleitforschung in den Modellregionen Technology Roadmapping at Fraunhofer ISI: Concepts-Methods-Project examples, vol 3. Fraunhofer Institute for Systems and Innovation Research (ISI), Stuttgart
Estima JO, Marques Cardoso AJ (2012) Efficiency analysis of drive train topologies applied to electric/hybrid vehicles. IEEE Trans Veh Technol 61(3):1021–1031
Federal Goverment of Germany (2009) Nationaler Entwicklungsplan Elektromobilität. Berlin
Finn P, Fitzpatrick C, Connolly D (2012) Demand side management of electric car charging: benefits for consumer and grid. Energy 42(1):358–363
Follmer R, Gruschwitz D, Jesske B, Quandt S, Lenz B, Nobis C, Mehlin M (2010) Mobilität in Deutschland 2008. Kurzbericht Struktur-Aufkommen-Emissionen-Trends. infas-institut für angewandte gmbh, Deutsches Zentrum für Luft- und raumfahrt e.V.-Institut für Verkehrsforschung, Bonn, Berlin
Franke T, Neumann I, Buehler F, Cocron P, Krems JF (2012) Experiencing range in an electric vehicle: understanding psychological barriers. Appl Psychol Int Rev 61(3):368–391
Fraunhofer-IAO (2010a) Strukturstudie BWe mobil. Baden-Württemberg auf dem Weg in die Elektromobilität. Fraunhofer-IAO, Stuttgart
Fraunhofer-IAO (2010b) Systemanalyse BWe mobil. IKT- und Energieinfra struktur für innovative Mobilitätslösungen in Baden-Württemberg. Fraunhofer-IAO, Stuttgart
Gallagher KS, Muehlegger E (2011) Giving green to get green? Incentives and consumer adoption of hybrid vehicle technology. J Environ Econ Manage 61(1):1–15
Gallardo-Lozano J, Milanés-Montero MI, Guerrero-Martínez MA, Romero-Cadaval E (2012) Electric vehicle battery charger for smart grids. Electric Power Syst Res 90:18–29
Gnann T, Plötz P (2011) Status Quo und Perspektiven der Elektromobilität in Deutschland Sustainability and Innovation, vol S14. Fraunhofer Institute for Systems and Innovation Research (ISI), Stuttgart
Götz K, Sunderer G, Birzle-Harder B, Deffner J (2012) Attraktivität und Akzeptanz von Elektroautos. Ergebnisse aus dem Projekt OPTUM-Optimierung der Umweltentlastungspotenziale von Elektrofahrzeugen. In: GmbH Ifs-öFI (ed). Institut für sozial-ökologische Forschung (ISOE) GmbH, Frankfurt am Main
Gould J, Golob TF (1998) Clean air forever? A longitudinal analysis of opinions about air pollution and electric vehicles. Transp Res Part D: Transp Environ 3(3):157–169
Greenpeace (2010) Energy [r]evolution world energy scenario. Greenpeace International, Amsterdam, Brussels
Hanselka H, Jöckel M (2010) Elektromobilität-Elemente, Herausforderungen, Potenziale. In: Hüttl RF, Pischetsrieder B, Spath D (eds) Elektromobilität-Potenziale und wissenschaftlich-technische Herausforderungen. Springer, Berlin Heidelberg, pp 21–38

Hartmann N, Özdemir ED (2011) Impact of different utilization scenarios of electric vehicles on the German grid in 2030. J Power Sources 196(4):2311–2318

Hawkins T, Gausen O, Strømman A (2012) Environmental impacts of hybrid and electric vehicles: a review. Int J Life Cycle Assess 1–18

He L, Chen W, Conzelmann G (2012a) Impact of vehicle usage on consumer choice of hybrid electric vehicles. Transp Res Part D: Transp Environ 17(3):208–214

He Y, Chowdhury M, Ma Y, Pisu P (2012b) Merging mobility and energy vision with hybrid electric vehicles and vehicle infrastructure integration. Energy Policy 41:599–609

Heinrichs M, Hoffmann R, Reuter F (2012) Mobiles internet-Auswirkung auf Geschäftsmodelle und Wertkette der Automobilindustrie, am Beispiel MINI Connected. In: Proff H, Schönharting J, Schramm D, Ziegler J (eds) Zukünftige Entwicklungen in der Mobilität. Gabler Verlag, Wiesbaden, pp 611–628

Held M, Baumann M (2011) Assessment of the environmental impacts of electric vehicle concepts. In: Finkbeiner M (ed) Towards life cycle sustainability management. Springer, Netherlands, pp 535–546

Hidrue MK, Parsons GR, Kempton W, Gardner MP (2011) Willingness to pay for electric vehicles and their attributes. Resour Energy Econ 33(3):686–705

Hiwatari R, Ikeya T, Okano K (2012) A Design system for layout of charging infrastructure for electric vehicle. In: Matsumoto M, Umeda Y, Masui K, Fukushige S (eds) Design for innovative value towards a sustainable society. Springer, Netherlands, pp 1026–1031

Huelsmann M, Colmorn R (2011) Strategische Perspektiven der Elektromobilität. Einige Überlegungen für die Automobilindustrie. Retrieved 17 July 2012, from Regionale Projektleitstelle der Modellregion Elektromobilität Bremen/Oldenburg www.modellregion-bremen-olden-burg.de/fileadmin/CONTENT_MBO/PDFs/Vortraege/Prof._Dr._Michael_H%C3%BClsmann.pdf

Hugosson MB, Algers S (2012) Accelerated Introduction of 'Clean' Cars in Sweden Cars and Carbon. In: Zachariadis TI (ed). Springer, Netherlands, pp 247–268

Hwang JJ, Chang WR (2012) Characteristic study on fuel cell/battery hybrid power system on a light electric vehicle. J Power Sources 207:111–119

Indra F (2012) Womit bewegen wir unsere autos morgen tatsächlich? Zeitschrift für Herz-, Thorax- Gefäßchirurgie 26(2):137–140

Johnson MW, Suskewicz J (2009) How to jump-start the clean-tech economy. Harvard Bus Rev 87(11) 52

Johnson G, Scholes K, Whittington R (2011) Strategisches management-Eine Einführung: Analyse, Entscheidung und Umsetzung. Pearson Studium, Munich

Kasperk G, Drauz R (2012) Kooperationsstrategien von Automobilproduzenten entlang der sich neu ordnenden Wertschöpfungskette. In: Proff H, Schönharting J, Schramm D, Ziegler J (eds) Zukünftige Entwicklungen in der Mobilität. Gabler Verlag, Wiesbaden, pp 391–403

Kendall G (2008) Plugged in: the end of the oil era. WWF European Policy Office, Brussels

Kley F (2011) Neue Geschäftsmodelle zur Ladeinfrastruktur Sustainability and Innovation, vol S5. Fraunhofer Institute for Systems and Innovation Research (ISI), Stuttgart

Kley F, Wietschel M, Dallinger D (2010) Evaluation of European electric vehicle support schemes sustainability and innovation, vol S7. Fraunhofer Institute for Systems and Innovation Research (ISI), Stuttgart

Knoppe M (2012a) E-mobility generates new services and business models, increasing sustainability. In: Subic A, Wellnitz J, Leary M, Koopmans L (eds) Sustainable automotive technologies 2012. Springer, Berlin Heidelberg, pp 275–281

Knoppe M (2012b) E-mobility will change automotive retailing: A strategic approach. In: Subic A, Wellnitz J, Leary M, Koopmans L (eds) Sustainable automotive technologies 2012. Springer, Berlin, Heidelberg, pp 283–287

Krutilla K, Graham JD (2012) Are green vehicles worth the extra cost? The case of diesel-electric hybrid technology for urban delivery vehicles. J Policy Anal Manage 31(3):501–532

Kudoh Y, Ishitani H, Matsuhashi R, Yoshida Y, Morita K, Katsuki S, Kobayashi O (2001) Environmental evaluation of introducing electric vehicles using a dynamic traffic-flow model. Appl Energy 69(2):145–159

Kulkarni A, Kapoor A, Ektesabi M, Lovatt H (2012) Electric vehicle propulsion system design. In: Subic A, Wellnitz J, Leary M, Koopmans L (eds) Sustainable automotive technologies 2012. Springer, Berlin, Heidelberg, pp 199–206

Kushnir D, Sanden BA (2012) The time dimension and lithium resource constraints for electric vehicles. Resour Policy 37(1):93–103

Leitinger C, Litzlbauer M (2011) Netzintegration und Ladestrategien der Elektromobilität. e i Elektrotechnik Informationstechnik 128(1):10–15

Li Y (2007) Scenario-based analysis on the impacts of plug-in hybrid electric vehicles (PHEV) penetration into the transportation sector 2007 In: IEEE international symposium on technology and society, Technology and Society (ISTAS), pp 145–150

Lieven T, Mühlmeier S, Henkel S, Waller JF (2011) Who will buy electric cars? An empirical study in Germany. Transp Res Part D: Transp Environ 16(3):236–243

Link C, Sammer G, Stark J (2012) Abschätzung des Marktpotenzials und zukünftigen Marktanteils von Elektroautos. e i Elektrotechnik Informationstechnik 129(3):156–161

Lucas A, Silva CA, Costa Neto R (2012) Life cycle analysis of energy supply infrastructure for conventional and electric vehicles. Energy Policy 41:537–547

Lunz B, Yan Z, Gerschler JB, Sauer DU (2012) Influence of plug-in hybrid electric vehicle charging strategies on charging and battery degradation costs. Energy Policy 46:511–519

Nischler G, Gutschi C, Beermann M, Stigler H (2011) Auswirkungen von Elektromobilität auf das Energiesystem. e i Elektrotechnik Informationstechnik 128(1) 53–57

Orbach Y, Fruchter GE (2011) Forecasting sales and product evolution: the case of the hybrid/electric car. Technol Forecast Soc Change 78(7):1210–1226

Özdemir ED, Hartmann N (2012) Impact of electric range and fossil fuel price level on the economics of plug-in hybrid vehicles and greenhouse gas abatement costs. Energy Policy 46:185–192

Pearre NS, Kempton W, Guensler RL, Elango VV (2011) Electric vehicles: how much range is required for a day's driving? Transp Res Part C: Emerg Technol 19(6):1171–1184

Pehnt M, Höpfner U, Merten F (2007) Elektromobilität und erneuerbare Energien. Arbeitspapier nr 5. Retrieved from http://www.bmu.de/files/pdfs/allgemein/application/pdf/elektromobilitaet_ee_arbeitspapier.pdf

Pehnt M, Helms H, Lambrecht U, Dallinger D, Wietschel M, Heinrichs H, Behrens P (2011) Elektroautos in einer von erneuerbaren Energien geprägten Energiewirtschaft. Zeitschrift für Energiewirtschaft 35(3):221–234

Peters, W. (2010). Der schwere Weg zur E-Mobilität Frankfurter Allgemeine Zeitung

Pierre M, Jemelin C, Louvet N (2011) Driving an electric vehicle. A sociological analysis on pioneer users. Energ Effi 4(4):511–522

Pohl H (2012) Japanese automakers' approach to electric and hybrid electric vehicles: from incremental to radical innovation. Int J Technol Manage 57(4):266–288

Pohl H, Elmquist M (2010) Radical innovation in a small firm: a hybrid electric vehicle development project at Volvo Cars. R&D Manage 40(4):372–382

PwC AG (2012) Elektromobilität Normen bringen die Zukunft in Fahrt. Berlin

Raich U, Sammer G, Stark J (2012) Gesamtwirtschaftliche Bewertung von Elektromobilität. e i Elektrotechnik Informationstechnik 129(3):162–166

Ramezani M, Graf M, Vogt H (2011) A simulation environment for smart charging of electric vehicles using a multi-objective evolutionary algorithm. In: Kranzlmüller D, Toja A (eds) Information and communication on technology for the fight against global warming, vol 6868. Springer, Berlin, Heidelberg, pp 56–63

Raykin L, MacLean HL, Roorda MJ (2012) Implications of driving patterns on well-to-wheel performance of plug-in hybrid electric vehicles. Environ Sci Technol 46(11):6363–6370

Reichert C, Reimann K, Lohr J (2012) Elektromobilität: Antworten auf die fünf entscheidenden Fragen. In: Servatius H-G, Schneidewind U, Rohlfing D (eds) Smart energy. Springer, Berlin, Heidelberg, pp 453–461

Richter T, Schreiber A, Schreiber M (2012) Planung eines Ladeinfrastrukturnetzes für Elektrofahrzeuge in Berlin. In: Proff H, Schönharting J, Schramm D, Ziegler J (eds) Zukünftige Entwicklungen in der Mobilität. Gabler Verlag, Wiesbaden, pp 549–561

Rudolph C (2012) Die Rolle der Kommunen bei Marktdurchdringungsszenarien für Elektromobilität. In: Proff H, Schönharting J, Schramm D, Ziegler J (eds) Zukünftige Entwicklungen in der Mobilität. Gabler Verlag, Wiesbaden, pp 81–89

Sammer G (2012) Wirkungen und Risiken einer City-Maut als zentrale Säule eines städtischen Mobilitätskonzepts. In: Proff H, Schönharting J, Schramm D, Ziegler J (eds) Zukünftige Entwicklungen in der Mobilität. Gabler Verlag, Wiesbaden, pp 479–491

Sammer G, Meth D, Gruber CJ (2008) Elektromobilität-Die Sicht der Nutzer. e i Elektrotechnik Informationstechnik 125(11):393–400

Sammer G, Stark J, Link C (2011) Einflussfaktoren auf die Nachfrage nach Elektroautos. e i Elektrotechnik Informationstechnik 128(1):22–27

Schill W-P (2010a) Electric vehicles: charging into the future. Weekly Report (27) 207–214

Schill W-P (2010b) Elektromobilität in Deutschland: chancen, Barrieren und Auswirkungen auf das Elektrizitätssystem. Vierteljahrshefte zur Wirtschaftsforschung/Q J Econ Res 79(2): 139–159

Schill W-P (2010c) Elektromobilität: kurzfristigen Aktionismus vermeiden, langfristige Chancen nutzen. Wochenbericht 77(27/28)

Schlick T, Hertel G, Hagemann B, Maiser E, Kramer M (2011) Zukunftsfeld Elektromobilität. Chancen und Herausforderungen für den deutschen Maschinenund Anlagenbau. Roland Berger Strategy Consultants, Düsseldorf, Hamburg, Frankfurt

Schroeder A, Traber T (2012) The economics of fast charging infrastructure for electric vehicles. Energy Policy 43:136–144

Schuster A, Leitinger C (2011) Fahrzeug- und Lademonitoring der ersten Generation von Elektromobilen in der Modellregion Vorarlberg. e i Elektrotechnik Informationstechnik 128(1) 2–9

Shinnar R (2003) The hydrogen economy, fuel cells, and electric cars. Technol Soc 25(4): 455–476

Thoma B, O'Sullivan D (2011) Study on Chinese and European automotive R&D-comparison of low cost innovation versus system innovation. In: Chiu ASFTJMLWGKJ (ed) International conference on Asia Pacific business innovation and technology management, vol 25

Ustaoğlu M, Yıldız B (2012) Innovative green technology in Turkey: electric vehicles' future and forecasting market share. Procedia Soc Behav Sci 41:139–146

van Wee B, Maat K, De Bont C (2012) Improving sustainability in urban areas: discussing the potential for transforming conventional car-based travel into electric mobility. Eur Plann Stud 20(1):95–110

VDE (2010a) E-mobility 2020. Frankfurt VDE Verband der Elektrotechnik Elektronik Informationstechnik e.V

VDE (2010b) Elektrofahrzeuge-Bedeutung, Stand der Technik, Handlungsbedarf. Energietechnischen Gesellschaft (ETG) des VDE Verband der Elektrotechnik Elektronik Informationstechnik e.V, Frankfurt

Wang Z, Liu P, Han H, Lu C, Xin T (2011) A distribution model of electric vehicle charging station. In: Chen R (ed) Frontiers of manufacturing and design science, Pts 1–4, vol 44–47. pp 1543–1548)

Wang Z, Wang L, Dounis AI, Yang R (2012) Integration of plug-in hybrid electric vehicles into energy and comfort management for smart building. Energy Buildings 47:260–266

Werther B, Hoch N (2012) E-mobility as a challenge for new ICT solutions in the car industry. In: Bruni R, Sassone V (eds) Trustworthy global computing, vol 7173. Springer, Berlin, Heidelberg, pp 46–57

Yabe K, Shinoda Y, Seki T, Tanaka H, Akisawa A (2012) Market penetration speed and effects on CO_2 reduction of electric vehicles and plug-in hybrid electric vehicles in Japan. Energy Policy 45:529–540

Yang C-J (2010) Launching strategy for electric vehicles: lessons from China and Taiwan. Technol Forecast Soc Change 77(5):831–834

Zhang H, Shi Y (2011) Understanding cultural impacts upon the development of business ecosystems in the electric vehicle industry

Zhang Y, Yu Y, Zou B (2011) Analyzing public awareness and acceptance of alternative fuel vehicles in China: the case of EV. Energy Policy 39(11):7015–7024

Zimmer R, Rammler S (2011) Leitbilder und Zukunftskonzepte der Elektromobilität. Unabhängiges Institut für Umweltfragen e.V., Institut für Transportation Design der Hochschule für Bildende Künste, Berlin, Braunschweig

Author Biographies

Christian Hanke is member of the research staff at the Institute for Project Management and Innovation (IPMI) at the University of Bremen and was also research associate at the workgroup "System Management" at the School of Engineering and Science at Jacobs University Bremen. His research interests are in the field of innovation, strategic management, future studies and urban and regional development in the framework of mobility and energy systems.

Michael Hülsmann holds the chair of "System Management" at the School of Engineering and Science at Jacobs University Bremen. He focuses on Strategic Management of Logistics Systems. Additionally he leads the sub-projects "Business model concept/product idea" and "Sustainable Innovation and technology strategies" in the framework of the electric mobility model region Bremen/Oldenburg.

Dirk Fornahl leads the Centre for Regional and Innovation Economics (CRIE) at the University of Bremen. His work is settled on cluster analysis, analysis of regional supply chains and potential, innovation processes and networks. He also coordinates the socio-economic research in the electric mobility model region Bremen/Oldenburg.

Mobility Scenarios for the Year 2030: Implications for Individual Electric Mobility

Tobias Kuhnimhof, Irene Feige and Peter Phleps

Abstract As other mobility innovations individual electric mobility will presumably have a significant market share only in decades from now. Scenario planning technique provides a framework for exploring the mobility market of the future with a long term perspective. Against this background the chapter at hand has a twofold objective: (a) It promotes scenario planning as an alternative approach to explore the distant future of mobility to establish market conditions. (b) The chapter presents results from a scenario planning study for Germany as an example. Three scenarios for the future of mobility emerged from an elaborate scenario planning process: (1) Matured Progress, (2) Global Take-Off, (3) Frantic Standstill. In all three scenarios, the German population ages significantly, concentrates in urban areas and is increasingly aware of environmental issues with consequences for consumer choice. The scenarios differ specifically regarding immigration to Germany, the global and national economies, and the farsightedness of political decision making. Conforming to expectation, these heterogeneous external developments lead to different projections for future travel. Hence, the presented study can assist in identifying future developments that are very likely while others are subject to specific conditional processes. This helps explore the future market for electric mobility among other mobility innovations.

T. Kuhnimhof (✉) · I. Feige · P. Phleps
Institute for Mobility Research (ifmo), Petuelring 130, 80788 Munich, Germany
e-mail: Tobias.Kuhnimhof@ifmo.de

I. Feige
e-mail: Irene.Feige@ifmo.de

P. Phleps
e-mail: Peter.Phleps@ifmo.de

1 Introduction

For decades the development of travel in Germany—as in other industrialized countries–was largely characterized by growth (BMVBS 2011). This was mainly driven by increasing levels of car ownership and car travel (The World Bank 2010). During these decades of growth, policy and planning as well as suppliers of transport services and products primarily were faced with increasing demand. Against this background existing approaches for forecasting and appraisal are specifically tailored to situations of overall growth.

After the aging of the German population has set in decades ago, the overall population has started to decline in 2003. Moreover, as regards everyday passenger travel there is no significant growth of per capita travel anymore since the late 1990s (Zumkeller et al. 2004; infas and DLR 2010). There are multilayer reasons for these trend reversals that include the economic development as well as the country's changing demography (FGSV 2006).

In this situation a look into the future of mobility (here: the physical movement of persons and goods) is much more characterized by uncertainty and heterogeneity than in the past. Growth and recession occur at the same time in different places and different segments of travel. Also, the extension of the infrastructure has slowed down.

Nevertheless, public and private stakeholders still have to make investment decisions which need to be informed by projections about future developments of mobility. However, in an environment of increasing uncertainty and heterogeneity the requirements for making informed decisions change. In such an environment robustness of decisions in the sense that they withstand a wide range of possible scenarios is increasingly important. This raises the question if conventional approaches of modeling and forecasting need to be supplemented by foresight instruments which allow for a broader perspective when drafting policy measures or appraising them.

In this chapter we present a possible alternative view to the future. In doing so, the chapter has a twofold objective: On the one hand, the chapter presents the findings from the latest up-date of the ifmo (Institute for Mobility Research) "Future of Mobility" scenario series for Germany. On the other hand and in a much broader sense the chapter presents scenario planning as an alternative way of thinking about the future of mobility in order to support robust decisions. Hence, the chapter is also about why and how scenario planning methods have an additional value when envisioning the future of mobility.

The chapter is structured as follows: At the outset, we discuss the motivation to go beyond conventional model-based projections. Thereafter we briefly outline the methodological basis of scenario planning in order to acquaint readers who are not familiar with scenario planning methods with the approach. Subsequently, the chapter presents the three resulting scenarios from a scenario study for Germany. The chapter discusses important findings from the study, highlighting insights which would not have been obtained from model-based projections.

2 Why Go Beyond Model Based Projections?

Model based projections pretend an exactness which they are in principle not able to achieve. First, most models–specifically those that are composed of regression models–produce point projections even though they are based on stochastic data. In statistical modeling there is fair consideration of error, however, this usually gets lost when implementing the models. Error propagation through transport models is an important issue discussed among academia but this has very little impact on model applications in practice (De Rham et al. 2006; Cools et al. 2011).

Second, most models are not even designed to predict reality but to inform about certain aspects of reality in hypothetical situations. This can be best exemplified with equilibrium seeking models which represent a significant share of models used for projections: While it is acknowledged that equilibria, e.g. in transportation networks or housing markets, hardly ever exists in reality, equilibration is an important criterion in order to produce comparable model results (de Dios Ortúzar and Willumsen 2011).

Third, models tend to build on relationships that were observed in the past and extrapolate these—even if doing so in a sophisticated manner–into the future. However, relationships which were observed in the past might not exist in the future. We might even anticipate that historic relationships change in the future– e.g. fuel price elasticity after specific benchmark values are exceeded. Most models would still fail to incorporate such anticipation (Kutter 2003).

In an environment where planning or corporate strategies do not solely administer growth but increasingly have to deal with uncertainty, new requirements for projections arise: There is rising awareness that alternative projects should be evaluated as to how they perform under different scenario conditions. This is the first step into thinking about different possible futures which represents the basis for scenario planning (Gausemeier et al. 1995).

3 Outline of Scenario Planning Methodology

The future is unknown. Scenarios support "what-if"-thinking and lay the foundation for alternative strategies to reduce uncertainties in mid- and long-term planning. They are applied in corporate, product or technology development to enhance robust and sustainable decision making for complex and long-term issues (Gausemeier et al. 1995; Fink and Siebe 2006). Two main characteristics have decisively supported the dissemination of scenario techniques: a cross-linked thinking and simultaneously multiple-future approach (Gausemeier et al. 1995).

In the case of the future development of mobility, scenario techniques help to identify driving forces or influential factors which today have important impact on how mobility develops (Shell 2008). In many cases there is pre-existing knowledge or plausible anticipation how these driving forces interrelate with specific dimensions of mobility, e.g. the relation of fuel prices and car use.

From today's perspective, the future development of some of these influential factors appears to be quite certain such as the natural demographic development of a country. Others are less certain such as migration or the economic development. For such factors we have to consider multiple possible developments or projections when imagining the future (Gausemeier et al. 1995).

Finally, there are plausible combinations of different projections for relevant influential factors. For example, extensive immigration is more plausible if a prospering economic environment generates demand for labor force while the combination of high immigration and economic despair appears less likely (ifmo 2010).

Such knowledge about the mutual interrelationship of relevant influential factors, the certainty of their future development, and their impact on mobility essentially represents the toolkit for a scenario planning study on mobility (Schoemaker 1995; Yuva 2011). In the case of our study we followed a standard scenario approach (Fig. 1) relying on knowledge from experts which is gathered and consolidated in expert workshops.

Our scenario study builds on 90 influential factors ranging from the demographic development to outsourcing strategies in logistics (ifmo 2005). These 90 factors are categorized into ten domains (Society, Economy, Environment and Policy, Technology, Water Transport, Air Transport, Rail Transport, Road Transport, Urban Transport, Logistics). Each of these domains is dealt with in a two-day workshop. In these workshops experts for the respective field develop up to three alternative projections for each influential factor that the domain

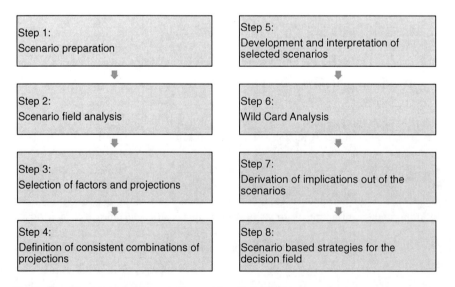

Fig. 1 Scenario approach (compare Gausemeier et al. 1995)

comprises of. Overall, 85 multidisciplinary experts contributed to the development of these single factor projections.

In a next step, likely combinations of different single factor projections have to be identified, respectively projections which appear to be rather mutually exclusive. This again is subject of expert workshops which are contributed to establishing the consistency of alternative projections for the different influential factors by filling in a consistency matrix.

Subsequent tool based analysis of this matrix identifies combinations of projections which are characterized by a high degree of consistency. In our study we used the INKA 3 software to identify the scenarios (Geschka and Partner 2012). Taken together, such combinations of alternative projections represent a scenario framework which features a high degree of consistency. From this set of consistent scenarios specific clusters are selected for further elaboration, which fulfill the following requirements: First, the degree of consistency within each scenario needs to be high, i.e. single scenarios do not contain contradictory developments of different influential factors. Second, the different selected scenario frameworks should cover the range of different imaginable future developments as broadly as possible.

Each scenario framework is described within stories covering the most relevant developments und interconnections in the regarded scenario field. Additional Wild Card analysis can be performed to cope with low-probability but high-impact events before specific implications can be derived from the scenarios to prepare robust decisions.

In summary, our scenario planning approach can be thought of as a structured voting of experts on likely future developments based on their longstanding expertise in the field. As a result it leads to consistent stories of what the future could be like and draws possible pathways to these future scenarios. These pathways inform us which decisions or events are likely to be influential for the future development of mobility.

4 The Future of Mobility—Three Scenarios for Germany

The following three scenarios imagine possible futures of mobility in Germany in 2030 (ifmo 2010). When presenting them in the following we fast-forward to the year 2030—and look back to the two decades after 2010 in retrospect. In doing so, the scenarios not only depict the situation in 2030 but also tell the story how this situation has emerged from the 2010 status-quo. For each scenario the key developments are presented in Figs. 2, 3, 4.

Fig. 2 Key developments in the matured progress scenario

MATURED PROGRESS

- Population decrease and aging
- Moderat economic development
- Export remains driver for economy
- Education policy averts social mismatch

Fig. 3 Key developments in the global take off scenario

GLOBAL TAKE OFF

- Immigration extenuates population decrease
- Prospering economic development
- Booming exports
- Reforms within the education, social and labor sector

Fig. 4 Key developments in the frantic standstill scenario

FRANTIC STANDSTILL

- Population decrease and aging
- Global crises slowdown economic development
- Exports stagnating
- Political focus on crisis abatement

4.1 Scenario 1: Matured Progress

Compared to 2010, the population has shrunk by 5 million and has aged significantly. There has been moderate economic development during the past twenty years with an average GDP growth of 0.7 % per year (1 % per capita per year) which–in light of the demographic decline—is still remarkable. The key prerequisite for this development was a significantly better exploitation of the labor force potential in Germany, mainly by raising the retirement age and increasing the employment rate of women. The social gap has not widened further, and social mobility has been maintained, mostly thanks to investments in education. The share of skilled workers in the labor force has increased.

Germany continues to be highly integrated into the global economy. Compared to the 1990s and 2000s, there has been slower growth in foreign trade, but the development of the foreign trade volume, with an average of 3.25 % per year (in values), is still significant. Export trade continues to be the key driver of economic growth in Germany.

During the last two decades, politics have focused on social and economic key challenges, not just in education and social policy, but also in other sectors. Transport policy, too, has concentrated on a clear strategy, and coordination between transport-policy actors at the federal, state and municipal levels has improved. Investments in transport infrastructure, which could essentially be kept at a constant level, have been characterized by a close integration of economic policy and transport policy objectives. Across all modes of transport, most investments are made to maintain and locally develop the infrastructure in highly frequented areas of the transport network. This includes the infrastructure in the booming metropolitan areas but also the important cross-border rail services, the German sea and inland ports, and air traffic hubs.

While freight traffic volume has further increased across all modes of transport (plus 20 % during the past two decades), passenger kilometers traveled have decreased by 5 % compared to 2009. In 2030, the car continues to be the main mode of transport. The significance of owning and using a car has not much changed since 2010.

The high oil price—averaging $200 per barrel throughout the last twenty years—has had a dampening effect on the development of both motorized individual transport and passenger air traffic. In combination with a marked increase in emissions-related taxes, flying has become significantly more expensive. Despite further growth in international business travel, the growth rates in air traffic have been nowhere near the growth rates of the 2000s. The slump in demand was especially marked in the price-sensitive customer segment. As for motorized individual transport, rising costs due to user fees, levies and taxes had an additional dampening effect on the traffic volume. In sum, there is a slight shift from motorized individual transport towards public transport.

4.2 Scenario 2: Global Take Off

The global economy has experienced a long, dynamic period of growth during the past two decades. Germany was able to consolidate—in some areas even improve—its competitive position internationally. The German export economy is strongly integrated into global value-generating networks based on a division of labor. The export trade thus was a key driver of economic growth, which averaged 1.5 % per year during the last two decades. Averaging 5.5 % per year (in values), the growth in German foreign trade has turned out to be significant.

Because of high immigration, the German population has declined only slightly by 1.4 million to 80.2 million people. All in all, the social gap has not widened. Social and education policy reforms, labor market policy measures but also transport policy measures were of crucial importance for these positive developments. Growing cities and metropolitan areas in particular were able to benefit from these favorable conditions, while the disparities between them and structurally weak rural areas have increased. Due to education policy measures especially, large parts of the population were able to benefit from better education opportunities and thus better employment opportunities. The percentage of skilled workers in the labor force has increased.

The growing industrial division of labor, the booming foreign trade, the positive economic development, and the concomitant increase in the mobility budgets of private households led to growing transport volumes specifically in freight traffic (plus 50 %). Passenger travel has also increased, however, compared to previous decades the 4 % increase during the period under observation constitutes an enormous slow-down of growth. This development suggests a long-term stagnation in passenger traffic volume.

In general, the mobility behavior in society has changed markedly and has become significantly more pragmatic. In many parts of the population, the mobility behavior, which was previously very much defined by routine, has become considerably more flexible when it comes to choosing a mode of transport. Depending on travel costs, travel time, comfort and lifestyle, the best mode of transport for a particular route is chosen.

Public transport as a whole was able to benefit from the concentration of the population in the booming cities and metropolitan areas, while the share of motorized individual transport in the modal split has slightly decreased. The number of passengers has increased in both public transport and high-speed rail traffic between the big cities. Because of the overall positive economic development, air traffic, too, has continued to grow.

4.3 Scenario 3: Frantic Standstill

The past two decades were defined by major business cycle fluctuations, with alternating periods of recession and growth. This was due to several global, crisis-laden developments whose effects reinforced one another. The consequences of another financial and economic crisis and an oil supply shock—successive individual events–, were enormous and manifold. Moreover, the effects of climate change have become more and more noticeable in Germany.

Germany was affected both directly and indirectly from the impact of these temporary crises and continuous processes. The political actors were often powerless in the face of these developments, but they always responded with measures. However, these measures often had a short-term effect only and in many cases turned out to be largely ineffective in the long run. The options available to the political actors have become increasingly limited, as long-term developments, e.g. the continually growing national debt as a result of the government's crisis intervention as well as demographic changes, place a heavy burden on public budgets.

Taken together, these developments caused the economic output in Germany to stagnate over the past two decades. For Germany, it has been the longest sustained period without economic growth since the end of World War II. With a growth rate of 1.25 % (in values), growth in foreign trade is also rather slow. This is largely due to the fact that the global economy is lacking momentum.

Because of economic and demographic developments, there has been a sharp decline in population (−5 million) as well as in the number of gainfully employed people. The increasing population concentration in urban and metropolitan areas exacerbated the consequences for structurally weak rural areas and resulted in growing regional inequality in economic performance. The social gap has widened, mostly because of insufficient investments in education and integration.

While freight traffic stagnated due to the economic stagnation, the passenger kilometers traveled decreased by 8 %. There was an especially sharp decline in motorized individual transport. The high oil price of $200 per barrel as well as slightly decreasing household incomes reinforced a cost-oriented, pragmatic mobility behavior that had been on the rise already. Here, shrinking household consumer budgets and environmental arguments were also prominent factors. Public transport and rail transport experienced a significant increase in their modal share, whereas air traffic's modal share declined; in part, this is also due to tougher regulations for climate protection and a concomitant rise in costs.

4.4 Discussion of Scenarios

The presented scenarios are story lines about how the future might look like and how evolutionary paths of development might lead to this distant future. Figure 5 depicts the projected development of passenger travel in the three scenarios.

Table 1 shows the relative development of mode shares in the three scenarios. It are predominantly varying economic and regulatory conditions which cause long distance and consequently air travel to diverge in the three scenarios. Mode choice in everyday travel between the car and public transport on the other hand is subject to the prevailing mobility culture. This is most car oriented in the case of the

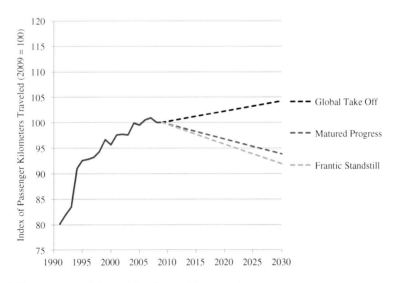

Fig. 5 Passenger travel demand development by scenario

Table 1 Mode share development by scenario

Mode \ Scenario	Matured Progress	Global Take Off	Frantic Stand still
Car	→	↘	↘
Public Transport	→	↗	↗
Air	→	↗	↘

matured progress scenario. In the Frantic Standstill scenario it are predominantly economic conditions which induce travelers to reduce their automobile travel. In the Global Take Off scenario, it are inspiring stimuli from the vibrant and innovative economic environment which open travelers' minds for a new and less car oriented mobility culture.

The question arises which implications can be derived from these scenarios for making decisions about the future, for example with regard to new mobility patterns which represent the mobility market for individual electric mobility. At the outset it makes sense to identify characteristics which all three scenarios have in common as these are developments with a high likelihood of occurrence. Increasing concentration of population in urban agglomerations is one common denominator. Moreover, the organization of everyday life is characterized by increased time efficiency in all scenarios. Also increasing environmental awareness with consequences for consumer choice is a common characteristic. In all three scenarios, public transport evolves as an important instrument for pursuing environmental, transportation, economical, and land use objectives. An increase of car sharing and cycling is also likely under all three scenario conditions. Digging deeper into the scenarios reveals the impact of certain regulations on the future of mobility, such as the introduction of road pricing on motorways in Germany which is foreseen in all scenarios.

Overall, these conditions which represent common denominators in all three scenarios are relatively favorable with regard to new mobility services and the adoption of electric mobility. However, it is also important to note that scenarios do not differ much with regard to the proliferation of specific automobile vehicle technology. The share of electricity as a source of energy for road transport is between three and 10 % in the three scenarios (ifmo 2010). Hence, electric energy is foreseen to remain a niche source of energy for road transport while fossil and bio fuels continue to dominate. Moreover, in none of the three scenarios vehicle propulsion technology tips the scales for the future mobility to develop in any specific direction. In all three scenarios other driving forces such as social, economic, cultural or environmental conditions are more relevant for the development of personal mobility patterns. Hence, vehicle propulsion technology will most likely not induce new mobility patterns but needs to serve mobility demand which is shaped by other external conditions.

5 Conclusions

This chapter presented findings from a scenario planning study about the future of mobility in Germany. Even though the three scenarios differ significantly with respect to their implications for travel demand an important finding from the scenarios is that passenger travel in the future will most probably not exhibit the high growth rates of the past decades. Hence, the scenarios reinforce the stagnation which has been observed in the last couple of years. Nevertheless, the findings from the scenarios also show that the development of passenger travel is also very much subject to external developments, specifically the global and national economic development, and the farsightedness of political decision making. An advantage of the scenarios is the ability to identify the impact of qualitative factors such as consumer choice or regulation on mobility.

With regard to individual electric mobility the implications from our scenarios are heterogeneous: On the one hand, mobility in Germany continues to be dominated by the car. Moreover, innovative mobility products have a fair chance if they are cost-efficient and in line with environmental awareness which increasingly shapes consumer choice. On the other hand, total mobility demand in Germany will most likely continue to stagnate and electricity as a source of energy for road transport remains a niche. Even though the current public discourse about electric mobility appears to suggest otherwise, according to our scenarios we will not see pervasive penetration of electric mobility in the next two decades.

References

BMVBS (Bundesministerium für Verkehr Bau und Stadtentwicklung) (2011) Verkehr in Zahlen 2010/2011. Deutscher Verkehrs-Verlag, Hamburg

Cools M, Kochan B, Bellemans T, Janssens D, Wets G (2011) Assessment of the effect of microsimulation error on key travel indices: evidence from the activity-based model feathers. In: Presentation at the 90th transportation research board annual meeting. Washington D.C., Jan 2011

De Rham C, Schwarz R, Schaufelberger W (2006) Error propagation in macro transport models. Forschungsauftrag SVI 2004/015

de Dios Ortúzar J, Willumsen LG (2011) Modelling transport. Wiley, Chichester

FGSV (Forschungsgesellschaft für Straßen- und Verkehrswesen) (2006) Hinweise zu verkehrlichen Konsequenzen des demographischen Wandels. FGSV-Verlag, Köln

Fink A, Siebe A (2006) Handbuch Zukunftsmanagement—Werkzeuge der strategischen Planung und Früherkennung. Campus Verlag, Frankfurt

Gausemeier J, Fink A, Schlake O (1995) Szenario-management. Carl Hanser Verlag, München Wien

Geschka & Partner (2012) Szenariosoftware INKA 3. Darmstadt. http://www.geschka.de/?id=62. Accessed 20 Mar 2012

ifmo (Institut für Mobilitätsforschung) (2005) Zukunft der Mobilität—Szenarien für das Jahr 2025. Erste Fortschreibung: Institut für Mobilitätsforschung (ifmo). Berlin

ifmo (Institut für Mobilitätsforschung) (2010). Zukunft der Mobilität Szenarien für das Jahr 2030. Zweite Fortschreibung infas (Institut für Angewandte Sozialwissenschaft GmbH), DLR (Deutsches Zentrum für Luft- und Raumfahrt) (2010) Mobilität in Deutschland 2008, Ergebnisbericht, Struktur –Aufkommen—Emmissionen—Trends. Bonn, Berlin: infas, DLR

Kutter E (2003) Modellierung für die Verkehrsplanung. Theoretische, empirische und planungspraktische Rahmenbedingungen. ECTL Working Paper 21. Hamburg

Schoemaker PJ (1995) Scenario planning: a tool for strategic thinking. Sloan Manage Rev 36:24–40

Shell International BV (Hg.) (2008) Scenarios: an explorer's guide

The World Bank (2010) World development indicators (WDI). Washington D.C. http://data.worldbank.org/data-catalog/world-development-indicators/wdi-2010. Accessed 8 Aug 2011

Yuva, J. (2011). Plan tomorrow's scenarios today. Inside Supply Management (2.2011)

Zumkeller D, Chlond B, Manz W (2004) Infrastructure development in germany under stagnating demand conditions: a new paradigm? Transp Res Record: J Transp Rese Board 1864:121–128

Author Biographies

Tobias Kuhnimhof is a research manager at ifmo, Munich. A civil engineer and transportation planner by trade, he obtained his doctoral degree from the Karlsruhe Institute for Technology in the field of travel demand modeling. Before taking up the position at ifmo, Tobias Kuhnimhof has also worked for INRETS, Paris and STRATA GmbH, Karlsruhe on research and consulting projects.

Irene Feige is head of ifmo, Munich. She holds a masters degree in Economics and Business Administration from the Universities Innsbruck, Vienna and Verona. Irene Feige obtained her doctoral degree from the University of Innsbruck on the relationship of transport, trade and economic growth.

Peter Phleps is a research manager at ifmo, Munich. He has a doctoral degree in Aeronautical Engineering from Technische Universität München (TUM). Before joining ifmo he worked as a Research Associate at the TUM Institute of Aircraft Design, where he focused on the combination of scenario analysis and technology evaluation, applied for different aspects of the air transport system.

How Does the Actual Usage of Electric Vehicles Influence Consumer Acceptance?

Uta Schneider, Elisabeth Dütschke and Anja Peters

Abstract Electric vehicles are being intensively discussed as a possible sustainable and energy-efficient means of transport. Throughout Europe, broad programmes have been launched to support electric vehicle research, field trials and market diffusion. However, for a successful diffusion of electric vehicles, their acceptance by consumers is crucial. So far this issue has not been analysed sufficiently involving actual users of recent electric vehicle models. What do electric vehicle users and those intending to use an electric vehicle in the near future really think about electric vehicles? How do these perceptions change if they actually use an electric vehicle in everyday life? In order to provide answers to these questions, a longitudinal set of survey data has been analysed of participants in field trials in the eight pilot regions for electric mobility in Germany. These findings are compared to the survey data of nearly 1,000 German car drivers classified into four groups (current electric vehicle users, non-users with a concrete purchase intention, electric vehicle-interested people and consumers with no interest in electric vehicles). The analyses and the comparison between the two studies indicate that gaining real experience with electric vehicles has a positive influence on some predictors of the acceptance of electric vehicles according to the diffusion of innovation model by Rogers (2003). This indicates the relevance of the visibility and observability of electric vehicles. For example, providing test drive opportunities allows consumers to experience electric vehicles themselves and might help to increase consumer acceptance.

U. Schneider (✉) · E. Dütschke
Competence Center Energy Technology and Energy Systems, Fraunhofer Institute for Systems and Innovation Research ISI, Breslauer Strasse 48, 76139 Karlsruhe, Germany
e-mail: uta.schneider@isi.fraunhofer.de

E. Dütschke
e-mail: elisabeth.duetschke@isi.fraunhofer.de

A. Peters
Competence Center Sustainability and Infrastructure Systems, Fraunhofer Institute for Systems and Innovation Research ISI, Breslauer Strasse 48, 76139 Karlsruhe, Germany
e-mail: anja.peters@isi.fraunhofer.de

1 Introduction

Electric vehicles[1] are a much discussed topic today, especially in the context of climate change, air quality and energy security. In several countries, governments are promoting the use of electric vehicles. The German federal government wants Germany to become a lead market for electric mobility by 2020. By then, 1 million battery-electric and plug-in hybrid vehicles are expected to be driving on German roads.

At present, the share of electric passenger cars in Germany is still very low: Only 0.01 % of the total fleet of passenger cars are electric. Thus, there is very little up-to-date research on actual consumer experiences and their acceptance of electric vehicles.

"Second generation" electric vehicles (i.e. with lithium batteries) have just started to enter the market and several models are now available or will be available soon from renowned international car manufacturers as well as from new enterprises all over the world, for example Mitsubishi, Tesla, Luis, Peugeot, Nissan, e-Wolf GmbH, Ford and Think. German automotive companies are also developing their own battery-electric or plug-in hybrid vehicles. In order to prepare for an electric future, the German Federal government is supporting several series of field trials of electric vehicles (e.g. "Electric Mobility in Pilot Regions", funded by Germany's Transport Ministry).

Driving an electric vehicle still implies having to deal with some challenges which could be critical from a consumer's point of view, for example, a limited driving range, a high purchase price and a long charging duration. However, there are also electric vehicle characteristics which offer advantages compared to conventional vehicles, for example low engine noise, fast acceleration, no local emissions or low running costs. Nevertheless, driving electric vehicles requires significant changes in consumer behaviour (Anable et al. 2011).

As consumer acceptance is a crucial precondition for the successful market penetration of electric vehicles, it is necessary that electric mobility concepts meet consumers' real needs. Up to now, most consumers have hardly any real experience with electric vehicles; many consumers have never even seen an electric vehicle. Skippon and Garwood (2011) state that new technologies with which consumers have not yet had any experience are "psychologically distant" (Liberman et al. 2007) and abstract. Asking consumers to evaluate such new products can hardly lead to valid predictions of actual future behaviour; consequently, surveys which focus mainly on non-users who are not yet familiar with the idea of electric mobility may not be useful for studying consumers' expectations of electric vehicles. Consumers themselves have even indicated in market research studies that they feel they do not know enough about electric vehicles (Fraunhofer IAO and PwC 2010). As other recent market research studies have

---

[1] In this paper electric vehicles include battery-electric and plug-in hybrid vehicles as well as four- and two-wheeled vehicles.

shown (e.g. ADAC 2009; Roland Berger 2010; TÜV SÜD and Technomar 2009), expectations of the attributes of electric vehicles by non-users are therefore strongly influenced by the attributes of conventional vehicles which serve as the main frame of reference. The question is how these expectations and perceptions might change if consumers are given more information. A study of Anable et al. (2011) provides evidence that receiving more information about a specific new technology such as electric vehicles could influence consumer acceptance, i.e. purchase intentions. This indicates the relevance of conducting studies involving consumers who have already gained more knowledge and experience with the particular technology, in particular actual users, in order to draw conclusions about the future acceptance of this technology. Since electric vehicle owners are still rare, it is important to analyse user acceptance in field trials in which consumers can test electric vehicles for a longer period of time.

In this chapter, we explore how actually using an electric vehicle influences individuals' acceptance, in particular their perception of the specific characteristics of electric drive vehicles. To this end, we analyse questionnaire data from two studies. The first data set (*study 1*) presents longitudinal data from user surveys conducted within the German electric vehicle pilot regions: participants in these field trials were asked about their perception of electric vehicles before and after using the vehicle for a certain period of time (one week to three months). A second data set (*study 2*) provides cross-sectional data on the perceptions of electric vehicles by several consumer groups, for example, individuals who expressed a high interest in buying an electric vehicle and respondents who have actually acquired and used an electric vehicle. The results of both studies are comparatively discussed on a qualitative basis against the background of Rogers' diffusion of innovation model and of prior empirical studies on battery-electric and plug-in-hybrid vehicles.

2 Theoretical Framework and Prior Empirical Findings

In this chapter, acceptance is defined as the willingness to adopt, i.e. to regularly use an electric vehicle in everyday life. The adoption of innovations (i.e. ideas, applications or objects that are perceived as new) by individual consumers is analysed using Rogers' diffusion of innovation model (Rogers 2003). According to this model, besides socio-economic characteristics, the decision to adopt or reject an innovation is influenced by its attributes as perceived by the individual: (1) the *relative advantages* (and disadvantages) of an innovation compared to conventional alternatives on the market; (2) the *compatibility* with the adopter's values, experiences and needs; (3) the *complexity*, i.e. how easy it is to understand and use the innovation; (4) the *trialability*, i.e. the possibility to test the innovation before the decision to adopt; and (5) the *observability* or visibility of an innovation and its consequences (Fig. 1).

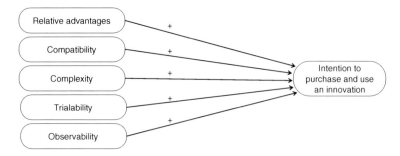

Fig. 1 Theoretical model to explain the intention to purchase and use an innovation, adapted from Rogers (2003)

Applied to electric vehicles, this implies that (1) electric vehicles will be evaluated based on their *relative advantages* when compared with conventional vehicles as well as other modes of transport. As outlined above, when comparing electric vehicles to conventional vehicles, consumers might perceive their high purchase price as well as their low driving range in combination with long charging durations as major disadvantages. On the positive side, electric vehicles could score highly from a consumer viewpoint with regard to driving pleasure and noise, positive image and zero exhaust emissions. Regarding the *compatibility* of electric vehicles (2), evaluating mobility panel data for Germany (Kley 2011) shows that the majority of weekly travel profiles (roughly 60 %) can be covered by the current range of electric vehicles (24 kwH battery) and overnight charging. However, the consumers' perception might differ with regard to electric vehicles matching their travel profiles as they might not keep track of their exact driving profile, or might still prefer to be flexible concerning spontaneous longer trips. Besides individual needs, values play an important role when evaluating whether electric vehicles are perceived as compatible. For example, consumers who believe strongly in the protection of the environment and resources may perceive electric drive vehicles as an attractive and sustainable means of transport which allows them to reduce the environmental impact of their car use (Skippon and Garwood 2011). With regard to the *complexity* of use (3), electric vehicles are sometimes said to be easier to use than conventional cars (e.g. Knie et al. 1999) as they do not have e.g. a gearshift. However, consumers who have not had the opportunity to actually drive an electric vehicle might expect them to be more complex. With regard to the remaining factors of the diffusion of innovation model (Rogers 2003) (4 and 5), as stated before, there are only a few electric vehicles already on the roads, so that *trialability* and *observability* might be perceived as relatively low.

In order to promote the acceptance of electric vehicles, it is important to analyse the perception of these attributes and their role in the acceptance of electric vehicles; particularly, how the evaluation of attributes changes when consumers gain more information and experience, with electric vehicles and expectations become more realistic. That changes do indeed take place is shown by a study by

Carroll and Walsh (2010), who surveyed 42 participants of a public drive event with electric vehicles (with the SmartFortwo electric drive) both before and after a test drive. Participants indicated a higher intention to use an electric vehicle as a regular car after the trial and their ratings based on their actual experience exceeded their previous expectations of several vehicle attributes (acceleration, top speed, braking performance, comfort, range, operation of controls). In other short field tests of electric vehicles, the participants were similarly enthusiastic about the vehicle's performance and in some cases acceptance, i.e. usage and purchase intentions, increased (Skippon and Garwood 2011; CABLED 2010). Other recent studies have pointed out that some participants of field trials (Graham-Rowe et al. 2012) or early users (Kurani et al. 2007) adapted themselves to the special characteristics of plug-in-hybrid vehicles and electric vehicles, for example, they changed their driving behaviour or rethought their lifestyles.

In addition to these recent studies, several authors have analysed data from field trials conducted in the late 1990s or the early 2000s, when electric vehicles experienced their first boom (e.g. Knie et al. 1999). Their results are basically in line with the findings described in more recent studies. Both Gould and Golob (1997) and Gärling (2001) surveyed participants of electric vehicle trials before and after using the vehicles.

The respondents (n = 53) in the study by Gould and Golob (1997) participated in a 2-week long trial of electric vehicles (prototypes). After the trial, the participants were more positive about electric vehicles as a key technology to solving air pollution than beforehand. Regarding the reasons for buying an electric vehicle, the environmental benefit was cited the most frequently before the trial. After the trial, the most frequently selected reason for acquiring an electric vehicle were the lower running costs.

Gärling (2001) analysed the perceptions of 42 families as well as of 32 owners of a conventional Renault Clio who participated in field trials with the Renault Clio Electrique (range 60–70 km) and came to different conclusions. In the family-sample, significant differences were found between the ratings for buying intentions, safety and usefulness before and after the trial. After the trial, purchase intentions were lower than before; the electric vehicle was perceived as less useful for shorter and longer leisure trips and was perceived as less safe than before. The author assumes one reason might be that this make of electric vehicle is not very popular in Sweden, so some respondents might not have distinguished between the make and the technology. Another reason could be the fact that the respondents—mostly families with children—compared the electric vehicle to their conventional car, which was usually larger than the electric vehicle. In the second group of consumers, the owners of a conventional Renault Clio, no significant effects were found of the trial on their ratings. However, more than half of the participants reported problems with the range.

Thus, these early studies present different effects for acceptance before and after use. However, these results relate to the first generation of electric vehicles and are not fully transferable to the current situation, as they refer to vehicles and charging infrastructures which were less developed than those currently on the market or soon to be available.

3 Research Questions

The presented outline points out the need for studies which analyse acceptance of electric vehicles of the current generation by actual users who integrated the electric vehicle into daily routines. Thus, we focus on analysing how the preconditions of acceptance identified by Rogers are perceived and how they differ or develop, respectively, when consumers have gained experience on using electric vehicles. These questions are explored by means of two studies providing both longitudinal data (*study 1*) to analyse possible changes due to the experience of driving electric vehicles and cross-sectional data (*study 2*) to analyse differences in acceptance between consumers using electric vehicles with consumers still intending to adopt electric vehicles for regular usage.

4 Study 1

The first two authors of this chapter are members of the team which coordinated the social scientific accompanying research of the pilot regions for electric mobility, in which the users of electric vehicles are surveyed on their expectations as well as their actual experiences with the vehicles. This project created the dataset for the first study.

4.1 Methodology

Sample and procedure. The first study presented is conducted within the programme "Electric Mobility in Pilot Regions" which was implemented by the German Federal Ministry of Transport, Building and Urban Development. In eight pilot regions several kinds of electric vehicles (two-wheelers, transporters, and passenger cars) are tested by several types of users (private as well as commercial). Additionally, the vehicles are used in various business models: car-sharing or hired car, as company or fleet vehicles or in exclusive private use. Every region conducts several projects on electric mobility, including several field trials. For all of these projects, identical surveys were developed and—if possible—distributed to the participants of the field trials. A longitudinal design was applied in order to identify possible changes in consumer acceptance over time:

- Survey T0 assessed consumer expectations of electric mobility prior to vehicle delivery.

- Survey T1 assessed impressions of the vehicles and electric mobility after a few weeks of usage.
- Survey T2 assessed impressions after several months of usage.

The survey includes questions on the vehicle types, planned usage, demographics, expected advantages and disadvantages of the vehicles, and item batteries with general aspects of acceptance as well as more detailed questions about specific attributes of the electric vehicle and the infrastructure. The survey was available online as well as in a paper version.

In this chapter, we analyse changes in perceptions between the T0 and T1 survey by comparing selected items based on a sub-sample of 145 participants who took part both in the T0 and T1 survey.

Table 1 shows that the majority of this sub-sample uses a battery-electric car. Most of the cars are used very frequently and mainly for private purposes. As far as data was available, the respondents are mostly male, highly educated, middle-aged and married or in a relationship. This corresponds to sample characteristics of similar studies (e.g. Gould and Golob 1997).

Measures and analyses. In order to measure the acceptance of electric vehicles, i.e. the willingness to adopt, participants were asked to indicate to what extent they

Table 1 Characteristics of the sample, the vehicles, frequency and mode of use

Attribute	Share
Vehicle	
Battery-electric car	56 %
Pool of different electric vehicles	16 %
Two-wheelers	20 %
Battery-electric transporter	6 %
Plug-in-hybrid car	2 %
Mode of use	
Predominantly private use	40 %
Predominantly commercial use	22 %
Duration of use	
1 week	35 %
2–4 weeks	15 %
5 weeks to 2 months	11 %
More than 2 months	9 %
Frequency of use	
(almost) every day	50 %
1–3 days per week	26 %
1–3 days per month	15 %
Less often	7 %
Gender	
Male	70 %
Female	29 %
Age	
Range	21–72 years
Mean	39 years

are willing to "substitute a conventional vehicle by an electric vehicle". As for most other items, evaluations were provided on a six-point Likert scale (1 = "do not agree at all"; 6 = "fully agree").

With regard to Rogers' (2003) diffusion of innovation model, several items refer to the first three factors as described in the following, while (4). trialability and (5). observability were assumed to be given by embedding the survey within the fleet trials.

1. With regard to *relative advantages* (and disadvantages) of electric vehicles, participants were asked for a general evaluation of the possibility to save money by using the electric vehicle (instead of a conventional vehicle) and, in more detail, about the perception of the purchase price, the operating as well as the service and maintenance costs. Furthermore, the participants were asked to evaluate charging time, availability of public charging stations, driving pleasure and driving comfort, acceleration, maximum speed, and driving noise. Moreover, the participants were to evaluate the range and their confidence in the indicated range and if there are any positive reactions of others towards the electric vehicle.
2. With regard to the *compatibility* of the electric vehicle with the individual values, experience and needs, the participants were asked about the perceived usefulness in everyday life, the environmental friendliness and if the participants are enthusiastic about the electric vehicle.
3. In relation to the perceived *complexity*, participants rated the ease of using as well as of charging the vehicle.

For these indicators of Rogers' (2003) diffusion of innovation model, we analyse how evaluations change before and after using an electric vehicle in a field trial. To test for statistically significant changes in attitudes, paired t-tests are conducted.

4.2 Results of Study 1: How Do the Predictors of Acceptance Change When Consumers Have Gained Experience in Using Electric Vehicles?

In the following, the results of the survey are presented for each diffusion factor of the innovation model.

4.2.1 Relative Advantages of Electric Vehicles

Table 2 summarises the results of the items referring to the relative advantages of electric vehicles. The general potential to save money by using the electric vehicle is rated positively on average. When looking at answers to more detailed

Table 2 Evaluations of the relative advantages of electric vehicles concerning the aspects costs, range and infrastructure and performance

	T0 mean	T1 mean	n	Interpretation
Costs				
Saves money	4.51	4.69	118	No difference
Low purchase price	2.00	1.81	31	No difference
Low operating costs	4.79	4.76	58	No difference
Range and infrastructure				
Sufficient range	3.92	3.65	103	No difference
Confidence in indicated range*	3.53	3.93	85	Perception more positive in T1
Short charging time**	3.64	3.17	101	Perception more negative in T1
Availability of public charging stations***	2.85	2.14	95	Perception more negative in T1
Performance				
Driving pleasure***	4.47	5.11	92	Perception more positive in T1
Good acceleration**	4.60	5.02	94	Perception more positive in T1
Adequate maximum speed	4.01	4.09	103	No difference
Good driving comfort	4.13	4.27	106	No difference
Agreeable driving noise	4.93	4.96	106	No difference
Image				
Positive reactions of others**	4.70	5.19	69	Perception more positive in T1

The items were rated on a 1–6 scale, where "1" means "does not apply at all" or "not at all" and "6" means "applies perfectly" respectively "fully". Mean values from 1 to 2.5 are interpreted as negative, mean values of 2.51–4.5 as undecided and mean values from 4.51 to 6 as positive. Significance level: *$p < .05$, **$p < .01$, ***$p < .001$

questions, it can be seen that the participants know about the specific cost structure of electric vehicles: purchase prices are rated negatively, operating costs, in contrast, positively. None of the evaluations regarding costs changes significantly between T0 and T1.

The range of the electric vehicles is rated neither positively nor negatively and is the same for T0 as for T1. Similarly, there is no strong confidence in the indicated range, but the evaluations become more positive in T1. The charging time is a critical issue for the respondents: the evaluations are in the middle spectrum and become more negative in T1. The evaluations regarding the public charging infrastructure are neutral in T0 and become (significantly more) negative in T1.

Before using the electric vehicles (T0), the participants expect a high level of driving pleasure which is rated even more positively after using the vehicle for some time (T1). Equally, in T1, the participants are positively surprised by the acceleration of their vehicles. Maximum speed and driving comfort are rated neither positively nor negatively in the T0 and the T1 surveys and ratings do not change over time. Driving noise is rated positively in both surveys and, again, perceptions do not change between T0 and T1.

Participants expect positive reactions of others to the vehicle (T0) and actually perceive them as even more positive after having used the vehicle for some time (T1).

4.2.2 Compatibility with the Adopter's Values, Experiences and Needs

As shown in Table 3, the usefulness of the vehicles as well as their environmental friendliness are positively rated. Perceptions do not change between T0 and T1. The statement that the respondent is enthusiastic about the electric vehicle is rated positively by the average participant and becomes more positive at T1.

4.2.3 Complexity

The participants in the field trials expect that using their vehicle will be easy in the T0 questionnaire. Their ratings are even more positive while using it. Similarly, expectations concerning the handling of the vehicle are positive and become better in the course of usage.

Charging was not expected to cause any problems according to the statements of the participants in the T0 survey and the evaluations become even more positive when using the vehicle (Table 4).

4.3 Summary and Discussion of Study 1

Even though the majority of the participants do not have concrete purchase intentions (57 % in T0 and 46 % in T1 "do not agree at all" resp. "tend not to agree" with the statement "substitute a conventional vehicle by an electric vehicle"), for most of the items, the respondents evaluate electric vehicles and their attributes positively: twelve item ratings are in the positive range of the Likert

Table 3 Evaluations regarding the compatibility of the electric vehicle

	T0 mean	T1 mean	n	Interpretation
Useful in everyday life	4.98	5.17	117	No difference
Environmentally friendly	5.16	5.10	126	No difference
I am enthusiastic about the electric vehicle**	4.64	5.03	132	Perception more positive in T1

The items were rated on a 1 to 6 scale, where "1" means "does not apply at all" or "not at all" and "6" means "applies perfectly" respectively "fully". Mean values from 1 to 2.5 are interpreted as negative, mean values of 2.51 to 4.5 as undecided and mean values from 4.51 to 6 as positive. Significance level: *p< .05, **p< .01, ***p< .001

Table 4 Evaluations regarding the complexity of using the electric vehicle

	T0 mean	T1 mean	n	Interpretation
Ease of use***	4.84	5.34	134	Perception more positive in T1
Handling easy to learn***	5.12	5.69	93	Perception more positive in T1
Easy handling of charging**	5.07	5.45	108	Perception more positive in T1

The items were rated on a 1 to 6 scale, where "1" means "does not apply at all" or "not at all" and "6" means "applies perfectly" respectively "fully". Mean values from 1 to 2.5 are interpreted as negative, mean values of 2.51 to 4.5 as undecided and mean values from 4.51 to 6 as positive. Significance level: *p< .05, **p< .01, ***p< .001

scale, six in the neutral range, while two are rated negatively. All items measuring the compatibility as well as the complexity of the electric vehicle received mean ratings in the positive range. By contrast, the respondents are still sceptical towards some aspects regarding the relative advantages of the vehicles.

Looking at the changes, it transpires that of all the items integrated into the analyses, most ratings do not significantly change between T0 and T1; eight ratings become more positive (spread over all factors of Rogers' (2003) diffusion of innovation model), and two items are rated more negatively (both assessing relative advantages). In the following, the results are discussed in detail.

4.3.1 Relative Advantages

The costs of electric vehicles are assessed sceptically by the participants: the purchase price is evaluated as unsatisfactory—for future as well as actual users. Graham-Rowe et al. (2012) found similar results. On the other hand, the participants appreciate the low operating costs of electric vehicles. No change in evaluations could be observed between T0 and T1 with regard to costs.

Aspects of range and infrastructure receive mixed reviews: participants seem undecided regarding the aspects range, confidence in the indicated range and charging time, i.e. the ratings are in the middle of the spectrum. While perceptions of range do not change between T0 and T1, confidence in range increases. In contrast, the evaluations of the charging duration and the availability of charging stations become more negative.

It can be assumed that the short range of electric vehicles compared to conventional vehicles is perceived by users as a limitation of their personal mobility and autonomy. This interpretation corresponds with the results of other studies: Gould and Golob (1997) found that participants in field trials with electric vehicles have high requirements regarding the range (more than 100 miles). Although they kept a travel diary and thus knew that most of their trips were less than fifty miles per day, the tolerance towards the limited range is low. The experience with the electric vehicles did not influence the perceptions of desired range. Similarly, Gärling (2001) found that 70 % of the participants in a 9-week field trial perceived the range of their electric vehicle (65 km) as too short. They required a range of at least 130 km. Graham-Rowe et al. (2012) who interviewed battery-electric vehicle users reported that users became more aware of their driving profiles because they had to plan their journeys because of limited range and long recharging times. Knie et al. (1999) found similar results.

However, confidence in the indicated range obtained better ratings in the group of actual users in our study. Consequently, the participants perceive they are better able to handle the limited range. The evaluation of the CABLED (2010)-trial came to similar results: it turned out that the longer people used them, the vehicles were driven more miles, i.e. they were driven more frequently and for longer journeys. Thus confidence in the vehicles may have increased and 'range anxiety', i.e. worries that the vehicle has insufficient battery performance or charge to reach the destination may have decreased. The fact that range is a complex issue from the user perspective has also been pointed out by a recent study by Franke et al. (2011) which showed that the perception of what is a comfortable range varied substantially among the participants in a field trial.

Medium and negative ratings of charging time and public charging infrastructure respectively in T0, which become even more negative in T1, indicate that the already low expectations of the participants were not met by actual usage experience and that these aspects might be critical issues.

The variables concerning performance aspects received medium (maximum speed, driving comfort) to positive (driving pleasure, acceleration, and driving noise) ratings. The evaluation for acceleration and driving pleasure in T1 became more positive than in T0. In the 1990s, Knie et al. (1999) found that such perceived advantages can induce that technical constraints of electric vehicles were not necessarily considered as a barrier to adopting an electric vehicle. Our first analyses point to the direction that this conclusion may still be valid.

Positive reactions of others are rated positively and evaluations become better between T0 and T1. Axsen and Kurani (2009) showed that social interactions in households and social networks shape the evaluations of plug-in-hybrid vehicles and probably also of electric vehicles generally. The majority of the networks named at least one social interaction that they had perceived as being highly influential on the evaluation of plug-in-hybrid vehicles. For example, these interactions included asking others about their potential motives for buying a plug-in-hybrid vehicle or seeking for help to understand technical aspects. Thus, a perceived increase of positive reactions of third parties towards the electric vehicle might lead to a better acceptance of the electric vehicle.

Taking an overall look at the items which assess perceived *relative advantages* before and during actual use, it turns out that most of the ratings remain basically unchanged, while they become more positive for some aspects and more negative for others. Thus it is difficult to draw a general conclusion as to how users appraise the relative advantages of electric vehicles after usage experience.

4.3.2 Compatibility

The *compatibility* of the electric vehicle with the users' values, experiences and needs, i.e. enthusiasm, is rated positively and assessments became better in the course of usage.

Usefulness in everyday life is rated positively as well and there is no difference between T0 and T1. In contrast, in the study of Graham-Rowe et al. (2012), the utility of the electric vehicles was perceived as limited since the interviewees hesitated to make longer journeys with their vehicles because of their scepticism regarding range.

Evaluations regarding environmental friendliness were very positive and did not change between T0 and T1. This is in line with findings of Skippon and Garwood (2011). In addition, Graham-Rowe et al. (2012) found that some users appreciate the "feel-good factor" related to environmental benefits when driving an electric vehicle. The perception that the car is environmentally friendly might, however, lead to rebound effects (Fuji 2010), i.e. higher consumption related to energy-efficient technologies.

In sum, based on these aspects, the evaluations of compatibility with values, experiences and needs seem to be in a positive range, at least for the issues under study.

4.3.3 Complexity

Handling the vehicle did not cause any problems: learning how to use it, the ease of use as well as handling of charging are all rated positively and the ratings further increase in T1. The results of the survey show that, although some of the vehicles driven are not yet ready for the market, this does not seem to be a challenge for the participants. They do not expect difficulties nor do they indicate that they experienced any. Using an electric vehicle is not perceived as complex.

4.4 Limitations of Study 1

The generalisability of these findings is certainly limited due to the special characteristics of the participants: the sample is quite homogeneous regarding gender, age and education. Thus it is not possible to draw conclusions for other potential users of electric vehicles. In addition to this, field trials can be regarded as a special case of adoption, because the participants are part of special programmes and often do not have to pay the full costs of vehicle use. Additionally, as the usage phase is declared an official test and is limited in time, it can be assumed that the participants display relatively positive attitudes and a higher degree of tolerance towards possible disadvantages of the electric vehicles. When the electric vehicle is used as an additional vehicle instead of replacing an existing one in the household or the company fleet, this may lead to similar effects. Further, the vehicles might be highly promoted; possibly more than in the case of purchasing a conventional car. Thus the final or actual adoption has to take place after the field trials.

However, 20 % of the respondents in the first study used their vehicle for more than one month. The participants were also able to integrate the vehicles in their everyday life, since the majority used it at least once per week. A wide variety of battery-electric or plug-in hybrid vehicles were field-tested in the pilot regions. Based on these arguments, we assume that valid conclusions can be drawn from our results for the actual adoption of electric vehicles in the current market phase.

5 Study 2

In order to compare consumer groups who differ in their affinity towards electric vehicles and to identify promising target groups for electric mobility, as well as relevant factors for adoption, one of the authors of this chapter together with her colleagues conducted a large online survey in Germany in 2010 with potential car buyers (cf. Peters et al. 2011).

5.1 Methodology

Similar to *study 1*, the questionnaire of this online survey includes items assessing perceived attributes of electric vehicles according to Rogers' (2001) diffusion of innovation model, and the participants' likelihood to purchase and use an electric vehicle, as well as socio-demographic items. Moreover, the affinity towards electric vehicles, i.e. the knowledge and general interest in electric vehicles, was measured.

The survey sample ($N = 969$) contains 81.4 % men. The mean age was 40.9 years, the modus of monthly household income was € 2,001–3,000, the average household size was 2.48 persons, and the average number of cars owned by a household was 1.43.

The participants were divided into different consumer groups according to their use of an electric vehicle as well as their interest in electric vehicles: (1) actual users of electric vehicles (n = 92), (2) consumers intending to adopt electric vehicles in the future (n = 244), (3) consumers interested in electric vehicles, but without concrete purchase intention (n = 352), and (4) consumers who are not interested in electric vehicles ($N = 281$).

Analyses of variance were conducted in order to analyse differences in the perceived characteristics of electric vehicles between these consumer groups. In the following, we summarise the results of these analyses based on Peters et al. (2011).

5.2 Results of Study 2: How Do Consumers with Different Levels of Affinity Towards Electric Vehicles Differ in Their Acceptance of Electric Vehicles?

The analyses generally show that the more interested respondents are in purchasing or using an electric vehicle, and the more experience they have, the more they tend to evaluate relevant characteristics and aspects of electric vehicles in favour of these vehicles.

Comparing consumers with purchase intentions with those who are interested but have not (yet) decided to adopt an electric vehicle, compatibility with own habits and needs, driving characteristics, operational costs and environmental consequences as well as social norms are evaluated significantly more positively by the consumers with intentions to purchase. The respondents who are not interested in electric vehicles rate these aspects even more negatively than the respondents who are interested but without intention to purchase.

While these groups thus show many differences in their perceptions of electric vehicles, consumers with concrete purchase intentions and users hardly differ significantly from each other in the assessed perceptions. Only the trialability of electric vehicles is rated clearly lower by the first group, as well as by all other non-users when compared to the group of actual users. With regard to the other aspects which were assessed, users and consumers with intention to purchase perceive electric vehicles as nearly equal (or slightly superior) to conventional vehicles in terms of driving characteristics, and slightly lower in terms of basic characteristics, like security or storage capacity. Operational costs and environmental consequences are perceived to be clearly in favour of electric vehicles. Also, the perception of compatibility of electric vehicles with own habits and needs, ease of use and social norms is clearly in the upper, i.e. positive range of the scale. Infrastructure, not surprisingly, is perceived as highly superior for conventional vehicles. For more details on the survey and its results, we refer to Peters et al. (2011).

5.3 Discussion of Study 2

The prevailing correspondence between actual users and consumers with concrete intention to purchase indicates that consumers who already intend to adopt an electric vehicle in the near future have come to almost the same conclusions regarding the properties of electric vehicles as actual users, and thus may have rather realistic expectations. However, they have not yet implemented their intention into real action. The results suggest that, in fact, a perceived or objective lack of possibilities to try out and evaluate electric vehicles in use and to compare different models in real life could be one significant barrier to actual adoption.

Such a pronounced decision-making and evaluation behaviour is typical of the group of early adopters, according to Rogers (2003). Moreover, the currently still limited availability of electric vehicles and the expectation of future price reductions could inhibit the final purchase decision. Consumers with intention to purchase could still be waiting for a larger variety of models to enter the market at a lower price, which is a reasonable forecast of market development.

6 Final Discussion and Conclusion

In this chapter, we surveyed participants of heterogeneous field trials and analysed how increasing experience with electric vehicles influenced the predictors of individual acceptance in applying a longitudinal design. Further, perceptions of electric vehicles by both individuals highly interested in buying an electric vehicle and respondents who have acquired and used an electric vehicle are compared in order to identify potential differences between these different consumer groups.

The results of the survey in the pilot regions (*study 1*) suggest that the participants are already very aware of the advantages and disadvantages of electric vehicles since the majority of the ratings of the items do not change between the T0 and T1 survey. Some relative advantages, for example, regarding performance or image, are more positively evaluated in T1. All of the items in the complexity factor received better assessments in the T1 survey, thus the users were positively surprised by the easy handling of the vehicles, although ratings before using them were already very positive. However, purchase intentions are still very limited—in the T0 as well as in the T1 survey. Thus there may be other reasons for these low intentions to adopt an electric vehicle than those analysed in this chapter.

The results of the cross-sectional online survey (*study 2*) concerning differences between the various consumer groups generally indicate a clear relation between a more positive perception of the relevant characteristics of electric vehicles with more experience and interest in electric vehicles and are therefore in line with the findings from the pilot regions.

When comparing the results of these two studies, it has to be considered that they differ in some aspects. The survey with participants in the field trials applied a longitudinal design: the same respondents were questioned at two different times whereas the cross-sectional survey compares different consumer groups. Moreover, the samples are different: in the first study participants of field trials and in the second one German car drivers were questioned, including electric vehicle owners who have regularly used an electric vehicle. However, both studies point out that increased information about and experience with electric vehicles positively influence the evaluation concerning some aspects of the vehicles. These results correspond to previous studies which indicate that it is crucial to expose customers to electric vehicles or to give them detailed information, on the one hand, to study acceptance of electric vehicles, and on the other hand to promote actual adoption of electric vehicles (cf. for example Anable et al. 2011; Carroll 2010; Martin et al. 2009). This finally points to the great importance of enhancing the trialability and observability of electric vehicles.

Acknowledgments We gratefully acknowledge funding from the German Federal Ministry of Transport, Building and Urban Development (BMVBS) and the German Federal Ministry of Education and Research (BMBF). Special thanks go to the participants of our surveys and the projects which provided the data.

References

ADAC (2009) ADAC-Umfrage Kaufbereitschaft Elektroautos. ADAC, Landsberg a Lech
Anable J, Skippon S, Schuitema G, Kinnear N (2011) Who will adopt electric vehicles? A segmentation approach of UK consumers. In: Proceedings to ECEEE summer study, June 2011, Belambra Presqu'île de Giens, France
Axsen J, Kurani KS (2009) Interpersonal influence within car buyers' social networks: five perspectives on plug-in hybrid electric vehicle demonstration participants. UC Davis ITS Working Paper. UCD-ITS-WP-09-04.ö
CABLED (2010) Electric vehicle drivers enjoy increased confidence and low 'refueling' costs. Aston University, Birmingham
Carroll S, Walsh C (2010) The smart move trial. Description and initial results. Cenex, London
Franke T, Neumann I, Bühler F, Cocron P, Krems JF (2011) Experiencing range in an electric vehicle: understanding psychological barriers. Appl Psychol: Int Rev. doi:10.1111/j.1464-0597.2011.00474.x
Fraunhofer IAO (2010) Elektromobilität—Herausforderungen für Industrie und öffentliche Hand. Fraunhofer IAO, Stuttgart PwC (PricewaterhouseCoopers)
Fuji S (2010) Does purchasing an "eco-car" increase the vehicle distance travelled? In: Abstracts of the 27th international congress of applied psychology, Australia. http://icap2010.eproceedings.com.au. Accessed 16 Dec 2011
Gärling A (2001) Paving the way for the electric vehicle. VINNOVA Rapport 2001:1
Gould J, Golob TF (1997) Clean air forever? A longitudinal analysis of opinions about air pollution and electric vehicles. Transp Res Part D 3:157–169
Graham-Rowe E, Gardner B, Abraham C, Skippon S, Dittmar H, Hutchins R, Stannard J (2012) Mainstream consumers driving plug-in battery-electric and plug-in hybrid electric cars: A qualitative analysis of responses and evaluations. Transp Res Part A 46:140–153
Kley F (2011) Ladeinfrastrukturen für Elektrofahrzeuge. Entwicklung und Bewertung einer Ausbaustrategie auf Basis des Fahrverhaltens. Stuttgart, Fraunhofer Verlag
Knie A, Berthold O, Harms S, Truffer B (1999) Die Neuerfindung urbaner Automobilität. Elektroautos und ihr Gebrauch in den U.S.A. und Europa. Berlin, edition sigma
Kurani K, Heffner R, Turrentine T (2007) Driving plug-in hybrid electric vehicles: reports from U.S. drivers of HEVs converted to PHEVs, circa 2006–2007. University of California, Davis
Liberman M, Trope Y, Stephan E (2007) Psychological distance. In: Kruglanski AW, Higgins ET (eds) Social psychology: handbook of basic principles. Guilford Press, New York
Martin E, Shaheen SA, Lipman TE, Lidicker JR (2009) Behavioral response to hydrogen fuel cell vehicles and refueling: results of California drive clinics. Int J Hydrogen Energy 34:8670–8680
Peters A, Agosti R, Popp M, Ryf B (2011) Electric mobility—a survey of different consumer groups in Germany with regard to adoption. In: Proceedings to ECEEE summer study, June 2011, Belambra Presqu'île de Giens, France
Rogers EM (2003) Diffusion of innovations, 5th edn. Free Press, New York

Roland Berger (2010) Powertrain 2020. Electric vehicles—voice of the customer. Munich, Roland Berger

Skippon S, Garwood M (2011) Responses to battery electric vehicles: UK consumer attitudes and attributions of symbolic meaning following direct experience to reduce psychological distance. Transp Res Part D 16:1–7

Tüv SÜD, Technomar (2009) Kurz- und mittelfristige Erschließung des Marktes für Elektroautomobile Deutschland—EU. Technomar, Munich

Author Biographies

Uta Schneider studied social sciences and psychology at the universities of Mannheim, Giessen and Brussels. She gained experience in international automotive market research. From March to July 2010 she worked as an intern at Fraunhofer ISI in the field of consumer acceptance of alternative fuels. Since January 2011 she has been working as a (junior) researcher and PhD student at Fraunhofer ISI. Her research work focuses on: acceptance of electric mobility, qualitative and quantitative methods.

Elisabeth Dütschke studied psychology, business administration and marketing at the TU Darmstadt and RWTH Aachen. She was a research associate at RWTH Aachen and then a research associate and lecturer at the University of Konstanz; she gained a PhD in 2010 from the University of Konstanz; and further work experience in consulting for private and public organizations and journalism. Since June 2009 she has been employed as a researcher at Fraunhofer ISI. Her research focuses are: technology acceptance (e.g. electric vehicles, carbon capture and storage), chances of and barriers to energy efficiency, evaluation studies, qualitative and quantitative methods.

Anja Peters studied psychology at the University of Trier from 1997 till 2003. From 2004 till 2008 she was a researcher at ETH Zurich in the Institute for Environmental Decisions; in 2008 she gained her doctoral degree at the University of Zurich on psychological factors which influence the purchase of fuel-efficient new cars. Since February 2009 she has been employed as a researcher at Fraunhofer ISI. Her main work focuses are: factors and measures influencing and promoting sustainability behavior, acceptance of innovations, mobility behavior, qualitative and quantitative methods.

Identifying Consumer Groups with Satisfactory Characteristics for Electric Mobility Usage

Dominik Santner and Dirk Fornahl

Abstract Who will use an electric car? This question will be addressed by identifying those societal groups which show socioeconomic characteristics, mobility patterns and attitudes that are compatible with a future usage of electric cars. To answer this question a survey in the model region Bremen/Oldenburg is conducted to gather the necessary information. We group the more than 700 respondents according to selected socioeconomic characteristics and then analyze whether their needs, resources and attitudes are in favor of electric mobility. We come to the conclusion that urban singles, urban seniors and rural families are three user groups with the highest likelihood to employ electric mobility in the near future.

1 Introduction

Electric mobility is widely seen as a potentially core development towards a paradigm shift in car engine technology and individual mobility behaviour. The German government expressed its intention to foster electric mobility as a key future technology in mobility in a national development plan ("Nationaler Entwicklungsplan Elektromobilität der Bundesregierung"; Bundesregierung 2009). One main goal of this plan is the development of Germany to become the leading market for electric vehicles in the world. Still, until now electric mobility's portion of the existing German fleet remains very low: At January 1st 2011 only 2,307 of a

D. Santner (✉) · D. Fornahl
Centre for Regional and Innovation Economics (CRIE), University of Bremen, Bremen, Germany
e-mail: dsantner@uni-bremen.de

D. Fornahl
e-mail: dirk.fornahl@uni-bremen.de

total of more than 42 million registered passenger cars (0.005 %) were electric ones (KBA 2011).

The identification of markets' characteristics is crucial for promoting electric mobility and individual mobility patterns need to be investigated in order to design vehicles as well as public and private promotion schemes. Until now technical development often takes place without ample knowledge on potential users' needs and their mobility patterns (Sammer et al. 2008). The investigation of such mobility patterns has to deal with the fact that the mobility behaviours of different persons strongly diverge from each other. Different circumstances of life, determined by aspects such as the place of living, the stage of life and personal preferences, lead to those diversified patterns of mobility. A large proportion of the population in developed post-modern societies use their private motorized vehicles as a central mode of transport that is often not easily and fully replaceable by other means of transport. However, not the whole population is equally dependent on using a car, e.g., urban dwellers can employ a whole range of public transport and are able to satisfy their mobility needs by using this public transport. Electric mobility as a new form of individual motorized transport should be most interesting for those groups in society that already show a high usage of private cars today.

Electric vehicles as a developing and emerging form of individual motorized transportation devices have to deal with manifold requirements of potential users if they want to compete with conventional fuel-driven cars. Even despite a likely increasing advance in technology, electric cars might keep a lower operating distance and relatively long charging spans for the next years to come compared to traditional vehicles. Until electric cars will be able to fully compete with conventional ones producers of electric cars have to focus on certain groups of people that show characteristics making them likely to become early users. Such characteristics might be certain mobility patterns or a certain way of life and the personal appreciation of cars in general or electric cars in particular. Additionally those early adopters have to be able to afford the still higher purchase costs of an electric vehicle up to the point where economies of scale strongly reduce the production costs and result in an increasing rate of diffusion of the technology in the population. Besides this consumer and firm-level considerations there exists also the demand of political decision makers to get advice where and how much investments in public infrastructure for electric mobility should take place. Finally, information strategies of car producers need to be focussed since information, public relations and marketing have a significant influence on purchase potential and adaptation frequency of electric cars (Sammer et al. 2008).

The underlying investigation deals with these topics. Based on a survey on mobility patterns and individual attitudes towards mobility in general and electric mobility in particular conducted in spring 2011 in the model region Bremen/

Oldenburg[1] of the federal research programme "Modellregionen Elektromobilität" of the German Federal Ministry of Transport, Building and Urban Development we identified specific groups of persons with different likelihoods to adapt electric mobility. We will show that certain relatively wealthy urban groups like those living in single person households as well as seniors have a high potential to be first movers or early adopters but at the medium-term scale other groups with a high dependence on cars to serve their personal mobility needs (e.g., groups living in rural areas) are likely to be the main target group for electric mobility.

The research questions of this chapter are: Which societal groups showing which socioeconomic characteristics have mobility patterns and attitudes that are compatible with a future usage of electric cars? Which groups can be main user groups in the long run? Which groups can be potential early users of the technology to form a critical mass for market diffusion?

The remainder of the chapter is structured as follows: Sect. 2 deals with theoretical aspects of group specific mobility needs and patterns. Section 3 introduces the methodology of the investigation while Sect. 4 presents the most significant results of our user group centred analysis. Section 5 draws some conclusions.

2 Mobility Needs and Patterns

Our analysis deals with individual mobility patterns as well as intentions and expectations towards electric mobility. Therefore, it has to be compared to similar studies in these two fields. The recently most recognized studies on mobility patterns in Germany are those in the series "Mobilität in Deutschland" (MiD) (Infas and DLR 2003, 2010) which are successors of the KONTIV series ("Kontinuierliche Erhebung zum Verkehrsverhalten") which only focused on West Germany. These studies investigate mobility patterns within Germany according to used transport modes, travelled distances, travel reasons and many other aspects. The most recent version (Infas and DLR 2010) also describes differences between regions of different structural type like rural and urbanized areas based on the regional merger of urban and rural districts ("Stadtkreise/kreisfreie Städte" and "(Land-)Kreise"). Additionally ten specific regional versions of the study were developed (see, e.g., for the Bremen case Infas 2009). All these studies are characterized by a set of investigations based on household surveys. The KONTIV/MiD methodology was adapted in several other studies (see, e.g., for the Austrian Salzburg case IMAD 2003). Similar studies with a regional focus on mobility within cities and municipalities in Germany are developed in the SrV series ("System repräsentativer Verkehrsbefragungen") (latest version: Ahrens

---

[1] The model region Bremen/Oldenburg consists of the urban districts of Bremen, Bremerhaven, Delmenhorst, Oldenburg and Wilhemshaven as well as the rural districts of Ammerland, Cloppenburg, Cuxhaven, Diepholz, Friesland, Oldenburg, Osnabrück, Osterholz, Vechta, Verden and Wesermarsch.

2009). The latest version has been conformed to the MiD methodology and indicators resulting in a higher degree of comparability. The studies of this series ask for individual mobility behaviour on precise dates to get a general picture on mobility patterns within German cities. A rotating panel analysis is offered by the MOP ("Mobilitätspanel") commissioned by the German Federal Ministry of Transport, Building and Urban Development (latest publication: Zumkeller et al. 2011). MOP and MiD are comparable to a high degree according to indicators and methodology.

The part dealing with mobility pattern of our investigation is based on these studies, especially on questions employed in the MiD. Questions and categories on transport modes, travel reasons and car ownership have been adapted. However, some adjustments have been made to fit our research questions. Hence our study is focussed on a specific regional population from the metropolitan region Bremen/Oldenburg and in contrast to MiD studies, children younger than 18 years are excluded.

Studies on the acceptance of electric mobility are scarce and it has to be distinguished between those studies investigating the acceptance of persons with no experiences with electric cars and those studies working with test users. Our study is of the former type. A recent study on user acceptance of electric cars based on mobility patterns is presented by Sammer et al. (2008). One problem of this study is that current mobility patterns are interpreted as acceptance of electric mobility if electric mobility can serve mobility needs without really asking whether users would change to electric mobility or not. Even if the conclusion is drawn that current mobility behaviour hardly jars with current technological realities of electric vehicles, this does not mean that electric mobility is accepted or not. However, the study gives good hints for the necessity of keeping in mind that consumer needs and mobility behaviour are important aspects in technological diffusion processes. A more general analysis on consumer acceptance of electric mobility offer the Aral studies (for example Aktiengesellschaft 2011). These studies investigate consumer expectations and desires for potential car purchases and include an analysis on attitudes towards electric cars. Even if it has to be kept in mind that these studies have been commissioned by the respective industry, it offers some interesting key questions in the survey on car purchase and acceptance.

According to Rogers (2003) the diffusion of innovations within society takes place successively. Individuals decide to adapt an innovation in respect to several aspects like relative advantages of the innovation, compatibility to the adopter's values, experiences and needs, the possibility to test the innovation and an imagination of the innovation by its observability at different times. Rogers (2003) subdivides all potential consumers into five categories (innovators, early adopters, early majority, late majority, laggards) according to the relative point in time of adaption and all characterised by a specific set of attributes. Electric cars in their current technological form are a relatively new technology to the market. Uncertainty about market development and the establishment of a technological standard combined with still high costs for electric cars make a current acquirement not attractive to most potential users. Market diffusion has just started. The few current users have to deal with this uncertainty and therefore can be characterized as innovators. As this is only a marginal part of society the influence of innovators on the successful diffusion is limited. According to Rogers (2003) a much more significant position is held by the early adopters which play a dominant role in forming majority opinions and being the basic stock for the critical mass the innovation has to overcome. Hence, the identification of possible early adopters for the diffusion process of electric mobility is crucial for a successful and efficient establishment of the technology. However, as the development is still emergent and current users are still characterized as innovators, the categorization of all potential adopters of the technology can only be an approximation. The main goal of our study is to identify such early adopters of electric cars.

Certain socioeconomic features affect people's behaviour. Aspects like age, gender or domestic location within urban or rural areas are crucial to a person's needs, consume behaviour and mobility patterns. Due to practical reasons of marketing the identification of relevant consumers for a product such as an electric vehicle should take a focus on these socioeconomic characteristics. In that way potential early adopters of the technology can be identified.

Demographic change towards an aging society is important when investigating future sales of a new product. Elderly people aged 50 years or older, often labelled as "best agers", "woopies" or similar marketing categories (Lilienthal 2007), show specific characteristics which make them an interesting consumer group. Especially "younger elderly" of the age between 50 and 65 are often characterized by a mixture of an active and curious lifestyle with an emerging need for security (Häusel 2008). Certain aspects of aging, especially after entering retirement, such as becoming a widow or health problems are a high risk of impoverishment. While most elderly are relatively wealthy, another part of this group is dependent on additional social welfare (Lilienthal 2007). However, most elderly people in Germany are relatively wealthy compared to younger people (Strauch 2008) and even became wealthier since the early 1990s (Lilienthal 2007). High income, health issues and the demand for security and comfort also form mobility patterns of elderly people because, for example, income and car ownership are positively correlated (Infas and DIW 2003). Seniors in advanced societies have a high affinity towards cars nowadays. Especially those who started driving at an early age and, hence, have a long history of driving are very likely to use cars as their preferred transport mode up to a quite advanced age with increasing importance of public transport and walking for older elderly (OECD 2001). Hence, seniors are an important user group for public transport in rural areas (Steinrück and Küpper 2010). While remaining a central transport mode for elderly people, the usage of cars successively declines after retirement and seniors avoid driving under unfamiliar or risky conditions (ibid). Electric vehicles could be of high interest to elderly people, if the specific needs of this group are addressed. High incomes, a high stickiness to car usage combined with a declining number of tours made with a car might fit well with electric mobility's reality. Still it has to be kept in mind that elderly are a quite heterogeneous group. While a large group of people older than 50 is very wealthy, free-spending and active, especially very old people are relatively immobile. Thus the interesting consumer group is more the former than the latter.

Areas with different settlement patterns are characterized by different mobility behaviours of their residents. Densely inhabited areas are in general better accessible by different transport modes than peripheral ones (Infas and DLR 2010). In contrast to urban areas many destinations for rural residents are only accessible by car (Steinrück and Küpper 2010). The study "Mobilität in Deutschland" (Infas and DLR 2010) shows that in German districts ("Landkreise") of the categories "Verdichtete Kreise" (urbanized districts) and "Ländliche Kreise" (rural districts) 82 % and accordingly 81 % of all traffic kilometres are travelled by private motorized vehicles. In urban central towns this degree is only at 70 %. Even if the use of private motorized vehicles is an important and dominant transport mode in all spatial categories, the study reveals a high dominance of motorized individual mobility outside of large urbanized centres. This is due to the fact that public transport systems can be run much more efficient in urban areas than in rural ones (Steinrück and Küpper 2010).

3 Methodology

The basic analysis has been done in spring 2011. It is based on a standardized web-based questionnaire on personal mobility patterns and attitudes towards mobility in general and electric mobility in particular. The target group consists of adults which represent a cross section of the population in the model region Bremen/Oldenburg. Participants have been recruited in the context of ten important traffic locations[2] within the whole model region including urban and rural areas. The locations where chosen based on the traffic simulation system VENUS developed by the IVV Aachen and due to the recognition of the most important traffic reasons (working, education, shopping and recreation) identified in the mobility study "Mobilität in Deutschland" (Infas and DLR 2010). In total 706 people participated in the survey. The questionnaire consists of a set of questions divided in five major categories "mobility in respect to the location of the survey", "general mobility patterns and behaviour", "personal opinions on mobility and electric mobility", "intentions for a car purchase" and "socioeconomic parameters".

For the identification of appropriate user groups we decided to use compiled socioeconomic parameters. Most studies on mobility patterns and consumer

---

[2] The locations are the business parks "Airportstadt" and "Technologiepark" in Bremen, the universities of Vechta and Oldenburg, the shopping malls "Dodenhof" in Ottersberg and "Waterfront" in Bremen, the leisure destinations "Havenwelten" in Bremerhaven and "Südstrand" in Wilhelmshaven as well as the pedestrian zones of the two most significant cities of the region Bremen and Oldenburg. At these locations possible participants received a flyer with the respective web address or where contacted by email or via specific internet portals.

acceptance follow a similar approach and base their results on single socioeconomic aspects like gender, age or occupation. Infas and DLR (2010) integrate an analysis of mobility patterns in respect to these aspects as well as the domestic location in regions of different settlement types. Biere et al. (2009) take a focus on occupation and municipality size to form 20 different user groups. For our analysis we decided to adapt this approach on the identification of user groups that are characterized by a set of different socioeconomic features. After a first analysis we identified the parameters "age", "number of people in the household", "existence of children in household", "occupation" and "type of municipality of domicile" as relevant. Since an analysis of potential user should identify groups based on the most influential characteristics, we exclude "gender" and "highest educational achievement" because these were no significant parameters to differentiate persons. Attempts to apply a cluster analysis on the data led to no satisfying results. In return we choose a manual classification. The identified user groups are depicted in Table 1. The distinction between "urban" and "rural" is based on the classification of municipalities developed by the "German Federal Bureau for Building and Spatial Planning" (Bundesamt für Bauwesen und Raumordnung). Municipalities classified as "large town" or "medium-sized town" are treated as "urban" whereas those classified as "small town" or "rural municipality" are treated as "rural". The categorisation of the participants in "urban" or "rural" is oriented on the stated postal code of the domicile in the survey. Students are the only group where no distinction between "urban" and "rural" has been applied as this group in almost completely of urban character in our analysis. The identified groups are partly overlapping which means that one person can be part of different groups.

Table 1 User groups for analysis

Label	Urban	Rural	Criteria
Students	266	(University) students	
Working singles	56	18	Freelance, privately employed or civil servant
			Only one person in household
Families	89	75	At least one person in household younger than 18
Seniors[a]	76	50	Age 50 or older
Working couples without children	69	29	Freelance, privately employed or civil servant
			Exactly two persons in household
			No person in household younger than 18

[a] Seniors in our analysis are to a large extent younger elderly (65 % age 50–59; 30 % age 60–69; 5 % age 70 or older)

4 Empirical Results

Before we present those results that show significant differences between the identified user groups one has to realize that there is a large number of aspects where there seems to be a high degree of conformity. Whereas we will show that mobility patterns differ to a larger extent between our identified groups, personal attitudes towards certain aspects of mobility and the expectations concerning the characteristics of an own car are very similar. Automobility and electric mobility are both subjects that are of a high interest to people of all groups. In total 82 % are at least a bit interested in automobility and 88 % in electric mobility. The valuation of general topics like environmentalism (99 % think this is an important issue) and, a bit more cautious, being personally technologically up to date (69 % think this is important) is in general high.

Additionally all groups show similar patterns when asked if they expect electric vehicles to be satisfying with respect to selected aspects. The central criticism is the purchase price of electric cars. The majority of 56 % does not think that electric cars are satisfactory in respect to this aspect (19 % yes, 25 % do not know). On the other hand operating costs are evaluated much more positive. Only 12 % said no, whereas 65 % believe electric vehicles to be satisfying (23 % answered with "do not know"). The operating distance is seen cautiously critical. Almost half (49 %) of all participants do not expect electric vehicles to be satisfying, 31 % say yes and 20 % do not know. A very optimistic evaluation can be observed with respect to environmentally friendliness: The vast majority of 82 % belief electric cars are environmentally friendly. Only 5 % do not belief this and 12 % do not know.

An aspect of mobility behaviour that does not differ a lot between the user groups (including urban and rural groups) is the frequency of long distant journeys by car. For most participants of the survey these are quite seldom events. Overall, 76 % only use a car to drive distances longer than 200 km less than once a month or never. Even distances of a length between 100 and 200 km are travelled by car less than once a month or never by 54 % of the survey's participants. For most participants the majority of their travel needs could be coped with electric cars in respect to the current technical operating ranges.

A different picture can be drawn when comparing most other aspects of mobility patterns of the identified user groups. Figure 1 shows that the daily use of a car is a much more significant phenomenon observed for rural groups than for urban ones, where bicycles and public transport are more often used. This goes along with empirical results identified in other studies (for example Infas and DIW 2010). Especially students and urban singles use cars rarely. It is expectable that individuals currently independent of cars are not likely to become significant users of electric cars in the future. From this point of view rural citizens have a much larger potential to become significant users of electric cars than urban singles or students.

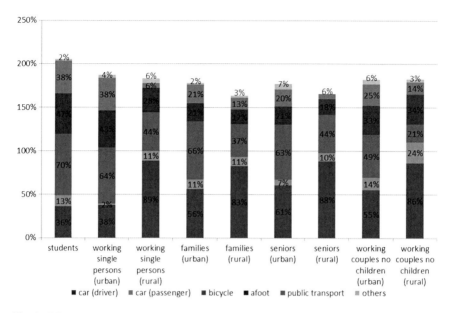

Fig. 1 What transport modes do you use on a normal day?

Another important aspect for the diffusion of electric mobility is the possibility for the potential user to charge the battery easily. Charging of the battery can occur at a socket in a privately used garage or parking space or at public accessible charge columns. While the first option can build on already existing or easily installable infrastructures, the second on requires costly and time-consuming investments first. Hence, electric vehicles are more attractive for users in the upcoming early years of market diffusion, if they can be easily charged at home with a private electricity supply. For those users without these possibilities the ownership of an electric car depends on a network of public or semi-public charge columns which is costly to install. Figure 2 depicts the domestic parking situation of car owners assorted by user groups. Most of our user groups are characterized by a high share of car owners with a satisfying parking situation at home. In most groups a high number of individuals could charge an electric car at home easily. This especially holds for inhabitants of rural areas, but also urban groups like families, seniors and even students can find appropriate conditions at home. Two groups are characterized by a much smaller degree of permanent parking space availability: Urban singles and couples are relatively often dependent on public or semi-public infrastructures. This situation is related to living conditions in central areas of cities characterized by apartment buildings. Especially urban singles are only to a small degree able to charge a potential electric car at home.

Car ownership is a most common phenomenon among most of our user groups (Fig. 3). All rural groups and the urban groups of families, seniors and working couples show a very large portion of car owners. The exceptions are students and urban singles which contain a larger portion of non-car owners. Accordingly, the

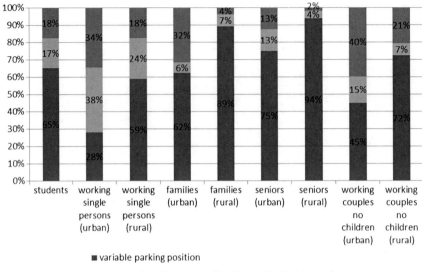

Fig. 2 What parking situation do you have at home?

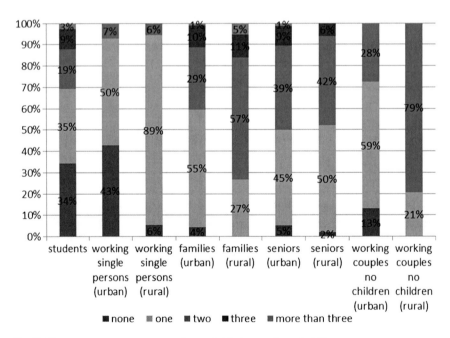

Fig. 3 How many cars are permanently available to your household?

usage of electric cars by a larger share of these user groups in the early phase of establishment of the technology is not very likely as many urban singles currently do not own or use cars very often, and are not likely to have charging facilities at home. Similar conclusions can be drawn for students. Rural families and rural working couples in most cases have even more than one car at home. Such a situation offers the possibility that one car employs conventional fuel and the other one is powered by batteries. Depending on the mobility need at hand, the matching car is used, e.g., for a long distance trip the conventional car is selected. Hence, the problem that the maximum range of electric vehicles is limited is reduced if several alternative cars are available. In contrast Aktiengesellschaft (2011) discovered that most potential users would buy an electric car to replace their first car.

Occasional demand for cars could be served by alternative mobility concepts such as car sharing which could bring electric cars into downtown areas. Figure 4 shows that the acceptance of electric cars in car sharing concepts is high within urban groups whilst very low within rural ones.

Seniors in rural, as well as in urban areas, show a high degree of car usage (Fig. 1), car ownership (Fig. 3) and domestic charging possibilities (Fig. 2). All these characteristics make them a target group for an early use of electric cars. Additionally, seniors are less often sensitive to higher prices of new cars (Figs. 5, 6). Urban seniors also stated to a higher degree than other groups that they will likely choose an electric car the next time they will purchase a car (Fig. 7).

Electric cars are until now much more expensive than fuel-driven cars of the same size and probably this situation will not change in the upcoming years. Electric cars affordable for a larger proportion of the population are nowadays mostly in the classes city cars or superminis. Potential users of electric cars should

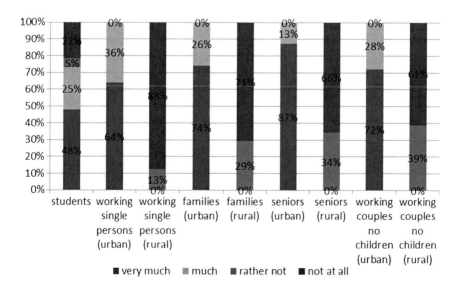

Fig. 4 Can you imagine using an electric car in car sharing?

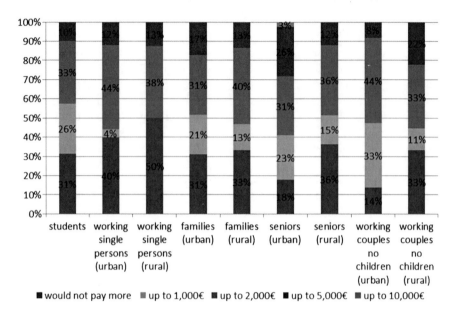

Fig. 5 Would you be willing to pay more for an electric car than for a conventional one and if yes, how much?

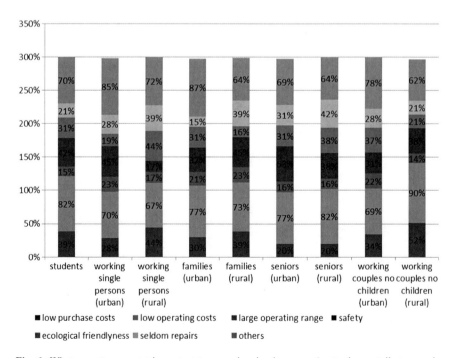

Fig. 6 What aspects are most important to you when buying a car (up to three attributes can be chosen)?

Identifying Consumer Groups with Satisfactory Characteristics

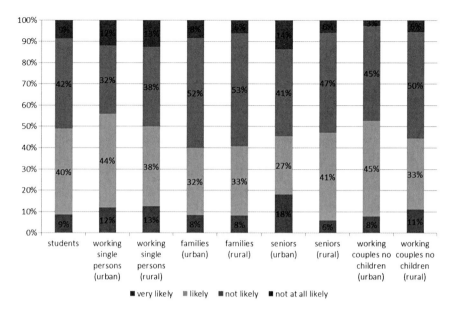

Fig. 7 How likely is it that you will choose an electric car at your next car purchase?

show a certain acceptance of these small cars. One question of our survey asks for the preferred class of a potentially purchased electric car. Figure 8 shows the results in respect to the investigated user groups. Families from rural areas alongside with students are often more affine to smaller types of cars than members of the other observed groups. Whereas urban seniors could be early adopters because of a high willingness to pay a premium for electric cars combined with a relatively high degree of car usage, rural families could be an important target group in the long run. As Fig. 3 shows and as already was mentioned above within this rural group in most cases more than one car is available which enables this group to select the car best matching to the current mobility need.

The presented analysis shows that our groups of investigation are characterized by different aspects making them more or less adaptable to electric mobility. However, it is unclear whether they really will adapt this technology and at which point in time this will take place. As already mentioned, diffusion of electric mobility has just begun and the identification of possible early adopters is crucial to the near future of the technology. Our investigation assigns users to specific categories of types of adopters based on specific statements given in the survey or by specific characteristics of the users. The categories and the underlying criteria are listed in Table 2. Individuals who are assigned to groups which occur in later periods of the diffusion process (early/late majority and laggards) are summed up within one category as it is currently very unclear how the development will take place. The picture drawn from this categorization of our user groups is depicted in

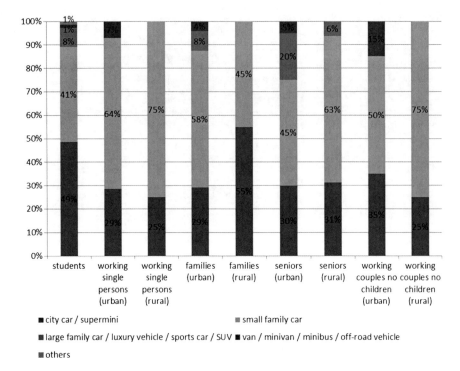

Fig. 8 Which class would you prefer if you would buy an electric car?

Table 2 Categories of diffusion process

Category	Criteria
Innovators	Current owner of an electric car
	Or
	Currently using an electric car frequently
Possible early adopters	Intention to buy a car within the next 5 years and it is very likely to be an electric car
Probable majority and laggards	Using a car as driver or passenger in the last 2 years but no or little intention to buy an electric car within the next 5 years
	Or
	Not using a car within the last 2 years but intention to buy a car within the next 5 years with no or little intention to be it an electric car
Probable non-users	No car usage within the last 2 years and no intention to buy any car within the next 5 years

Fig. 9. Even if all groups show similar patterns in general some aspects are interesting. Urban groups, especially urban singles, show a larger portion of probable non-users than rural groups. On the other hand urban seniors show the largest portion of possible early adopters.

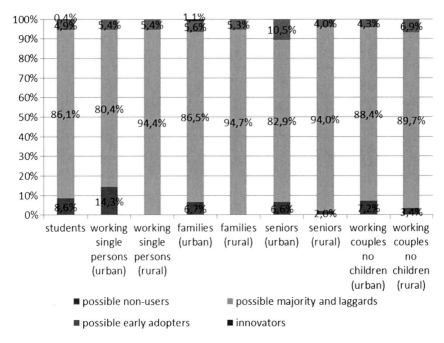

Fig. 9 Potential diffusion groups

5 Conclusions

Our investigation on mobility needs and attitudes towards (electric) mobility shows some patterns giving interesting hints for the identification of potential users of electric mobility. Personal attitudes, expectations towards electric mobility and the frequency of long-distant rides do not differ significantly among our user groups. Thus by these criteria it is not possible to identify potential user groups of electric mobility. However, in respect to many other aspects differences can be detected. One of the main conclusions is that the location of the domicile has a significant impact on mobility behaviours and car dependency. Persons living in rural areas are highly dependent on privately owned cars while city dwellers have often additional options to serve their mobility needs. Some user groups, especially typical inhabitants of densely populated urban centres like a person living in urban single-person households and students also often do not have appropriate charging facilities at home. This leads to the conclusion that the introduction of electric mobility in urban and rural areas requires different strategies. Private ownership of electric cars is most reasonable for people with possibilities to charge the battery at the domicile which is most common in rural and suburban areas. People without these possibilities in urban areas could be introduced to electric mobility by means of car sharing. Electric cars could be charged at selected public or semi-public locations and serve urban dwellers occasional needs of individual motorized mobility. The wide acceptance of electric mobility and car sharing concepts among urban dwellers might show good chances that electric mobility could be introduced to them by car sharing. Private ownership of electric among urban dwellers might be limited. Urban seniors could be one group with the potential to become early adopters of the technology through ownership. This group has, in contrast to other urban groups in our investigation, a high portion of owners of private parking facilities. Additionally, a large share of this group shows a high willingness to pay a premium for an electric car and is also likely to buy an electric car within the next 5 years. Hence, the introduction of electric mobility could be started in urban areas by the combination of car sharing for a larger number of citizens and private car ownership by selected wealthier urban inhabitants, especially seniors.

In the long run, when electric vehicles will become more established and prices will decline, the results of our investigation point to different target groups of electric mobility. Highly car dependent groups located in rural areas which normally own more than one car and have a high acceptance of smaller vehicles as well as domestic charging possibilities might become main users of electric cars. Especially families located in rural areas, but also, to a lesser extent, other groups like working couples without children, show these characteristics. If solutions for long distant rides can be offered to inhabitants in rural areas only owning an electric car or if an additional fuel-driven car is available to the household, it is very likely that electric mobility's final triumph will take place in rural areas. Summing up, Table 3 shows the most relevant user groups identified in our investigation.

A reduction in the rate of private car ownership in urban areas caused by the diffusion of car sharing concepts with electric cars leads to a decrease of urban traffic problems. Political decision makers could support this development by regulating the access to urban centres by fuel-driven vehicles. The development of

Table 3 User groups with high potential to adapt electric vehicles

Classification	Urban single person household	Urban seniors	Rural families
Characteristics	Low degree of car ownership and usage	Highest willingness to pay	High dependence on car
	Often no parking facilities on domestic lot	Often parking facilities on domestic lot	Good domestic charging possibilities
	Positive attitude towards car sharing	Positive attitude towards car sharing	Large degree of second car ownership
		Relatively high willingness to buy an electric car within the next 5 years	High acceptance of small vehicles
Conclusions	Possible target group for electric vehicles combined with car sharing	Possible early owner of electric cars (early adopters)	Possible main target group in the long-run

a good urban car sharing system, combined with additional efforts to improve public transport and bicycle networks leads to a cleaner, healthier and more silent city. Additionally, charging possibilities for commuters from the rural surroundings of the city are necessary. This holds even if this infrastructure is seldom used and the car owners always charge at home; the psychological effect that one could charge an electric car if needed, helps to reduce uneasiness and to increase the diffusion rate. But our findings imply that no encompassing and expensive charging infrastructure network is needed since most owners charge at home. Instead places for charging infrastructure should be selected very carefully by taking into account visibility as well as the likelihood that a driver really charges the car in a particular location.

Private owners of electric cars can reduce their personal pollution even further if renewable energy is used. If this energy is produced on the own lot this could lead to a significant cost reduction of their private mobility. Electric mobility could provide an important contribution to ecological improvements, if renewable energy is used and electric mobility is reasonably integrated in powerful public transport networks, bicycle infrastructures and car sharing models. Further scientific investigations should focus on this integration of electric mobility in future mobility systems as a possible driver of environmentally friendlier mobility behaviour. Our investigation gives first hints which groups have the potential to be powerful supporters of the establishment of electric mobility. Future research might identify even more precisely defined target user groups.

References

Ahrens G-A (2009) Endbericht zur Verkehrserhebung,Mobilität in Städten—SrV 2008 und Auswertungen zum SrV-Städtepegel, Dresden
Aktiengesellschaft A (2011) Aral Trends beim Autokauf 2011, Bochum
Biere D, Dallinger D, Wietschel M (2009) Ökonomische Analyse der Erstnutzer von Elektrofahrzeugen. ZfE Z Energiewirtschaft 02(2009):173–181
Bundesregierung D (2009) Nationaler Entwicklungsplan Elektromobilität der Bundesregierung
Häusel H-G (2008) Brainsights: Was das Senioren-Marketing von der Hirnforschung lernen kann. In: Meyer-Hentschel H, Meyer-Hentschel G (eds.) Jahrbuch Seniorenmarketing 2008/2009, pp 139–154
IMAD (2003) Mobilitätsanalyse 2002/2003. Innsbruck Stadt und Umlandgemeinden
Infas (2009) Mobilität in Deutschland 2008. Tabellenband Bremen, Bonn
Infas and DIW (2003) Mobilität in Deutschland 2002. Kontinuierliche Erhebung zum Verkehrsverhalten. Endbericht, Bonn/Berlin
Infas and DIW (2010) Mobilität in Deutschland 2008. Ergebnisbericht. Struktur—Aufkommen—Emissionen—Trends, Bonn/Berlin
KBA (2011) Emissionen, Kraftstoffe—Deutschland und seine Länder am 1. Jan 2011. http://www.kba.de/cln_031/nn_269000/DE/Statistik/Fahrzeuge/Bestand/EmissionenKraftstoffe/2011__b__emi__eckdaten__absolut.html. Accessed 20 Dec 2011
Lilienthal A (2007) Der Reisemarkt für Senioren. Salzwasser, Bremen
OECD (2001) Ageing and transport: mobility needs and safety issues. OECD Publication Service, Paris

Rogers EM (2003) Diffusion and Innovations, 5th edn. Free Press, New York
Sammer G, Meth D, Gruber CJ (2008) Elektromobilität—Die Sicht der Nutzer. Elektrotech Informationstechnik 125(11):393–400
Steinrück B, Küpper P (2010) Mobilität in ländlichen Räumen unter besonderer Berücksichtigung bedarfsgesteuerter Bedienformen des ÖPNV. Arbeitsberichte aus der vTI-Agrarökonomie, 02/2010
Strauch H-J (2008) Das neue Mittelalter—Die Zukunft einer werberelevanten Zielgruppe. In: Meyer-Hentschel H, Meyer-Hentschel G (eds.) Jahrbuch Seniorenmarketing 2008/2009, pp 159–178
Zumkeller D, Vortisch P, Kagerbauer M, Chlond B, Wirtz M (2011) Deutsches Mobilitätspanel (MOP)—wissenschaftliche Begleitung und erste Auswertungen. Bericht 2011: Alltagsmobilität. Karlsruhe

Author Biographies

Dominik Santner is research associate and PhD-student at the Centre for Regional and Innovation Economics (CRIE) at the University of Bremen. His work is centred on regional industrial clusters and networks. The thematic focus lies on the industries of agricultural engineering and electric mobility. He holds a diploma in geography of the Philipps-University of Marburg.

Dirk Fornahl leads the Centre for Regional and Innovation Economics (CRIE) at the University of Bremen. His work is settled on cluster analysis, analysis of regional supply chains and potential, innovation processes and networks. He also coordinates the socio-economic research in the electric mobility model region Bremen/Oldenburg.

What Do Potential Users Think About Electric Mobility?

Christian Hoffmann, Daniel Hinkeldein, Andreas Graff and Steffi Kramer

Abstract Whilst a growing number of agencies have begun to define individual electric vehicles as the key to sustainable transportation, others propose to use electric vehicles as a complement to public transport for intermodal and/or multimodal use (Canzler and Knie 2009; Beckmann 2010). However, little is known about the views of potential electric transport users. Who should offer such an integrated mobility product? Which integrated products should be offered? Which target groups should be addressed? What are their mobility needs? How do the peculiarities of electric vehicles (range, cost) match the transportation needs of users? This chapter reveals insights into these aspects by presenting four integrated mobility products to potential users in five focus groups. Five different product concepts were developed: pay-as-you-go electric bike (pedelec) rental for highly urbanised areas; e-carsharing for urbanised areas 'berlin mobil'; 'last mile' shuttle service, including free rides with e-cars for less densely urbanised areas; e-car-leasing including free local transit and a contingent of free long distance rail journeys; and pay-as-you-go car-rental (for further information see Scherf and Wolter 2011; see also Beckmann 2010). Earlier research identified four homogeneous user groups among users of integrated mobility products (Maertins et al. 2004a): pragmatic public transit users (Canzler and Knie 2009); eco-friendly cycle and public transport users (Beckmann 2010); pragmatic multi-modal users and fun-oriented car users (Maertins et al. 2004b, Maertins 2006). Members of each of

---

C. Hoffmann (✉) · D. Hinkeldein · A. Graff · S. Kramer
Innovationszentrum für Mobilität und gesellschaftlichen Wandel (InnoZ),
Torgauer Straße 12-15, 10829 Berlin, Germany
e-mail: christian.hoffmann@innoz.de

D. Hinkeldein
e-mail: daniel.hinkeldein@innoz.de

A. Graff
e-mail: andreas.graff@innoz.de

these groups were recruited for a two and a half hour focus group to discuss different aspects of the five above-mentioned mobility products: price models; business models; compatibility in every day use; and trust in potential operators (transportation; energy). Additionally, a group of car-sharing clients was invited to delve deeper into usage problems, since carsharing users are already experienced with the system of instant access.

1 Introduction

German Government plans to bring up to 1 million electric cars on the road until 2020 as part of Germanys Climate- and Traffic-Policy (BMWI et al. 2011). Generally, the ecologic effects of electric cars are discussed to be dependent an many variables (Hacker et al. 2009), such as energy production, energy demand per vehicle or market penetration with electric vehicles. Recent studies (Sperling and Gordon 2009; Zimmer et al. 2011) confirm that electric vehicles have an eco-friendly advantage over conventional cars if green energy is used, especially, if additional capacities of green energy production are added for this. Also, electric vehicles have high public acceptance (Die Bundesregierung 2009; TÜV Rheinland 2010) and high acceptance by users (Wolter and Faltus 2011; Dütschke et al. 2012).

This chapter takes a qualitative study as its subject of discussion, which examines the integration of electric mobility into the public transport system, and beyond that the concept of e-carsharing.[1] Electric and hybrid cars[2] were tested in several studies in the project BeMobility/Berlin elektroMobil (Scherf and Wolter 2011; Wolter et al. 2011) as part of the Flinkster-Carsharing. In the beginning, group discussions, which are described here, were organized initially to develop the basis for a more extensive field trail later. The results of the current quantitative research will be reported elsewhere.

---

[1] Carsharing is an organized form of jointly using one or more cars. The precondition is a membership of a cooperative or club or a contract with a carsharing-provider. In contrast with car rental it is planned for a longer period and a vehicle could be rented for a shorter time. The vehicles belong to the provider, who is responsible for the condition of the cars and receives payment from the users. If there is a written arrangement between neighbours about using a car together, it is carsharing in a wider sense, too (Öko-Institut 2004; Baum and Pesch 1994).

[2] There are two types of vehicles used in this project: firstly, electric vehicles which have an electric motor only, a range between 100 and 150 km, and a charging time between 4 and 8 h; and secondly, plug-in-hybrid (in the following called 'hybrid') vehicles, which have both an electric and combustion motor. The range is about 25 km if driven electrical only and like cars with combustion motor if driven with combustion motor. Charging time is about 1.5 h.

2 Theory and Current Research

To compensate for the compared to combustion engines limited range of electric vehicles (between 100 and 150 km) it would seem useful to combine them with other services such as local public transport and carsharing (Canzler and Knie 2011). Regarding possible users, there could be a focus on those living in metropolitan areas, since they would be more likely to use local transport services instead of owning a car (Canzler et al. 2007). Research carried out by the InnoZ (Innovation Centre for Mobility and Societal Change) shows us that very mobile people ('Multimodale') with high levels of education are especially open-minded about services including electric cars (Scherf and Wolter 2011).

Regarding these potential users of electric cars or mobility services based on electric vehicles, it is important to ascertain the influencing factors on travel mode choice. In the discipline of environmental psychology, several influencing factors on the choice of travel mode were found e.g., environmental attitudes and perceived behavioural control (Hunecke and Wulfhorst 2000; Harms 2003; Hoffmann 2010). Research on carsharing and Call a Bike (Hoffmann 2010) identified several influence-Factors on customer loyalty, e.g., ease of use, price-system or attractiveness, which served as input to the research-plans of the qualitative and quantitative studies in the project BeMobility/Berlin elektroMobil of which the first results are reported here.

Growing openness towards electromobility services
Mobile services like "Car2Go", where vehicles could be rented open-end and one-way, have recently become very popular (Solberg 2009); also, trends predict a growth of carsharing in the future (Frost and Sullivan 2010). But to fulfil the user needs of convenience and flexibility, new forms of integration and product concepts are needed (Mitchell et al. 2010). In this chapter, we explore some ideas for new electromobility services that were tested as the starting point of some field research, utilising real services in the Berlin mobility market (BeMobility, see below).

Typology of users as the empirical basis to carry out research on test users[3]
The distinction between mobility types leads though the analysis of the whole range of private customers. In the last 10 years, separate research on mobility types, with the intention to identify attitudes that are relevant to behaviour and to express them as specific target groups, has emerged (Hunecke and Haustein 2007; Maertins 2006; Hoffmann and Stolberg 2005; Stolberg and Hoffmann 2005). Development trends in the field of personal attitudes and preferences are described as target-group-specific and applied in those terms to the market of passenger transport. On the basis of these analyses, estimations of the expectable run-up of

---

[3] To analyse user needs it is necessary to distinguish private from institutional customers. This study will focus on private customers.

product alternatives, willingness to pay and necessity of local variation will be available.

Previous research provides the tools to distinguish between these different types of private costumers (Hunecke and Haustein 2007). Hereafter a summary table of the INTERMODI-mobility types is used as the basis of the study reported in this chapter. This typology bases on results of the INTERMODI-Project (for detailed descriptions see Maertins 2006; Hoffmann and Stolberg 2005; Stolberg and Hoffmann 2005):

1. *Pragmatic public transit users* decide in favour of the one or the other means of transport for pragmatic reasons, and values are scarcely of any importance. Mobility in everyday life is seen as secure because of the public transit system—they have low car ownership and high use of monthly tickets for public transport.
2. *Eco-friendly cycle and public transport users* view cars negatively, so everyday mobility could be realized through bicycle and public transport. These people have a high level of education, but a below average salary, low car ownership and high use of monthly tickets for public transport. Many distances are managed on foot.
3. *Pragmatic multimodal users* are very mobile business people, who travel frequently and for long distances. Their requirements regarding means of transport are very high. In general they have a positive attitude to public transport but their environmental values are not strong. Public transport is less suitable for securing mobility (refusal of bicycles). They have high car ownership and high use of monthly tickets for public transport, and use means of transport pragmatically.
4. *Fun-oriented car users* have low environmental values, and for them public transport means low autonomy. The most positive rating of cars can be found in terms of availability, especially in view of autonomy, flexibility and fun. Ownership of a car enhances social position (highest car ownership).

Based on these findings, the focus groups were developed for the identification of the qualitative dimensions of attitudes and evaluations on mobility services including electric driven vehicles. Also in these focus groups evaluations on the first product concepts of mobility services based on electric vehicles could be gained.

3 The BeMobility-Project and its Research Issue

At the moment there are several different projects (Bühler et al. 2011a, b), which examine electric vehicles; BeMobility is one of these projects with a special "[…] focus [on] the development of integrated concepts, their practical implementation as well as complimentary research of customer acceptance and traffic effects" (Wolter 2010). It is essential for the project to organize a service of a public and integrated fleet of 40 cars in Berlin for two years. Intensive studies on potential customer needs, user profiles and business concepts will be pursued simultaneously.

The approach of integration is consequently geared towards the private and commercial interests of customers and to the needs of urban development, so that the preconditions for the realization of an innovative, efficient public transit system should be created.

Further goals are to develop customer-friendly and -orientated electromobility offers, which convey innovations in the field of electromobility to users, improve fair prices for use and are more environmentally friendly.

4 Methods

The method of focus groups was used to reach the aims outlined above. The goal was to collect qualitative empirical data about the previous knowledge, opinions and evaluations of the interviewed target groups in reference to electric vehicles that are integrated in public traffic.

This was realized through five group discussions (GD), and each took approximately two and a half hours and had 7–10 participants. The first wave of research took place in December 2009 (GD 1–3), and the second wave in February 2010 (GD 4 and 5). The interviews were conducted in the studio of a market research institute.

Each focus group was mixed considering the INTERMODI-typology of the mobility and residence of the interviewees (according to the three tariff zones of urban public transport in Berlin). Intermodi-types are important in this study because the project INTERMODI discovered clear differences in behaviour regarding transport use, distance travelled and carbon footprint between the mobility types, based on the typification of DB (Deutsche Bahn)-Carsharing and -Call-a-Bike customers (Maertins 2006).

Each discussion was filmed and recorded, and afterwards two independent raters evaluated them. Finally the results were discussed and validated with experts from the BeMobility consortium.

Within the focus groups (described below) the following aspects were examined:

- Analysis of current mobility behaviour;
- Experience and attitude towards carsharing;
- Perception and estimation of electric and hybrid vehicles;
- Acceptance of the integration of electric vehicles into the public transport system;
- Evaluation of three market models and the collection requirements on their respective realizations;
- Appraisal of three specific model suggestions: pay-as-you-go rental; e-car-sharing for urbanised areas; and e-car-leasing, including free local transit and a contingent of free long distance rail journeys.

5 Results

The results of the focus groups will be described in this chapter. Before going into detail, an overview will be given.

5.1 Overview

In general, the combination of public transport and electric vehicles was rated very highly, but on closer examination, differences in willingness of use between the presented product concepts, mobility types, location and availability of means of transport could be found. Some of the price systems were generally evaluated poorly because of their expense. Multi-modal concepts were seen as suitable only by the few interviewed multi-modal users. The willingness of use is mainly influenced by the everyday suitability of the product concepts. In the following section, the results will be described in depth.

5.2 Attitudes and Evaluations of Carsharing

At the beginning of the focus groups there were some questions about carsharing in general. This approach was chosen to examine experiences, attitudes and reservations that may influence the presented product concepts structurally. During the analysis of the records, differences between carsharing customers and those, who don't use carsharing, were found.

The principle of carsharing is perceived as inexpensive by both carsharing customers and non-customers. Carsharing customers see carsharing as flexible, whilst the others expect limits if they use the service (above all if they own a car and had strong car use routines). From their own experience, carsharing customers know how simple the principle is—the others suspect that it is more complicated. The environmental benefit from carsharing is more seen by those who are not carsharing customers. Driving pleasure polarizes the customers of carsharing: one group of customers is very interested in new attractive models and fun-oriented driving. For the other group, carsharing vehicles are only means of transport without any attraction. For both groups, carsharing mainly seems to be innovative, and it fits into the transport needs of the city.

Furthermore, differences between automobile owners and those, who don't own a car, were found. Automobile owners think that they would have some mobility restrictions for smaller distances if they used carsharing because they wouldn't have a car in front of their door.

In summary, carsharing is seen as suitable for daily use. There were no extreme positions relating to carsharing that could influence the evaluation of the presented product concepts.

5.3 Alternative Motor Vehicles (Electric and Hybrid)

Electric and hybrid vehicles were of interest to the focus groups in general, but at the moment they produce polarizing feelings because of their highly innovative nature.

On the one hand, users and owners would gain high recognition (for a pioneering role), but on the other hand some of the interviewees had worries that they could make mistakes (e.g., putting vehicles into service) or overestimate technical conditions (in particular the range and availability of filling stations). With regard to the environment, new motors seemed on the whole to be more eco-friendly than conventional engines for the interviewees. But if they had to evaluate environmental-friendliness, most only thought of exhaust emissions. The complete lifecycle of car production, battery production, energy production and consumption was rarely recognized by most of the participants except some, who were extremely well informed.

Regarding the vehicles there were two response categories:

1. Responses for electric vehicles: participants considered performance parameters (above all speed); range and charging times were seen as important; range was only seen as sufficient for driving out of cities; and charging times were regarded as partially critical.
2. Responses for plug-in-hybrid vehicles: participants thought that a 25 km range of battery was rated sufficient for cities (recharging during journey), but regarding medium and long distances the long range of combustion engine was rated positively.

5.4 Requirements for Electric and Hybrid Vehicles

One group of questions concerned requirements for electric and hybrid vehicles. At the first question, where the interviewees were asked about their background knowledge of this topic, some people were well informed, but most only had rudimentary knowledge about electric and particularly hybrid vehicles.

Therefore, all interviewees were provided with basic knowledge of electric and hybrid vehicles via an information sheet after the first part of the focus group. In the following Table 1 the requirements of the participants are described.

Table 1 Requirements on electric and hybrid vehicles

Range	The responses covered the whole spectrum of potential ranges: city, outer conurbation area and longer distances (100–1,000 km; mode: 500 km). The influence of the expected range on the willingness of use always had a correlation with accomplishing everyday tasks
Booking	In this topic, there were two tendencies, which are similar to ownership of a private car. On the one hand the interviewees wanted flexible access to a car at any time, and on the other hand the opportunity of reservation was important
Return	Most of the interviewees wished there was an option of return throughout Germany, not just at the rental station
Equipment	The majority want the same equipment and safety as small and medium cars have
Locations/charging stations	It was important for most of the interviewees that charging stations are easily reachable. In addition, a large number want a location for rental and return at access junctions to public transport

5.5 Integration of Electric Vehicles into Public Transport: Requirements for an Integrated Concept

Overall, the integration of electric vehicles into public transport was seen as an appropriate addition to the public transport system. There were some notable characteristics of people with an affinity towards cars and of those who live in outer districts of the city: because they generally use cars in everyday life, a journey with several changes of means of transport was seen as complicated and exhausting:

> That's like a pursuit in a spy movie.

Public transport is taken for granted by most of the interviewees and therefore this field is hardly reflected. In the open setting of the group discussion they largely focused on characteristics of electric cars. The requirements of customers included service, availability, vehicles network and vehicles (Table 2).

5.6 Evaluation/Discussion Product Concepts

Within a differentiating evaluation of the focus groups, the particular service components were checked and the willingness to pay was queried, as well as identifying other requested additional services, which could increase the attractiveness of the product concepts.

5.6.1 Approach and Dimension of Evaluation

The willingness of use was also queried because if there was an affinity for a product concept there was not necessarily an automatic willingness of use

Table 2 Requirements for an integrated concept

Service	Interestingly, it emerged during the discussion on service that some interviewees had had positive experiences with the Deutsche Bahn carsharing hotline, and they set this as the standard for electric vehicle booking hotlines. Furthermore, a good price-performance ratio and the ability to book without staff (for example via the internet) were desired
Availability	Vehicles should be available and bookable 24 h a day. Furthermore, reserved parking space would be useful
Vehicles network	An area-wide system is required, that is to say optimal distances below 500 meters to the next location for rental and return, and a concentration of network points at access junctions to public transport
Vehicles	One striking requirement was diversity in car models provided: for different reasons of use (e.g., transportation, shopping, leisure time) and preferences, different vehicles should be available. Electric vehicles are recommended for city centres above all because of their short range

following from that (e.g., if one owns a car and routinely uses it). Along with general impressions, the following aspects were raised during the group discussions (according to Hoffmann 2010).

- Reasons to use: for which reason could the interviewees imagine using the presented product concept?
- Ease of use: do the described product concepts seem to be easy to use?
- Price system: is the price system transparent and are the prices fair?
- Attractiveness: is the presented product concept an attractive option for everyday life?

Before discussing the results it is recommended to have a look at the table to get an insight of the different product concepts (Table 3).

The given product concepts were evaluated very differently and the results are summarized in the following paragraphs.

Attractiveness: Product concepts like pay-as-you-go rental, Berlin mobil and pedelecs were more attractive to the interviewees than the other options. Pay-as-you-go rental and the opportunity to be most flexible are very attractive, and furthermore the option to use pay-as-you-go usable e-cars could be combined with an annual ticket for public transport. Pedelecs have very good ratings as an environmentally credible mode of transport for tourists and leisure time. In both models the prices fit best to the ideas of the interviewees. E-car-leasing and Last Mile ("sounds like the way to the grave") were not very attractive to the interviewees. Last Mile did not seem to fit well with their daily lives, and e-car-leasing was seen as too expensive.

Reasons to Use: Generally, the evaluations of product concepts were higher the better they fitted the everyday transport needs of the interviewees. Late journeys home, object transportation, business meetings and pay-as-you-go needs were named for berlin mobil/pay-as-you-go rental; leisure time and small errands were named especially for pedelecs. Only a few interviewees regarded pedelecs as

Table 3 Overview on product concepts

Product concept	Description
E-carsharing for urbanised areas ("berlin mobil")	Price per hour and per km and extended monthly ticket for public transport
	Cars at important connection points to public transport
	Return at the rental location only
	Rental of cars possible at every rental location
Last mile shuttle service including free rides with e-cars for less densely urbanised areas ("Letzte Meile")	Basic fee per month incl. power, km and collective transport to suburban train
	Overnight-rent
	Precondition: monthly ticket for public transport
	Return at the rental location only
E-car-leasing including free local transit and a contingent of free long distance rail journeys ("Langzeitmiete")	Basic fee per month incl. monthly ticket for public transport and long distance rail ticket
	Charging stations at main railway stations, at home, at work
Pay-as-you-go rental ("Spontanmiete")	Pay-as-you-go use
	Vehicles in public space of a defined area
	Price per minute
	Return possible on every street of defined area

adequate for everyday life in Berlin. In the case of e-car-leasing, the given reasons of use were similar to those of private vehicle traffic but the price and the low range were problematical.

Ease of Use: Berlin mobil and pay-as-you-go rental had good ratings because of the opportunity for use as required and for the flexibility. However, the limited area where you could rent and return an e-car was seen as a handicap. Pedelecs had good ratings particularly because of the opportunity for pay-as-you-go rental. Bad experiences with the quality and cleanness of hired bicycles (above all Call a Bike) were negative points, which are assumed for pedelecs also. Reachability and the notable query as to whether use with a guest is easy or even possible worried the interviewees. Last Mile and its concept of collective transport was seen very negatively due to the possibilities of delays and the unpunctuality of other users, fear of unpleasant passengers regarding the limited space, and missing privacy respectively (the last point was made especially by people with high car-using routines).

Price system: The view of the interviewees was that the price systems of the product concepts pay-as-you-go rental, berlin mobil and pedelecs were fair, but that the other product concepts were too expensive. Interestingly, the majority of people calculated prices per hour or per ten minutes for each product concept before rating them. Some interviewees would feel stressed if there were a price per minute (clock ticking).

5.6.2 Evaluation of Potential Providers

The question of potential providers produced different impressions. An overall view of the ratings is given in the following Tables 4 and 5.

In summary, if providers' logistic system is seen as being good and if they have experience with logistics and mobility, they receive a good rating. Consortiums, which unite several competences, for example mobility (carsharing provider) and generation of energy (electricity company), are evaluated more positively too.

In January/February of 2010 Deutsche Bahn and S-Bahn Berlin (a local public transport provider in Berlin) had some problems and many cancellations, so the ratings for these providers were much better in the first wave of group discussions than in the second one.

5.6.3 Evaluation Differentiated by Mobility Types

Generally there was more openness towards and higher ratings of the product concepts from the interviewees with an affinity towards public transport than from interviewees with an affinity towards private cars. In addition, there is a clear interaction between the availability of public transport, which varies between residential areas, and the ratings: if there is less availability of public transport (which goes together with an almost 100 % ownership of private cars) product concepts based on carsharing were mostly seen as not useful for everyday life. Costs of the product concept e-car-leasing were evaluated as much too high by

Table 4 Potential providers

Provider	Competence	Environmental credible
Local public transport (e.g., BVG)	Existing logistic system, but unreliable	Credible (e.g., eco ticket)
Car rental firm (e.g., Sixt)/ carsharing	Logistic system and competence in renting	Carsharing environmentally credible
Deutsche Bahn (German Rail)	Existing logistic system, but: unreliable	Credible, good environmental understanding
Electricity company (e.g., RWE, Vattenfall)	Fit in with e-cars, cross-selling, but low mobility competence	Unclear whether the power is clean or not
Provider of green energy (e.g., Greenpeace, LichtBlick)	Fit in with eCars, cross-selling, but low mobility competence, often unknown	Credible as provider of green energy
Car manufacturer (e.g., VW, Mercedes)	Technical competence, missing logistic competence and system	Continual further development of cars

Table 5 Potential providers—consortiums

Provider	Competence	Environmental credibility
Large store (e.g., IKEA)	Nothing special, but more suitable (transporting goods)	Credible with limits
ADAC (German automobile club)	Professional, fair with many competences, e.g., competence in renting	Credible with limits (automobile lobby)
Car manufacturer and electricity company	Competence in production, cross-selling, no rental logistics	Credibility higher if green energy
Carsharing provider and electricity company	Competence with eCars, cross-selling, rental logistics	High credibility
Deutsche Bahn and electricity company	Competence with eCars, cross-selling, rental logistics	Credibility higher if green energy
Local public transport and electricity company	Existing logistics system, cross-selling	Credibility higher if green energy

nearly all interviewees. Only a few business people understood the high prices, but nevertheless this concept was too expensive for them.

6 Discussion

Overall, the results of the interviews indicated strong fundamental acceptance of electric vehicles. At this stage, some of the interviewees have a high expert knowledge that will in all likelihood continue to develop because of the high media coverage of this topic. One further observation is that it was essential for almost all interviewees that the operating power is obtained from renewable sources.

The calculated prices of many product models are seen as too expensive. Many customers compare prices and services with their options and costs today, especially those who are professionally mobile. If the costs are too high, attractive offers are often rejected.

Both product concepts, pedelecs and pay-as-you-go rental, get a good rating from those with an affinity for public transport and who live in the city centre. A huge willingness to use these models with the presented price structure could be observed. However, user engagement is very small if the product concept is being seen as inappropriate (e.g., Last Mile shuttle service) or as too expensive (e.g., e-car-leasing including free local transit).

Generally, the results of the group discussions show that both practicability and price structure could be decisive factors for a drop in the willingness to use, even if there is a general approval of the product offer. In the case of private car ownership and very frequent use, the willingness to use these products is generally low proportional to how well interviewees evaluate their connection to the public transit system.

In principle, the results indicate that mobility types with an affinity towards public transport, who have a good connection to the public transport system, are easier to win over to the concept of electric vehicles as a supplement to public transport. The product concepts investigated here—pay-as-you-go rental, pedelecs and Berlin mobil—constitute a good basis, but need to be matched to customer requirements in detail.

This qualitative study was the first step to generate dimensions of evaluation of the presented services and by this the empirical basis of evaluating the project BeMobility; those that follow are different waves of quantitative studies—here the statistical details of the evaluation of mobility concepts will be in focus. The first results were published in this volume (see Hasse 2011, for further results see Wolter and Faltus 2011; Scherf and Wolter 2011).

Acknowledgments This study was funded by the German Federal Minister of Transport, Building and Urban Development (BMVBS—Bundesministerium für Verkehr, Bau und Stadtentwicklung), whose support made this research possible.

References

Baum H, Pesch S (1994) Untersuchung der Eignung von Car-Sharing im Hinblick auf Reduzierung von Stadtverkehrsproblemen. Köln

Beckmann K (2010) Elektromobilität. Hoffnungsträger oder überschätzte Chance des Stadtverkehrs? Difu-Bericht. Retrieved from http://www.difu.de/publikationen/difu-berichte-22010/elektromobilitaet.html. Accessed 29 July 2011

Bühler F, Neumann I, Cocron P, Franke T, Krems JF (2011a) Usage patterns of electric vehicles: a reliable indicator of acceptance? Findings from a German field study. Retrieved from http://www.tu-chemnitz.de/hsw/psychologie/professuren/allpsy1/pdf/Buehler%20et%20al.,%202011.pdf. Accessed 11 Dec 2011

Bühler F, Neumann I, Cocron P, Franke T, Krems JF (2011b) Enhancing sustainability of electric vehicles: a field study approach to understanding user acceptance and behaviour. Retrieved from http://www.tu-chemnitz.de/hsw/psychologie/professuren/allpsy1/pdf/Franke%20et%20al.,%202011.pdf. Accessed 11 Dec 2011

Bundesministerium für Wirtschaft und Technologie (BmWi), Bundesministerium für Verkehr Bau und Stadtentwicklung (BMVBS), Bundesministerium für Umwelt Naturschutz und Reaktorsicherheit (BMU), Bundesministerium für Bildung und Forschung (BmBF) (2011) Regierungsprogramm Elektromobilität. Berlin

Canzler W, Knie A (2009) Grüne Wege aus der Autokrise. Vom Autobauer zum Mobilitätsdienstleister. Heinrich-Böll-Stiftung. Retrieved from http://www.boell.de/downloads/wirtschaftsoziales/Autokrise_Endf.pdf. Accessed 29 July 2011

Canzler W, Hunsicker F, Karl A, Knie A, König U, Lange G, Martins C, Ruhrort L (2007) DB mobility: Beschreibung und Positionierung eines multimodalen Verkehrsdienstleisters. InnoZ-Baustein Nr. 1, herausgegeben vom Innovationszentrum für Mobilität und gesellschaftlichen Wandel (InnoZ) GmbH, Berlin

Canzler W, Knie A (2011) Einfach Aufladen. Neue Beweglichkeit durch Elektromobilität. München

Die Bundesregierung (2009) Nationaler Entwicklungsplan Elektromobilität der Bundesregierung. Retrieved from http://www.now-gmbh.de/uploads/media/Nationaler_Entwicklungsplan_Elektromobilit_07.pdf. Accessed 29 July 2011

Dütschke E et al (2012) Roadmap zur Kundenakzeptanz Elektromobilität. Technologie-Roadmapping am Fraunhofer ISI: Konzepte—Methoden—Praxisbeispiele Nr.3. Karlsruhe

Frost and Sullivan (2010) Carsharing: a sustainable and innovative personal transport solution. Retrieved from http://www.frost.com/prod/servlet/market-insight-top.pag?Src=RSS&docid=190795176. Accessed 29 July 2011

Hacker F et al (2009) Environmental impacts and impact on the electricity market of a large scale introduction of electric cars in Europe. Critical review of literature. European topic centre on air and climate change (ETC/ACC). Retrieved from http://air-climate.eionet.europa.eu/docs/ETCACC_TP_2009_4_electromobility.pdf. Accessed 29 July 2011

Harms S (2003) Besitzen oder Teilen. Sozialwissenschaftliche Analyse des Carsharing. Zürich

Hasse S (2011) Customer needs and acceptance of electric vehicles as part of integrated transport concepts. Presentation at the final conference of the socio-economic research of the "Model region electric mobility Bremen/Oldenburg", 18th and 19th August 2011, Jacobs University in Bremen

Hoffmann C (2010) Erfolgsfaktoren umweltgerechter Mobilitätsdienstleistungen: Einflussfaktoren auf Kundenbindung am Beispiel DB Carsharing und Call a Bike. Dissertation an der Universität Osnabrück < urn:nbn:de:gbv:700-201011046672>

Hoffmann C, Stolberg A (2005) INTERMODI—Wirkungsbilanz Carsharing: Kundensegmentierung auf der Basis von Mobilitätsorientierungen und soziodemografischen Merkmalen. Wissenschaftszentrum Berlin. Retrieved from http://www.wzb.eu/gwd/mobi/projects/closed/projects_closed.de.htm. Accessed 29 July 2011

Hunecke M, Haustein S (2007) Einstellungsbasierte Mobilitätstypen: Eine integrierte Anwendung von multivariaten und inhaltsanalytischen Methoden der empirischen Sozialforschung zur Identifikation von Zielgruppen für eine nachhaltige Mobilität. Umweltpsychologie 11(2):38–68

Hunecke M, Wulfhorst G (2000) Raumstruktur und Lebensstil—wie entsteht Verkehr? Internationales Verkehrswesen 52:556–561

Maertins C (2006) Die Intermodalen Dienste der Bahn: Mehr Mobilität und weniger Verkehr? Wissenschaftszentrum Berlin. Retrieved from http://bibliothek.wzb.eu/pdf/2006/iii06-101.pdf . Accessed 29 July 2011

Maertins C, Hoffmann C, Knie A (2004a) Automobil mit der Bahn. Bilanz zur Markteinführung von Call a Bike und DB Carsharing. Internationales Verkehrswesen 56:38–40

Maertins C, Knie A, Hoffmann C (2004b) Die automobile Bahn. Erfahrungen und Potentiale von Auto- und Fahrradbausteinen bei der Deutsche Bahn AG. In: Zanger C, Habscheid S, Gaus H (eds) Bleibt das Auto mobil? Mobilität und Automobil im interdisziplinären Diskurs. Frankfurt am Main, Peter Lang

Mitchell WJ, Borroni-Bird CE, Burns LD (2010) Reinventing the automobile. personal urban mobility for the 21st century. Cambridge, MA

Öko-Institut (2004) Bestandsaufnahme und Möglichkeiten der Weiterentwicklung von Car-Sharing. Retrieved from http://www.oeko.de/oekodoc/247/2004-032-de.pdf. Accessed 29 July 2011

Scherf C, Wolter F (2011) Multimodales Mobilitätsmanagement. Internationales Verkehrswesen 63:53–57

Solberg P (2009) Adieu ÖNVP—Car2go-Projekt in Ulm. Süddeutsche Zeitung. Retrieved from http://www.sueddeutsche.de/auto/cargo-projekt-in-ulm-adieu-oepnv-1.452002. Accessed July 29 2011

Sperling D, Gordon D (2009) Two billion cars. Driving toward sustainability. Oxford University Press, New York

Stolberg A, Hoffmann C (2005) INTERMODI—Wirkungsbilanz Call a Bike: Kundensegmentierung auf der Basis von Mobilitätsorientierungen und soziodemografischen Merkmalen.

Wissenschaftszentrum Berlin. Retrieved from http://www.wzb.eu/gwd/mobi/projects/closed/projects_closed.de.htm. Accessed 29 July 2011

TÜV Rheinland (2010) Ergebnisse der repräsentativen Befragung zur Akzeptanz von Elektroautos. Retrieved from http://www.tuv.com/media/presse_2/pressemeldungen/Ergebnisse_Umfrage_E-Mobilitt.pdf. Accessed 29 July 2011

Wolter F (2010) BeMobility—Berlin elektroMobil: The Future of Urban Transport. Retrieved from http://bemobility.de/site/bemobility/zubehoer__assets/de/dateianhaenge/bemobility__englisch.pdf. Accessed 29 July 2011

Wolter F, Faltus E (2011) BeMobility Berlin elektroMobil: Multimodal und elektrisch unterwegs. Retrieved from http://www.innoz.de/fileadmin/INNOZ/pdf/Brosch%C3%BCren/BeMobility_Apr_2012.pdf. Accessed 28 Mar 2012

Wolter F, Hasse S, Heinicke B (2011) Intelligent vernetzen. Internationales Verkehrswesen 63:16–19

Zimmer G et al (2011) OPTUM: Optimierung der Umweltentlastungspotenziale von Elektrofahrzeugen. Retrieved from http://www.pt-elektromobilitaet.de/projekte/begleitforschung/optum-1/copy_of_5BMUFKZ16EM0031_OPTUM.pdf. Accessed 30 Feb 2012

Author Biographies

Christian Hoffmann holds a degree in psychology. His research focuses on the relationship between human characteristics and mobility-behaviour, e.g., on customer loyalty in mobility-services. At the Innovation Centre for Mobility and Societal Change (InnoZ) he is concerned with method development and standardisation for mobility research and consumer acceptance. Dr.Christian Hoffmann has been co-editor of the german journal "umweltpsychologie" (environmental psychology) from 1997 to 2012.

Daniel Hinkeldein holds a degree in transportation engineering. He has worked as project manager at the German Aerospace Center (DLR e.V.) before he joined the Innovation Centre for Mobility and Societal Change. Dr. Daniel Hinkeldein is currently leading national and international OEM projects dealing with consumers 'acceptance of electric vehicles and publicly funded projects on consumers acceptance' of transport services that integrate public transport and electric vehicles.

Andreas Graff graduated from the Technical University Berlin (TUB) with a degree in Sociology and Technology Studies. He has worked as a research assistant at the chair of sociology at TUB where he taught methodology of scientific research and was involved in several projects concerning this topic. His work at InnoZ focuses on method development and standardisation for mobility research and consumer acceptance.

Steffi Kramer holds a degree in applied geography and geographical information systems. She has worked as transport co-ordinator and sustainability manager before she joined the Innovation Centre for Mobility and Societal Change (InnoZ) in 2011. Her research focuses on sustainable mobility from a social scientific perspective. Steffi Kramer has been involved in projects dealing with consumers acceptance of transport services that integrate public transport and electric vehicles.

Electric Car Sharing as an Integrated Part of Public Transport: Customers' Needs and Experience

Steffi Kramer, Christian Hoffmann, Tobias Kuttler and Manuel Hendzlik

Abstract The project BeMobility/Berlin elektroMobil aims to investigate the benefits and draw backs of electric vehicles as part of the public transport system. Therefore, about 40 electric vehicles (EV) were integrated into a public car sharing system in Berlin and research was conducted measuring users acceptance of this service. By utilising online surveys this service was tested for user friendliness, everyday usability and its modal integration. The study revealed unique user groups whose travel patterns are multimodal and dominated by public transport. These sample groups can be characterised as environmentally friendly and open-minded towards car sharing and mobility concepts in general. Positive expectations and experience with EVs are related to quality and pleasure of driving whilst aspects of charging and costs are considered in a critical way. In order to cater for everyday mobility an integration of electric car sharing into Berlin's public transport system is perceived as sensible. This will provide the opportunity to compensating for limitations of electric vehicles.

S. Kramer (✉) · C. Hoffmann · T. Kuttler · M. Hendzlik
InnoZ—Innovationszentrum f. Mobilität u. gesellschaftlichen Wandel GmbH,
Torgauer Str. 12–15, 10829 Berlin, Germany
e-mail: steffikramer.berlin@yahoo.de

C. Hoffmann
e-mail: christian.hoffmann@innoz.de

T. Kuttler
e-mail: tobias.kuttler@innoz.de

M. Hendzlik
e-mail: manuel.hendzlik@innoz.de

1 Background

Challenges in transport are linked to emissions from combustion engines, limited fossil fuel resources and provision of physical space for transport infrastructure. There is a worldwide trend towards a further increase in the number of vehicles of which about 90 % is due to road traffic. There is no single way to approach these challenges but there must be changes to travel behavior, mobility concepts and usage of fossil fuels in order to reduce CO_2-emissions from transport (Brake 2009). Integrated mobility concepts such as sharing concepts will play an important role in future urban transport. A greater share of fewer cars combined with public transport could lower traffic volumes and increase quality of urban life. These can be complemented by electric vehicles (EV), especially as they provide low carbon transport when using energy from renewable resources (Sperling and Gordon 2009).

Key benefits of electric cars for users have been widely discussed in media and can be summarized as "intersection of clean and fun" (Turrentine 2011). Limitations of electric vehicles are of less importance in a car sharing setting. Users have the opportunity to switch to alternative modes of transport if their needs are not met by the electric car.

E-Flinkster is the national brand for the car sharing concept of Deutsche Bahn (German Rail) that offers electric vehicles (Smart ED, Citroen C1, Citroen Zero, Toyota Prius Plug-In Hybrid) as part of Berlin's public transport.

"BeMobility—Berlin elektroMobil" is one of the projects in the Electric Mobility Pilot Region Berlin/Potsdam funded by the German government (German Federal Ministry of Transport, Building and Urban Development). Its aim is to combine electric vehicles and public transport, hence offering easy access to an integrated transport system. Usability is increased by providing associated services such as integrated information and offers. Successful cooperation between project partners from the energy sector, public transport, information technology, automotive industry, science and public services made best practice that could be adopted by other regions (Scherf and Wolter 2011).

Research was conducted as part of the project identifying costumer needs of electric vehicles, car sharing and car sharing with EVs. This chapter describes the socio-demographic characteristics of participants and their travel behavior in detail and presents the results of the surveys relating to expectations and experiences with EVs in car sharing.

2 Previous Research

Research has been done on both supply and demand side of new mobility concepts. Maertins et al. (2004) suggest that the combination of public transport and private motorisation is able to offer tailored modal choice to the customers. Each mode is applied according to its strengths and hence overall efficiency is increased.

Research on supply side has concentrated on testing new products and its market potentials. On the other side environmental research and modal choice has often focussed on personal attitudes. Canzler and Knie (2011) combine users' acceptance of new technology (i.e., electric mobility) with a known setting (i.e., car sharing). Potential customers are identified as multimodal persons. That means that they use different modes of transport, regarding a private car as only one option in a modal chain. This type of pragmatic users can primarily be found in cities like Berlin.

Based on empirical findings this chapter supports the approach of Canzler and Knie.

There are only few studies dealing with electric mobility from users' perspectives (e.g., Golob and Gould 1998; Turrentine 2011). The international MINI E trial involved consumer studies that primarily investigated the acceptance of the BMW Mini E as a substitute for privately owned cars. Therefore, a group of people has been leased an electric car for several months. Data were collected at three points of time utilizing different methods such as travel and charging diaries, questionnaires and conjoint analysis (Turrentine 2011).

3 Research Design

Research of the project BeMobility investigates the drivers and barriers that convince and withhold the public from combining different modes of transport including electric car sharing. Therefore, users' perspectives are integrated into the innovation process. A multi-methodical approach was applied and included both qualitative and quantitative methods (cf. Wolter et al. 2011).

This chapter presents the results of the quantitative evaluation of the field trials in Berlin that were conducted between November 2010 and September 2011. Questionnaires are a useful tool for gathering comprehensive information about users. Using online surveys data were collected at three points in time—before usage of electric car sharing, after a short period of usage, and at the end of the field trial. Table 1 provides an overview of dates of field trials, number of participants and key topics of the three online questionnaires.

Due to different settings in space and time the surveys do not present a panel situation. Each survey contains different respondents with 89 people having completed both T1 and T2.

Respondents to the surveys consist of customers of car sharing (Flinkster) and bike sharing (Call a Bike), both offered by Deutsche Bahn. Therefore, results presented here are based on the opinion of a unique group that might be considered as open-minded towards innovative mobility concepts. Furthermore, as car sharing is usually used for purposes not occurring every day (Intermodi 2004) it is assumed that participants use different modes of transport hence practicing multimodality.

Table 1 Details of the quantitative study concept

	T0—before usage	T1—after short period of usage	T2—at the end of the field trial
Field period	16.11.–05.12.2010	25.11.2010–20.04.2011	9.8.2011–5.9.2011
Participants	n = 311	n = 160	n = 178
Key topics and standardised tools (extract)	Expectations of EVs according to the committee for scientific evaluation of the model regions (NOW 2011)	Mobility patterns (Infas DLR 2010)	Mobility behaviour (Infas DLR 2010)
		Evaluation of user friendliness and everyday usability	Experiences with different aspects of electric vehicles in line with T0-questionnaire
	Attitudes towards integration of EVs into public transport	Usage patterns of electric car sharing including travel purpose and destination	Evaluation of everyday usability
	Evaluation of pricing structure	Evaluation of pricing structure in line with T0-questionnaire	Evaluation of pricing structure in line with T0 and T1 questionnaire
	Mobility patterns according to the representative study of MiD—Mobility in Germany, Mobilität in Deutschland (Infas DLR 2010)	Evaluation of integration into public transport	Usage patterns of electric car sharing including purpose and destination
	Attitudes towards environmental issues according to the representative study of environmental awareness (BMU 2011)	Socio-demographic characteristics (Infas DLR 2010)	Evaluation of integration into public transport in line with T0 and T1 questionnaire
	Evaluation of everyday usability		Preferences regarding renting locations
	Attitudes towards regenerative energy sources according to the MINI E study (Krems et al. 2010)		Car sharing routines in line with T1-questionnaire
	Socio-demographic characteristics according to MiD (Infas DLR 2010)		Socio-demographic characteristics (Infas DLR 2010)

Due to the design of the study this chapter will only give first impressions on car sharing with EVs but is not able to draw conclusions regarding a market potential of EVs in general.

4 Results

Results presented in the following chapter are based on questionnaires. They provide a useful source of information on the potential and real users of electric car sharing.

4.1 Socio-Demographic Characteristics of the Samples

The findings in this study confirm previous research relating to the socio-demographic composition of car sharing customers. In all three study groups male respondents outnumber female ones (Table 2). In comparison, 87 % of all E-Flinkster customers are male too. Further, respondents are highly educated, and often graduated from university. Rates of employment and self-employment are very high among participants, particularly in professional jobs. The age primarily ranges between 30 and 40 years whilst students and pensioners are represented below average.

The first survey (T0) included attitudes towards environmental issues and regenerative energy sources. In summary, the sample population agrees to concepts saving the environment and are acting accordingly. For example, nearly half of them (T0: 44 %) purchase electricity from renewable resources at home; in comparison, this applies to only three percent of the German population (BMU 2011). Seventy percent of respondents are willing to pay more for sustainable

Table 2 Sample characteristics

Characteristics		T0	T1	T2
		Percentages		
Gender	Male	87	94	87
	Female	13	6	13
Qualification	University degree	60	74	68
Employment type	(self-) employed	89	94	90
	Student	6	3	4
	Pensioner	3	1	1
		Mean		
Age		39	38	39
Household size		2.3	2.1	2.2

Comments. BeMobility T0—before usage of electric car sharing, n = 311, BeMobility T1—after a short period of usage, n = 160, BeMobility T2—at the end of the field trial, n = 178. Samples do not present the same persons products. Consequently, 69 % of the first sample regard the source of energy for the electric vehicles as important. This refers to renewable sources of wind, solar and water in particular. More than 90 % support their utilisation. Environmental concerns are an important reason for choosing sustainable modes of transport (T0: 63 %) and car sharing with electric vehicles (T0: 77 %) respectively.

4.2 Mobility Patterns of the Samples

Users' mobility routines were compared with the representative study of Mobility in Germany (Infas DLR 2010) and found to be very unique among the samples. Their modal choice represents their environmental attitudes described above. Compared to the Berlin population, car ownership and usage is below average and public transport has a great share in everyday travel (Table 3).

Whilst 61 % of the Berlin population travel by car at least once a week (Infas DLR 2010), this applies to less than 50 % of the participants. On the other side, up to 86 % (T2) use public transport on a daily or weekly basis as opposed to 59 % of the Berlin population. Cycling rates are above average too.

Results confirm the presumption that electric car sharing attracts mainly multimodal users who do not rely on a single mode of transport. This means they use different modes for different journeys often involving public transport.

Table 3 Modal split in comparison with the representative study of MiD

Mode of transport		T0	T1	T2	MiD
		Percentages			
Car as driver	Daily	21	14	10	31
	1–3 times a week	23	20	19	30
	1–3 times a month	25	24	33	16
	Less than once a month	21	24	21	8
	Never	10	20	17	15
Public transport (regional)	Daily	51	54	60	35
	1–3 times a week	23	24	26	24
	1–3 times a month	20	16	10	18
	Less than once a month	5	6	3	11
	Never	2	1	1	13
Bicycle	Daily	38	31	41	21
	1–3 times a week	23	20	21	19
	1–3 times a month	20	19	18	11
	Less than once a month	11	18	12	9
	Never	8	12	8	24

Comments. BeMobility T0—before usage of electric car sharing, n = 311, BeMobility T1—after a short period of usage, n = 160, BeMobility T2—at the end of the field trial, n = 178. Samples do not present the same persons

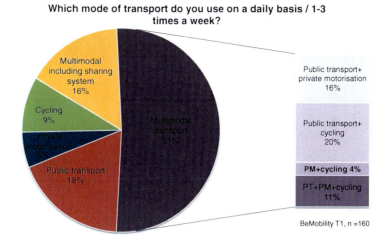

Fig. 1 Mode of transport (BeMobility T1)

Between 18 % (T0) and 24 % (T1) use more than one mode of transport every day. On a weekly basis the number of multimodal users rises to 67 % at the second point of data collection (Fig. 1). 85 % of these involve the use of public transport, mainly in combination with the bicycle.

4.3 Usage of Electric Car Sharing

During the project more than 40 electric and hybrid cars (Citroen C1, C-Zero, Smart ED, Toyota Plug-In Hybrid) were deployed in car sharing. In a two year period nearly 3,000 bookings were made by over 1,000 customers. Frequency of usage was low with over half the respondents having used the vehicles only once or twice. Apart from technical specifications of the electric vehicle (i.e., range, charging) reasons for modest usage are the availability of vehicles, accessibility of the stations, and greater charges. If no electric vehicle is available or not suitable for the purpose of travel respondents would use conventional car sharing instead. The hybrid vehicles provide the opportunity to extend the limited driving range. On average they are used for about 127 km per booking whilst battery electric vehicles have an average booking distance of 28 km.

4.4 Expectations and Experience of Electric Mobility in Car Sharing

The overall response to the electric vehicles was thoroughly positive. In each survey conducted during the study noise levels, driving pleasure as well as aspects of safety were evaluated positively. After a short period of usage (T1) open statements in the questionnaire included: "very silent", "no motor noise", "handling is intuitive", "not as complicated as expected" and driving is a "special event".

Negatively perceived aspects include costs for purchasing an EV, availability of recharging infrastructure at home or at the workplace, lack of flexibility and cargo space.

In a car sharing setting flexibility is constrained by the dependence on physical car sharing stations and limited driving range. The latter includes all aspects of the charging procedure that users face as they are required to recharge the vehicle at return. Both driving range and charging were criticised. However, electric car sharing is perceived as modern, useful and environmentally friendly.

4.5 Pricing Structure

In order to increase acceptance of new forms of mobility and possibly initiate changes in existing mobility patterns low entrance barriers are key (Maertins et al. 2004). A central barrier for electric vehicles to enter the market is the cost of the battery. As a result electric car sharing is more expensive than conventional car sharing. The evaluation of the price system shows that while it is easy to understand, its attractiveness has reduced during field period (Fig. 2). Although concerns about technical limitations of the vehicles predominate, current costs prevent their routine usage. It is possible that respondents of the first survey may not have been aware of the charges. In the final survey respondents were able to leave suggestions for improvement. Apart from reasonable charges they ask for cheaper charges per kilometer, more special offers and privileges for frequent users.

4.6 Integrated Mobility Concept

Integration into public transport is important for electric car sharing to be suitable for everyday use. In BeMobility three types of integration were implemented:

- Physical integration—car sharing that provides close links to public transport eases changing between means of transport.

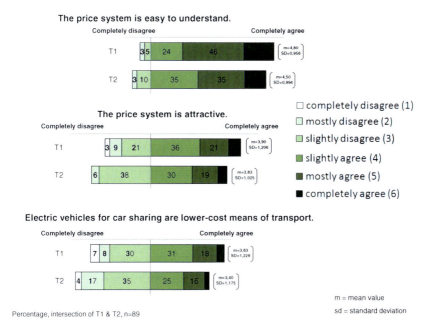

Fig. 2 Attitudes towards pricing system

- Integrated Information—travel information on different modes of transport was realized through a smartphone and web application called "BeMobility Suite".
- Integrated offers—one ticket ("mobility card") was introduced for a limited period of time that offered access to and usage of different means of transport (i.e., public transport, car and bike sharing).

Results of the surveys show that the majority of respondents agrees to the combination of different means of transport for their daily travel. They can benefit from flexible interchange, independence from parking space and time tables respectively. Efficient usage of different modes is seen as environmentally friendly and cost beneficial. These arguments tend to persist during the whole period of study; issues regarding independence from parking spaces and accessibility of car sharing stations are stressed more frequently after usage.

Visibility and spatial accessibility of car sharing increases demand. Nevertheless, 22 % of respondents need 30 minutes or longer from their place of residence to the next car sharing station. Due to the limited number of electric vehicles, this expenditure of time applies to even 38 % of respondents reaching an electric car sharing station. The majority of current car sharing stations is located at transport hubs and consequently, more than 60 % of users travel to and from them by public transport. 71 % of those who responded to the final survey are in favour of central locations with public transport accessibility. 66 % prefer locations in residential areas. A mixture of both concepts would make sense if the network of car sharing stations were to be extended.

In general, car sharing users ask for a dense car sharing network which extends over the borders of Berlin and is well connected with the public transport network. Availability of electric vehicles in residential areas, in rural areas and in other towns would improve the everyday usability of car sharing.

5 Discussion and Conclusions

This study presented results of user surveys that were conducted as part of the evaluation of the project BeMobility/Berlin elektroMobil. The research focused on user acceptance of new and innovative mobility services, i.e., electric car sharing.

The survey samples consist of people, who show a general interest in mobility services due to their membership with car sharing (DB Flinkster). They are characterised by being male, about 40 years old and highly educated. Respondents' environmental consciousness is expressed in their travel behavior that can be described as multimodal. A high proportion of them use public transport and bicycle on a daily basis. They agree that electric car sharing is able to cater for their daily mobility needs when offered as part of public transport.

Results of the conducted surveys both before and after usage of electric car sharing show that positive attitudes towards electric mobility predominate. Current users of e-Flinkster are quite relaxed in terms of technical limitations of EVs. They are aware of the disadvantages of EVs but in an urban context and in combination with other modes of transport driving range and cargo space are less of an issue. Consequently, vehicles are only used when travel distance and purpose of travel match vehicle abilities. Usage frequency of electric car sharing is accordingly modest.

Due to the unique characteristics of the sample groups it is not possible to draw conclusions for the German population. However, the study has shown that electric mobility is an environmentally friendly add-on to conventional car sharing that attracts primarily current users of the system. Whilst advantages and disadvantages of electric mobility are mostly related to the vehicle itself, i.e., quality of driving, driving range, charging aspects, and cargo space, further remarks comment on car sharing in general. These are, for example, the useful integration into the public transport system and the limited flexibility respectively.

Further research in the field of user acceptance of electric mobility in car sharing is necessary in order to learn more about motivations and drawbacks of the users. Some changes to the current setting are necessary if users other than early adopters shall be addressed and if usage of vehicles shall be increased. First of all, demand for a greater availability of vehicles and a dense network of car sharing stations was identified. Special offers like the "mobility card" must be established over a longer period of time for changes in behaviour to come into effect. Further research will include spatial and seasonal analysis and a study on effects on traffic.

References

Brake M (2009) Mobilität im regenerativen Zeitalter: Was bewegt uns nach dem Öl? Hannover, Heise
BMU (2011) Umweltbewusstsein in Deutschland 2008. Resource document. Bundesministerium für Umwelt, Naturschutz und Reaktorsicherheit. www.umweltdaten.de/publikationen/fpdf-l/3678.pdf. Accessed 21 June 2011
Canzler W, Knie A (2011) Einfach aufladen: Mit Elektromobilität in eine saubere Zukunft. München, Oekom
Golob T, Gould J (1998) Projecting use of electric vehicles from household vehicle trials. Trans Res Part B 32(7):441–454
Infas DLR (2010) Mobilität in Deutschland 2008. Resource document. Bundesministerium für Verkehr, Bau und Stadtentwicklung. Bonn und Berlin. www.mobilitaet-in-deutschland.de/pdf/MiD2008_Abschlussbericht_I.pdf. Accessed 11 June 2011
Intermodi (2004) Sicherung der Anschluss- und Zugangsmobilität durch neue Angebotsbausteine im Rahmen der "Forschungsinitiative Schiene". Gemeinsamer Schlussbericht von DB Rent und WZB. 6/2005
Krems J et al (2010) Research methods to assess the acceptance of EVs—experiences from an EV user study. In: Gessner (ed) Smart systems integration: 4th european conference and exhibition on integration issues of miniaturized systems—MEMS, MOEMS, ICs and electronic components. VDE Verlag, Como, Italy
Maertins C, Hoffmann C, Knie A (2004) Automobil mit der Bahn. Internationales Verkehrswesen 56(1+2):38–40
NOW (2011) National Organisation Hydrogen and Fuel Cell Technology. www.now-gmbh.de/de/elektromobilitaet/plattformen-elektromobilitaet/sozialwissenschaften.html. Accessed 3 Aug 2011
Scherf C, Wolter F (2011) Multimodales Mobilitätsmanagement. Internationales Verkehrswesen 63(1):53–57
Sperling D, Gordon D (2009) Two billion cars. Driving toward sustainability. Oxford University Press, New York
Turrentine T et al (2011) The UC Davis MINI E consumer study. Institute of Transportation Studies, University of California, UC Davis Institute of Transportation Studies Research Report 5
Wolter F, Hasse S, Heinicke B (2011) Intelligent vernetzen. Internationales Verkehrswesen 63(5):16–19

Author Biographies

Steffi Kramer holds a degree in applied geography and geographical information systems. She has worked as transport co-ordinator and sustainability manager before she joined the Innovation Centre for Mobility and Societal Change (InnoZ) in 2011. Her research focusses on sustainable mobility from a social scientific perspective. Steffi Kramer has been involved in projects dealing with consumers acceptance of transport services that integrate public transport and electric vehicles.

Christian Hoffmann works as psychologist at InnoZ and is interested in the relationship between human characteristics and ecological behaviour. He is member of a consultancy firm researching sustainable economy and co-editor of the journal "Umweltpsychologie".

Tobias Kuttler BA Geography, is an MA student in Urban and Regional Planning at Technical University Berlin. He assists in projects that are concerned with the acceptance of integrated mobility concepts at InnoZ. He is particularly interested in participation processes for sustainable urban transport developments, and is currently researching non-motorised transport in Indian megacities.

Manuel Hendzlik BA Geography, is a student assistant at the InnoZ too. He is currently writing his MA thesis in urban geography at Humboldt University Berlin researching the relationship between electric mobility and public transport in metropolitan areas.

Transforming Mobility into Sustainable E-Mobility: The Example of Rhein-Main Region

Birgit Blättel-Mink, Monika Buchsbaum, Dirk Dalichau, Merle Hattenhauer and Jens Weber

Abstract The probability that German mobilists are ready to make their everyday mobility more sustainable is still not very high, as individual mobility has a very high social standing. It has been the objectives of the social scientific accompanying research of 'Modellregion E-Moblität Rhein-Main'[1] to explore the acceptance of e-mobility as well as develop strategies to improve the applicability of concepts of e-mobility that are currently in a testing stage. Explorative in-depth interviews with selected people of the region, group discussions with users of e-vehicles and creative workshops with lead users of e-mobility identified the following facts and relationships: the central motive for choosing the means of

---

"Modellregion E-Mobilität Rhein Main" is one of eight regions in Germany, funded by the Federal Ministry of Transport, Building and Urban Development as part of an economic stimulus package. Specifically, a total of 15 pilot projects in Rhein-Main are supported.

---

[1] See also Schäfer et al. in this volume.

B. Blättel-Mink (✉) · D. Dalichau · M. Hattenhauer
Institut für Soziologie, Goethe-Universität Frankfurt am Main FB
Gesellschaftswissenschaften, Grüneburgplatz 1, 60323 Frankfurt am Main, Germany
e-mail: b.blaettel-mink@soz.uni-frankfurt.de

D. Dalichau
e-mail: dalichau@soz.uni-frankfurt.de

M. Hattenhauer
e-mail: hattenhauer@soz.uni-frankfurt.de

M. Buchsbaum
Institut für sozial-ökologische Forschung (ISOE), Hamburger Allee 45,
60486 Frankfurt am Main, Germany
e-mail: buchsbaum@isoe.de

J. Weber
Beratungsgesellschaft für Stadterneuerung und Modernisierung mbH, Uhlandstraße 11,
60314 Frankfurt am Main, Germany
e-mail: weber@bsmf.de

transportation is the convenience it offers to master every day demands of mobility. As a consequence the added value of e-mobility is not sustainability, but the chance to cope with everyday mobility demands. Secondly, e-mobility brings fun but is still much too expensive and under regulated. Thirdly, giving users the opportunity to participate in the development of technical and infrastructural aspects of e-mobility increases their readiness to change behavior.

1 Introduction

In order to evaluate the conditions under which a 'system change' towards sustainable e-mobility, i.e. beyond the universal usage of internal combustion engine and towards usage of regenerative energy, can be initiated, it is necessary to understand the complexity of the objective requirements and subjective desires in this field of needs (Bundesregierung 2009).

The research project's primary focus and central questions concern the acceptance of another type of individual mobility as well as possible strategies to improve the applicability of concepts of e-mobility that are currently in a testing stage. Therefore, two objectives ought to be achieved: The assessment of how a new culture of mobility can gain acceptance and how present offers can be optimized by potential users participating in the process. In order to achieve these goals research was divided in three segments.

1. Selected residents of the model region were asked to answer questions concerning their attitudes towards e-mobility.
2. Pilot project's participants discussed their experiences with e-mobility in focus groups.
3. By using an experimental concept of user participation during the developmental process, an attempt was made to improve the everyday usability of electrically driven vehicles in order to increase their acceptance as well as their chances on the market.

In terms of socio-demographics the majority of the participants of all three sub-projects (N = 93) are highly educated and between 30 and 49. There are slightly more men than women among the participants in total. Regarding the differences between the sub-projects it is not very astonishing that only in the prosuming workshops the majority were male ("lead-users"). The level of education is highest in this group as well and lowest among the people interviewed. Concerning age, the differences are not significant.

The following contribution briefly introduces the three segments, or subprojects, of the study (see also Fig. 1). The first part deals with the theoretical approach to the question of acceptance of e-mobility among the population of the Rhein-Main Region as well as among selected users of electrically powered

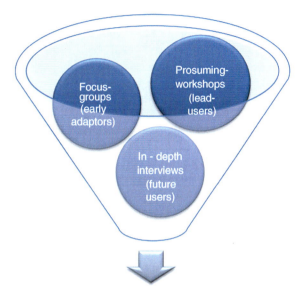

Fig. 1 Research design (Source: authors)

vehicles. The second part discusses the central results of the first two research segments. The third part which explores the theoretical concept of the working costumer, also known as the prosumer, presents the approach to the three creative workshops and finally discusses the results of the workshops. The article ends with a synthesis of all the findings.

2 Mobility and E-Mobility in the Perspective of Private Actors—A Theoretical Approach

The concept of 'lifestyle' is central to the project, whereas lifestyle, following Hradil (2005), is understood as the regularly reoccurring entirety of all behavior, knowledge, interactions, attitudes, and values of a human being. Lifestyle research has noted that different lifestyles each develop a highly specific style of mobility (Götz 1997; Deffner 2009; Götz and Deffner 2010; Lanzendorf 2003; Kramer 2005; Harms et al. 2007; Klöckner 2005; Preisendörfer and Rinn 2003). In this context, a recently emerging lifestyle: LOHAS (Ray and Anderson 2000) give reason for an optimistic outlook. The acronym LOHAS refers to people with a 'lifestyle of health and sustainability', that centers around a healthy, ecological and socially sustainable way of leading ones everyday life. People (or households) with such a lifestyle are usually well to do, from the center of society who value "pleasure without regrets" (Bilharz and Belz 2008). Also typical for this new generation of the responsible consumer is a strong identification with organic or eco, even though this is not necessarily substantiated by everyday behavior. So called 'green' products like organic food items or ecologically produced energy are in high demand (Weller 2008). Could it be that these types of consumers are pioneering for the acceptance and the use of e-mobility as 'lead users' (von Hippel 1986)?

So far, the social sciences have offered only small insights regarding the topic of e-mobility as such (Götz and Deffner 2010; Harms et al. 2007). It is expected that there is a pluralism of cultures of mobility in the region, not just one typical culture. The various cultures of mobility will in turn reflect the different lifestyles of people, their financial situation, their need to commute between home and work as well as the individual degree of awareness concerning ecological issues and so on, in short: Various cultures of mobility need to be adapted to individual motives and obstacles. Any empirical research strategy has to be aligned with this notion. Therefore, the following problem formulations and research hypotheses will be of purely descriptive nature and there will be no attempt to test or even verify them.

The question concerning the chances for acceptance of e-mobility can now be specified: How does the individual mobility in the model region present itself and what types of attitudes concerning e-mobility exist in the model region?

3 Acceptance of E-Mobility in Rhine-Mine

To answer these questions, two qualitative research methods were used: First, 30 selected residents of the model region who have never had any experience with e-mobility were interviewed with a manual guided open questionnaire. Second, five group discussions were conducted, again using manual guided open questions. The 6–8 participants of each focus group had been participants of selected pilot projects[2]. In the end, both segments should offer insights into the chances of the acceptance of e-mobility.

3.1 Acceptance of E-Mobility Beyond Any Experience

In the run-up of the empirical survey the following three guiding hypotheses were identified by analyzing former empirical findings as well as theoretical reflections.

1. **Mobility behavior is routine behavior**

Previous research on mobility behavior suggests that the individual choice of means of transportation is habitualized (Harms et al. 2007). In other words, this behavior turns into an integrated part of one's everyday action schemes, which will

---

[2] See chapter note on the first page.

not be easily altered or changed. The readiness to change such habits will most likely occur when other conditions change, such as alterations in the lifecycle. Thus, the sampling in the first segment of the study took the lifecycle into account.

2. **E-mobility has to offer a subjectively added value in contrast to conventional mobility**

Alterations in the lifecycle or changes in status will not lead automatically to a readjustment of individual mobility in the future. Any situation of radical change can in fact lead to a perpetuation of previously habitualized action schemata, since they tend to offer a sense of stability in everyday life (Klöckner 2005). Therefore, the question arises as to where the motives and obstacles for a switch to e-mobility would be if one takes the desire to maintain continuity in one's life into account. If necessary, e-mobility has to offer a subjective feeling of added value that can strengthen the motives and lower the barriers for a transition to e-mobility. Accordingly, problem centered interviews were conducted, focusing on the topics of the respondents general day-to-day mobility behavior and e-mobility as well as on motives and obstacles concerning a switch to e-mobility.

3. **Because e-mobility has strong ties to sustainability, such sustainability can signify a subjective sense of added value**

The political intention to introduce e-mobility in private transportation and its extension to the public transportation systems aims at a concept of mobility that meets the demands of an increasingly mobile and flexible world as well as the necessity to reduce the consumption of limited non-renewable resources and limit the impact on the climate and on the environment in general (Bundesregierung 2009). Because the topics of environmental protection and sustainability have a strong presence in the minds of people today and the concept of e-mobility has been linked closely with sustainability in the public discourse, this could possibly be the key to the subjective sense of added value of e-mobility. Therefore, the third group of questions in the interviews focused on topics concerning environmental protection and sustainability in general as well as more specifically on e-mobility.

The selection of the interviewees for the individual interviews followed along the lines of socio-demographic characteristics of lifestyle, variants of which define a particular life cycle (Deffner 2009; Dalkmann et al. 2004). Important variables here are gender and age, marital status and family income, education, occupational situation (employment and occupational status) as well as the location of the place of residence. This sampling procedure serves as a theoretical selection tool with simultaneous consideration of factors specific for the pertinent research project (Schnell et al. 2005). In the context of the present research project the availability of transportation and in particular, public transportation, is highly relevant. The location of the residence, therefore, is one of the most important criterion for selecting the various lifecycle groups. The location of the residence in relation to the place of employment will most likely have a decisive influence on mobility behavior. Thus, the differentiation was made between residence in the center of the city of Frankfurt, in the periphery of the city (commuting distance) or in the surrounding extended suburbs and in rural areas further away. The classification of these zones was made in accordance with the availability of public transportation and the theoretical accessibility of the Frankfurt city center using roads or highways, respectively (Regionalatlas Rhein-Main 2000).

The mobility behavior of the respondents varied considerably with the location of the place of residence. As expected, the use of an automobile in the city tended to be considered inconvenient because of the overall density of traffic and the lack of available parking space. Vice versa, the use of public transportation in rural areas was regarded as inconvenient and unpractical. Only those respondents living in the inner city area owned their own vehicle who either regularly needed to transport large items, who had to commute to the city's outlying areas to reach their place of employment or who frequented recreational areas outside of the city. In contrast, all of the respondents living in rural areas owned a car. Even if the willingness to use public transportation existed, it still is associated with a major effort because of the necessity of a long drive from the home to the train station. Surprisingly, all of the respondents living in the close periphery have their own cars available. The reason given is frequently the inconvenient access to the public transportation network. An attempt to reach one area in the periphery from another area in the periphery leads to a lengthy and time consuming triangular journey that is perceived to be inconvenient.

The central motive of the respondents for the selection of the means of transportation can be identified as convenience. In other words: The same motive, namely to master the demands of the mobility in every day life as comfortable, as convenient and as effortless as possible can lead to quite different mobility behavior depending on the location of the residence.

> And before I drive anywhere, when the route takes up 20 min of my time and then I have to drive around in circles for an additional 20 min to find a place to park, only to find a parking space on the wrong side of the street and only for residents with permits... No, this is clear, okay, the bicycle, I can lock it up anywhere. No bother, no searching, it's done. You know, it is very important to me that everything is really close by that I can get there quickly and easily, without any effort. (Z1_03)

> Yeah, the convenience. Because the bus connection from here goes only to X. From there, the S-Bahn only goes to XX and yes, to XXX there is no connection by bus, as far as I know. Way too difficult. If you have a car, you just use it. (Z3_02)

As expected, phases of change in the lifecycle proved to be particularly vulnerable to alterations of the choice of the means of transportation. The first job or the transition to retirement are examples of this. Even more so, if the change of status also entails a change of residence. Relocation is a major factor. Here, it seems to be most important where the relocation takes the respondent. If it is a relocation from one neighborhood on the periphery to another it can require a major adjustment in the mobility behavior. It might even require the purchase of a car. For parents, the development of their children is an influential factor. Again, the purchase of a car might be the only option.

> But, as I said: The recreational needs of the children will reflect their age and then, certainly, at one time or another, the car will be more important than ever. (Z2_05)

While regarding the acceptance of e-mobility it becomes clear that the only association here is with a car powered by electricity. And this electrical car is expected to have the same properties and can fulfill the same expectations that are met by a conventional car with a combustion engine. In other words: If an e-mobile is called an e-car, the expectation will be that it can at least replace a conventional car or better offer an additional added value to it. Accordingly, similar problems are being anticipated with the e-mobile, challenges that any conventional car already offered (parking, cost of purchase, taxes and insurance etc.). In the light of such requirements it is not surprising that the respondents do not consider the technology to be ready for the market. Electrically powered automobiles don't seem to be developed enough to meet the requirements, needs and demands of a conventionally motorized society of mobility with its highly diversified options of mobility as well as demands for mobility. Also the expectation of continuity of the respondents regarding their own mobility behavior when considering the e-vehicle is becoming evident.

> If the car offers the same driving comfort or what ever, if it doesn't lack anything in comparison to a regular car, if the driver has no major changes to manage: sure, why not. (Z1_08)

E-mobility is also connected to the 'Pedelec'[3]. While younger respondents have the opinion that the Pedelec is more suitable for the older generation, retired respondents are convinced that the Pedelec—because of the considerable speed it can reach—is much more suitable to the younger generation.

One motive for the transition to electrically powered vehicles in this sample is their environmental friendliness. However, many respondents questioned whether e-mobiles are really as environmentally sound as their image suggests. The reason for this is an ambiguity or a lack of clear information concerning the production process, the technology and the necessary resources as well as questions regarding the recycling of the batteries and the body of the car. Interestingly enough, the respondents do grant the e-mobile a better environmentally friendly rating because of its lower CO_2 emissions compared to gasoline or diesel powered cars. This environmentally friendly rating however, is considered as critical because of questions concerning the production process. In contrast, the production process of gasoline or diesel powered cars is not being questioned at all. Most of the respondents lament the lack of information about electric cars. In the light of this information deficit as well as the presumption that the electric car needs frequent reloading of its batteries, which again causes additional efforts, it is not surprising that the major obstacles to switching to an electric car are seen in a completely unreasonable relation between price and value.

---

[3] A bicycle, powered by a battery that does only work through pedaling, e.g. in contrast to e-scooters.

What is really important, the electric cars are outrageously expensive. I mean, the purchase price. You have to do some serious calculating whether or not this is worth it. You will have to drive an awful lot, to somehow get the cost of purchase back. (Zi_02)

The most important finding of this study is that convenience is the primary motive for selecting a means of transportation. E-mobility would have to offer an additional value regarding convenience, offer the opportunity to overcome the every day demands of mobility in more convenient way than any other alternative in order to achieve a higher acceptance and market saturation. Therefore, the following theses were developed that have to be investigated in further research:

- The easier the transition to an electrical car is possible, the higher the acceptance will be.
- The better potential users are informed about the technology, the higher the acceptance.
- If electrical cars are perceived as technological innovation with additional convenience, acceptance will increase.
- If electrical cars are not only fulfilling the average expectations one has of a car but offer a true advancement in the sense of technical progress, acceptance will increase.
- If electrical cars support the individual lifestyle of people and help facilitate daily demands, the acceptance will increase.

The question that now needs to be answered is: Where could the value added through technology, the innovation and the progress of e-mobility be located, that would help to facilitate daily demands, make daily tasks easier and more convenient and finally offer the motivation to switch to an electrically powered vehicle?

3.2 Acceptance of E-Mobility After a Testing Phase

The goals of the second segment of the research were to study the experiences and opinions of users of e-vehicles offered within the pilot project concerning mobility, e-mobility and new technologies, obstacles and chances to use e-mobility, to evaluate any potential for improvement and finally to describe the group dynamic processes evolving in the discussions.

The data were collected using the method of the focus group (Dürrenberger et al. 1999), and data analysis was conducted using qualitative data analysis.[4] The focus group as a method in qualitative social research is a focused, structured and manual guided group discussion on a particular topic. Focus groups offer the advantage that the participants engage in interaction with each other, therefore topics can be treated in a much more differentiated and realistic way in a

---

[4] In the analysis, all transcripts and codes were compared and summarized. Communalities as well as differences between the groups were recorded and described.

discussion compared to individual interviews (Mitchell and Olson 1981). Even very subtle verbal and non-verbal reactions in the group can be observed and taken into account (for example, the observation of gestures and mimic of the participants). As a rule, focus groups are being conducted with a relatively homogeneous group of participants (Waterton 1999).

Therefore, participants have to have similar background in regard to particular important criteria (for example concerning their interests, their membership in an organization, occupation or age). Homogeneity helps the participants to get acquainted, because they have common points of reference. Guided by the presumption that men and women differ in their approach to mobility, the groups were constructed according to gender (Flade and Limbourg 1999). Creating 'chains of mobility' is a standard attribute in gender oriented mobility research in differentiating the mobility behavior of men and women. Other authors (Hjorthol 2008) point to the different allocations of tasks and typical day to day activities to explain the various patterns of mobility of men and women. Yet another perspective is offered by Götz (2011) in which the behavior of individuals in the context of a specific household is determined by economic considerations. Such considerations could be the distance traveled or the time consumed by this distance. Even with such considerations there could be gender related issues.

The following questions were posed to the focus groups: What type of experiences and what kind of attitudes do the users of e-vehicles describe and what potentials or barriers exist concerning a system change in the Rhein-Main Region? The questions related to the present day to day mobility behavior, to their experience with electrically powered vehicles, to possible changes of individual mobility behavior, to the usage of e-mobiles as well as to attitudes towards sustainability and environmental protection.

The focus groups were based on the assumption that the group process would influence the opinion of the individual. The discussion was in part tightly structured so that the discussion would not be limited and in order to reduce outside influences. The discussion manual offered the possibility to alter the structure of the discussion in a flexible and situational manner to allow for the open and extensive discussion of certain issues.

The majority of the respondents who had previously experienced electrically powered vehicles first hand had enjoyed the straightforward use of the e-mobiles. They considered their individual and personal contribution to sustainability to be highly relevant, yet they questioned the economic viability of e-mobility (cost of purchase as well as cost of maintenance) and voiced their critique about the immature technology and the lack of standardized solutions.

> I'm a businessman, look at the costs, what will 10 years of mobility with this type of car cost me, right? I think about the risks with these rechargeable batteries, how long will they last, and then I think about how long would I drive such a car, and I always look at a timeframe of about 10 years. And during those years I would not want to spend more for an e-mobile, regardless of my environmental consciousness. (PublicOrg; male)

Most of the respondents had not changed their own mobility behavior in any significant way since they first started using e-vehicles. However, using them was perceived to be fun and there were reports of some deviation from the usual routes.

> Inevitably, the mobility behavior has changed somewhat when using e-mobility, because one has to map out different routes—always keeping the range in mind. On the other hand, one has less of a bad conscience when driving short distances because of the pollution—so, the short distance trips practically doubled. (Company; male)

Users of e-mobility see themselves as 'Lead-Users' and this self ascribed status is being reinforced by the resonance in their social network.

> Yes, same with me, even with an e-scooter, I feel like a pioneer, definitely. My friends and everywhere I go, I get reactions to that effect. There is surprise, start looking for the engine, wonder why this thing runs so quietly, but it is positive, you know... (Company; male)

Although there is a certain amount of interest in switching to new technologies, there are certain preconditions and there is a tendency to accept e-mobility more in an occupational context than in the context of private, individual mobility. However, there is a gender related difference: While women are primarily concerned with getting errands and shopping done conveniently, men focus on the fun of driving and their affinity for new technology. The main barriers to change for both men and women are the perceptions of the high price of purchase, of the limited reach of the e-vehicles and of possible high maintenance costs in comparison to conventional oil-based combustion vehicles. There seems to be a lack of political and institutional support that could offer a clear framework adapted to e-mobility as well as tax incentives in order to make e-vehicles more competitive in price and benefit relative to conventional vehicles. Overall, there is an awareness of limited fossil resources and the necessity to use other resources of power medium and long term. Subsequent ecological and social damages of the consistently increasing motorized individual traffic, e.g. the anthropogenic influence, emphasize the necessity of a system change.

> Yes, I think, there is no such thing as ecological individual mobility. Sure, you can try and calculate your own ecological footprint, and even if, when I continue doing what I do under present conditions, I will use up so many resources that I need more than one earth... The system as such is not sustainable and ecological. We, in the Federal Republic (of Germany), are still far away from using up more than one earth. You know, I believe there is this system that assumes that we will replace all cars with other cars in the next 10 or 20 years which will use up huge amounts of resources, can never be sustainable or environmentally friendly. In my mind that's impossible. If we in deed manage to switch to another system within the next 10 or 20 years, even if, yes, even if we in deed overcame capitalism, at which point we would have less of a need to get somewhere. Well, then I could imagine that maybe a development of e-mobility can contribute something. (Community; male)

There was consensus across all the focus groups that an electrically powered car with the present state of technology will not be able to compete with a conventional car, but that it might be suitable for daily use under certain inter modal criteria. Business plays a major role in the implementation of e-mobility. It would be seen as a positive sign if employers acted upon the opportunity to experience e-mobility first hand—particularly because of the present high cost of such vehicles for private use.

3.3 Adaptation of the System Change to Individual Needs Through Participation of End Users (Prosuming)

The term 'prosuming' or 'prosumer' originates from a mix of production and consumption outside of employment. The term was first introduced by Toffler (1980). Linguistically, 'prosumer' is a composite of the terms producer and consumer. According to Toffler, the idea of the prosumer is a post-industrial phenomenon, the so called 'third wave'. In pre-industrial times, the 'first wave', producers and consumers were not found to be clearly divided, as people did consume self produced goods. Only in the 'second wave', during the industrial revolution, the line between production and consumption became clearly defined and thus caused the expansion of the market, the exchange network in which goods and services make their way from producer to consumer.

Now, the return of the prosumer in the 'third wave' is evident, even if this takes place under different conditions. According to Toffler, the 'third wave' is characterized by dissolving the boundary between production and consumption. He uses the do-it-yourself-movement or self-service in grocery stores as examples. Voß and Rieder (2005) address the integration of the 'working customer' in the production process of the company. They use the terms prosumer or 'the new type of prosumer' as well. The focus here is how customer involvement causes consumption to be embedded within the producing company.

Therefore, boundaries between customer and company dissolve accordingly, which is also stated by Voß and Rieder (2005) by describing the role of the new type of Prosumer that is clearly converging the one of an employee. Reichwald and Piller (2009) offer an economical approach with the model of 'interactive value creation', that conceptualizes the relationship between customers and companies as a win–win-situation. Central to their analysis is the concept of voluntarism as well as the involved actors competence for interaction. The point of departure for the analysis is the identification of two central problems within the conventional arrangement of creation of value: First, the customer is usually seen by the company as a 'passive receiver of value' whose 'average' needs are commonly analyzed by market research. Secondly, the 'problem of searching locally' clearly confines the capabilities of the company to innovate, because only known solutions and approaches can be applied. In order to clarify these two issues, the authors introduce the concepts 'need information' and 'solution information'… ''need information relate to the needs and preferences of customers or users: this can be information on explicit as well as on latent needs […]. Solution information is (technical) knowledge on how to solve problems or fulfill needs through special product specifications or with a service." (Reichwald and Piller 2009: 47, own translation). According to the authors, both are important input factors for companies. While information of needs assures higher effectiveness during the creation of value process, because it allows the fulfillment of the customers desires. The information of solution focuses on the efficiency in the creation of value, because new solutions can be developed faster and more efficiently.

Furthermore, the work of von Hippel (1988, 2005) is of particular relevance for the present research project. He discusses the integration of consumers into the creative phase of product development and improvement. Primarily, the so called 'Lead Users' play a major role, as they are already familiar with new areas of work even at an early stage and drive innovations forward.

Also, von Hippel (2005) deals with the (increasing) desire of consumers for individualized products. According to his observations, more and more areas develop in which an individual is looking for the best suitable product while the readiness for compromises is rapidly decreasing.

In research on 'open innovation' the motives of participating prosumers in innovation processes are of importance. Motives are presumed to be quite variable: from societal improvement to work-life balance, from satisfaction of the need to individualize to economical optimization. From a psychological viewpoint, Schattke and Kehr (2009) conclude that central factors for open innovation projects to stimulate motivation is to focus on intrinsic motivation and the experience of flow (Csikszentmihalyi 1975). In the context of sustainable consumption, Codita et al. (2011) identified relevance of integrating intrinsically motivated consumers in the process of innovation and optimization of sustainable products in general as well as especially in context of sustainable (e-) mobility.

Because of the diversity of the different pilot projects in Rhein-Main, it is an absolute requirement to develop individual concepts of prosuming treatments, four of which have been conducted.

3.3.1 Prosuming Treatment 1: Pedelecs in the City

The preliminary discussion on the prosuming treatment revealed two aspects of particular interest. On the one hand, the issues of planning infrastructure were of importance and on the other hand, issues of improving the electric vehicle, specifically, optimizing the 'Pedelec' that is presently available. Because all the Pedelecs in the pilot study originated from the same manufacturer, a direct contact could be established for the second issue concerning the mutual innovative and creative development of the vehicle. Accordingly, a workshop for the prosuming treatment was designed that entailed both of the aspects. In January 2011, fifteen participants acted as their own experts and innovative creatives in discussing their own mobility. On two consecutive days both in groups and individually, issues concerning infrastructure as well as issues concerning the vehicles were identified and solutions were evaluated. An array of diverse issues were mentioned, such as the variety of options in the use of e-bikes and the possible arrangement of express lanes for pedelecs. In addition, questions concerning the improvement of the infrastructure were raised: whether such a bike could be transported on public transit, how to ensure safety when parking and whether useful sharing or leasing concepts could be developed. Lastly, there was a demand for clearly defined administrative regulations at the legislative level concerning the use of pedelecs. After the final discussion with the manufacturer of the pedelec at the end of the workshops, it was evident that many of the proposed ideas and visions of the participants could indeed be realized in the near future. Therefore, a list of ideas was drawn up for the experts to take back to the drawing board and to evaluate in depth, most specifically weather protection, weight reduction and the ability to transport the pedelec on public transit.

3.3.2 Prosuming Treatment 2: Light Weight Vehicle Pool

It became evident in the preliminary discussion on the prosuming treatment that the focus should not rest on the electrical power of the vehicles alone but should focus on their particular light-weight construction as well. After preliminary discussions, a two day workshop for the prosuming treatment was planned and conducted. During the workshop, twelve participants, half of which were members of the community and half of which lived close but outside of the community, identified the pros and cons of the light-weight construction of the e-mobiles in question. It was of primary importance to develop a concept of a sustainable carpool for the community and it was clear, that this carpool was supposed to consist only of electrically powered vehicles of light-weight construction. Considering the vehicle mix of the carpool, presently available models were discussed along side with visions of possible future types of vehicles. In the discussion with representatives of the pilot project problems linked with the present carpool could be addressed and possible solutions for the problems could be found that could be implemented at short notice. The concepts of model carpools developed in the workshop made it clear, that there had to be only one highly specific solution for one typical group of users. A concept concerning the ideal infrastructure for light-weight vehicles was one of the universally applicable findings of the workshop, yet it was a marginal area of discussion.

3.3.3 Prosuming Treatment 3: Pedelecs in a Hilly Region

Central to the project is the distribution of pedelecs as a new sustainable means of transportation in a region characterized by short and medium distances between several communities on hilly terrain. For this purpose, the producer of pedelecs extended its previous line of products by developing the pedelecs as a new type of vehicle with the goal to position them (exclusively) to a market of young and sports-orientated consumers. Therefore, the prosuming treatment had the intention of identifying the requirements the pedelecs must fulfill to satisfy potential customers. The energy provider facilitates the distribution to customers in the region in order to position themselves on the market as an energy provider with a sustainable profile and to explore new areas of business. Of particular interest has been the question concerning the necessity of loading stations for the pedelecs at suitable locations in the region. Furthermore, general ideas and problems in the context of the introduction of pedelecs should be addressed as well as additional needs in the development and planning of the necessary infrastructure. One result of the prosuming treatment proved to be that a number of desirable requirements of the pedelecs can not be realized at the present time because of current technical limitations. Of particular concern are questions of capacity and performance of the rechargeable batteries of the pedelecs that limit reach, the lack of the accuracy of information on the battery status and the resulting uncertainty about how far the battery might range. The need of using the pedelec as a multipurpose vehicle was also discussed: As far as the infrastructure was concerned, the central issues were battery change and battery loading stations as well as a possible leasing model for batteries because they were considered to be the most expensive spare part of the bike. Furthermore, there was a great concern for the development of infrastructure that would offer safe routes for pedelec drivers.

3.3.4 Prosuming Treatment 4: Pedelec Pool for Tenants

A renting company together with a regional power supplier started a demonstration project offering their tenants (and employees) to rent a pedelec at a reduced price. From their experiences the company decided to provide two large real estate properties with mobility stations for pedelecs. In the context of preliminary discussions with the project managers, main issues for the prosumer workshop were developed: mobility profiles from tenants in inner-city suburb residential communities, existing experiences with electrical mobility, requirements at driving equipment, renting system and infrastructure. A goal should be the production of concepts of one 'ideal mobility station' in real estate properties that eases the way of renting a pedelec. The invited participants did not have any experiences with e-mobility so far. One of the outcomes produced in the workshop was the insight, that tenants owing a bicycle conceive no additional value with the pedelec, since the electrical support is not necessary in city in question. Of interest for the tenants however, was the possibility to transport children or goods; here a pedelec would represent quite an additional value. Regarding the renting modalities wide spectrum of options have been identified, favored here has been the possibility to rent per hour or day, however, the system should hold ready also long-term renting solutions like renting a pedelec for three, six or twelve months.

4 Summary

The investigation of the level of acceptance of e-mobility in the "Modellregion Elektromobilität Rhein-Main" has been the primary objective of this study. The second objective has been to bring users and producers/service providers for electrically powered vehicles together to offer solution informations and by that optimise the product as well as the infrastructure, and: to increase the acceptance of e-mobility. To answer the question concerning acceptance, leading research hypotheses were constructed, according to which mobility behavior is for the most part routine behavior. E-mobility, so the assumption, would have to provide an additional subjective added value in contrast to conventional mobility. Because e-mobility is already politically associated with sustainability, the subjective added value is assumed to be found in a so called 'green' mobility. First, the empirical part of the study focused on the present choice of means of transportation by people living in the inner city area, in the periphery and those living in rural areas close to the city. Intentionally, only people were interviewed who had not participated in one of the projects in the model region. One of the most important finding was that the central motive for choosing the means of transportation is the convenience it offers to master the every day demands of mobility. This allows for an important thesis for further research on e-mobility: As previously assumed, people expect a subjective added value from e-mobility. This added value, however, is found to a lesser extent in the potential for sustainable mobility but in the expectation, e-mobility might help to cope with the demands for mobility and the chances of mobility in every day life, not just occupational life.

Next, attitudes towards e-mobility of people who had already experienced electrically powered vehicles within the model region were investigated using focus groups. The differences in attitudes towards e-mobility of people with previous experiences compared to people without such experiences are striking. Participants of the model projects enjoyed the ease of use and the chance to make a contribution to sustainability. Particularly noticeable was the reaction of respondents outside the pilot projects: They associated e-mobility with terms such as small, slow, unreliable etc. In contrast, participants of the pilot projects repeatedly reported, how much fun it is to drive an electric vehicle. They regard themselves as lead users and receive a lot of recognition in their personal as well as in their occupational environment. Thus, the conclusion: Anyone who has tested an electrically powered vehicle is delighted by the driving experience and handling, yet believe that action is required in regard of the efficiency and of the reliability of the technology as well as of the institutional conditions such as taxes, insurance, or other monetary benefits. However, there is consensus between the respondents of the focus groups that electrically powered vehicles will presently not be able to compete with conventional cars. They can, however, be quite suitable for daily use under certain conditions of increased multi modal transportation. In this context, it is important to note that women have different needs in mobility than men.

Finally, the prosuming treatments were designed to discuss advantages and problems in the use of specific types of electric vehicles with users, producers and service providers in a structured manner and to develop solutions together. On the one hand, the prosuming treatment proved to be quite an intensive marketing instrument with positive external impact. On the other hand, the prosumers took the role as lead users in a field of social innovation, whose experience and opinions were valued by the producers and service providers. Thus it became evident, that the prosuming treatment is a highly productive instrument, particularly for producers in the investigation of problems, in the usage of products as well as in the simultaneous identification of solutions. This way, the 'working customer' drives the development of products that will be successful on the market because their development is guided by experience and they are well aligned to everyday needs. Previously, the working customer was mainly included in the performing tasks of the production process (using ticket machines, using coffeemakers in fast food restaurants). Now, in a prosuming treatment like it has been done here, the working customer gains a new quality, and acquires analytical and conceptional competence. In a way, the user almost turns into a researcher and/or member of the R&D-department of a producer or service provider.

References

Bilharz M, Belz F-M (2008) Öko als Luxus-Trend. Rosige Zeiten für die Vermarktung "grüner" Produkte? Mark Rev, St. Gallen 25/4:6–10

Bundesregierung (2009) Nationaler Entwicklungsplan Elektromobilität der Bundesregierung. Resource document. http://www.elektromobilitaet.din.de/sixcms_upload/media/3310/Nationaler-Entwicklungsplan-Elektromobilitaet.pdf. Accessed 7 July 2011

Codita R, Belz F-M, Moysidou K (2011) Expected benefits of lead-users: a ethnographic study in the field of electric cars. Conference presentation: Sustainable consumption. Towards action and impact, Hamburg

Csikszentmihalyi M (1975) Beyond boredom and anxiety. Jossey-Bass, San Francisco

Dalkmann H, Lanzendorf M, Scheiner J (eds) (2004) Verkehrsgenese. Entstehung von Verkehr sowie Potenziale und Grenzen der Gestaltung einer nachhaltigen Mobilität. Mannheim: MetaGis-Informationssysteme

Deffner J (2009) Zu Fuß und mit dem Rad in der Stadt—Mobilitätstypen am Beispiel Berlins. Dortmunder Beiträge zur Raumplanung, Verkehr, vol 7. Dortmund: IRPUD

Dürrenberger G, Behringer J (1999) Die Fokusgruppe in Theorie und Anwendung. Akademie für Technikfolgenabschätzung, Stuttgart

Flade A, Limbourg M (eds) (1999) Männer und Frauen in der mobilen Gesellschaft. Leske+Budrich, Opladen

Götz K (1997) Mobilitätsstile. Ein sozial-ökologischer Untersuchungsansatz. Arbeitsbericht zum Subprojekt "Mobilitätsleitbilder und Verkehrsverhalten" im CITY:mobil-Forschungsverbund. Frankfurt am Main: Institut für Sozialökologische Forschung ISOE

Götz K (2011) Nachhaltige Mobilität. In: Groß M (ed) Handbuch Umweltsoziologie. VS Verlag, Wiesbaden, pp 325–334

Götz K, Deffner J (2010) Die Zukunft der Mobilität in der EU. Herausgegeben von Europäisches Parlament. Resource document. http://www.europarl.europa.eu/studies. Accessed 10 July 2011

Harms S, Lanzendorf M, Prillwitz J (2007) Nachfrageorientierte Perspektive. Das Verkehrsmittelwahlverhalten. In: Schöllgen O, Canzler W, Knie A (eds) Handbuch Verkehrspolitik. VS Verlag, Wiesbaden, pp 735–758

Hjorthol R (2008) Daily mobility of men and women: a barometer of gender equality? In: Uteng TP, Cresswell T (eds) Gendered mobilities. Ashgate, Aldershot, pp 193–209

Hradil S (2005) Soziale Ungleichheit in Deutschland. VS Verlag, Wiesbaden

Klöckner C (2005) Können wichtige Lebensereignisse die gewohnheitsmäßige Nutzung von Verkehrsmitteln verändern? Umweltpsychologie 9(1):28–45

Kramer C (2005) Zeit für Mobilität. Räumliche Disparitäten der individuellen Zeitverwendung für Mobilität in Deutschland. Franz Steiner Verlag, Stuttgart

Lanzendorf M (2003) Mobility biographies. A new perspective for understanding travel behaviour. IATBR Konferenz Luzern. Resource document. http://www.ivt.ethz.ch/news/archive/20030810_IATBR/lanzendorf.pdf. Accessed 6 Aug 2011

Mitchell AA, Olson JC (1981) Are product attribute beliefs the only mediator of advertising effects on brand attitude? J Mark Res 18(3):318–332

Preisendörfer P, Rinn M (2003) Haushalte ohne auto. Eine empirische Studie zum Sozialprofil, zur Lebenslage und zur Mobilität autofreier Haushalte. Leske+Budrich, Opladen

Ray PH, Anderson SR (2000) The cultural creatives: how 50 million people are changing the world. Three Rivers Press, New York

Regionalatlas Rhein-Main (2000). Resource document: http://www.planungsverband.de/index.phtml?mNavID=1.100&sNavID=1136.67&La=1#A1. Accessed 25 June 2011

Reichwald R, Piller F (2009). Interaktive Wertschöpfung. open innovation. Individualisierung und neue Formen der Arbeitsteilung. Gabler, Wiesbaden

Schattke K, Kehr HM (2009) Motivation zur open innovation. In: Zerfaß A, Möslein KM (eds) Kommunikation als erfolgsfaktor im innovationsmanagement. Gabler, Wiesbaden, pp 121–140

Schnell R, Hill PB, Esser E (2005) Methoden der empirischen Sozialforschung. Oldenbourg, Opladen

Toffler A (1980) Die dritte Welle. Zukunftschance. Perspektiven für die Gesellschaft des 21. Jahrhunderts. Goldmann Wilhelm GmbH, München

von Hippel E (1986) Lead user. A source of novel product concepts. Manage Sci 32(7):791–805

von Hippel E (1988) The sources of innovation. Oxford University Press, New York

von Hippel E (2005) Democratizing innovation. The MIT Press, Cambridge London

Voß GG, Rieder K (2005) Der arbeitende Kunde. Wenn Konsumenten zu unbezahlten Mitarbeitern werden. Campus, Frankfurt am Main

Waterton C, Wynne B (1999) Can focus groups access community views? In: Barbour RS, Kitzinger J (eds) Developing focus group research. Thousand Oaks, London, pp 127–143

Weller I (2008) Konsum im Wandel in Richtung Nachhaltigkeit? Forschungsergebnisse und Perspektiven. In: Lange H (ed) Nachhaltigkeit als radikaler Wandel. Die Quadratur des Kreises? VS Verlag, Wiesbaden, pp 43-69

Author Biographies

Birgit Blättel-Mink is Professor for Sociology with main focus on industry and organisation at the Department of Social Sciences at Goethe-University in Frankfurt/Main, Germany. She holds a PhD in Sociology from University of Heidelberg. Her main research fields: (open and social as well as economic) innovation, sustainable consumption. Gender (in)equality in higher education, transdisciplinarity and social accompanying research.

Monika Buchsbaum studied political science at the University of Zagreb, Croatia and is working since 2005 as research assistant on projects of sustainable development and its social-ecological aspects. In her PhD-Thesis she analyses biodiversity loss and management on water use efficiency as well as regional sustainability in a case study of Ohrid region. Since 2012 she is working at the Institute for Social-Ecological Research in Frankfurt/Main, Germany. Her research foci involve biodiversity conservation and sustainable use, integrated water resources management, knowledge transfer and inter- and transdisciplinary research process.

Dirk Dalichau studied sociology and is scientific assistant at the Department of Social Sciences at Goethe-University Frankfurt/Main, Germany. He is currently doing his doctorate about standardized working and consumption conditions in themed consumption settings. His main research interests: (sustainable) consumption; transdisciplinary research on sustainability; consumption and experience; prosumption and user integration; electromobility.

Merle Hattenhauer studied sociology, psychology and history at the universities of Frankfurt and Hagen. She was research assistant in Frankfurt and is now in charge of the unit microcensus, payments, prices, household surveys in the state statistical office of Rheinland-Pfalz. Her PhD-Thesis is about corporate legitimacy in economic crisis. Her research interests include economic sociology, sustainability and corporate social responsibility research, sociological theory and organizational sociology of power and legitimacy.

Jens Weber studied sociology, political science and psychoanalysis and specialized on sociology of organization and sociology of city and regional planning. He worked as scientific assistant at Goethe-University Frankfurt/Main, Germany, between 2004 and 2011 and is doing his doctorate about welfare capitalism and shareholder value-oriented restructuration in big German companies. Since summer 2011, he builds up the new department of research and development at BSMF, a city planning office based in Frankfurt. There, he works on several projects that connect urban planning with sustainable e-mobility and regenerative energies.

Validation of Innovative Extended Product Concepts for E-Mobility

Jens Eschenbaecher, Stefan Wiesner and Klaus-Dieter Thoben

Abstract Trends, technologies and developments in the e-mobility sector to reinforce the role of electric vehicles are currently subject of scientific cross-disciplinary discourses. Consequently, the implementation of e-mobility is currently promoted by a number of countries, such as the US, France or China, as well as companies and other organizations, to participate in this technology shift of higher sustainability and profitable business opportunities. Due to the serious challenges to overcome, the implementation of e-mobility applications, services and products requires proactive political, scientific, and technological cooperation. In this process, resources, competencies and demands need to be matched by global actors and regional coordination. In this chapter, our specific focus lies on the analysis of different Extended Product (EP) concepts for e-mobility, in order to identity and analyze the most valuable and feasible ones for implementation. Resulting chances, challenges and the positioning options of the involved organizations, such as mobility service providers, energy supply companies, automotive and supplier industry, along with battery manufacturers are depicted. The authors try to demonstrate the immense dynamic and inter-linkage of this evolving business sector and show that new EP concepts need to be introduced. Therefore, promising EP concepts are exemplified and modeled by the Business Canvas method as a possible approach to future business planning. An example for car sharing demonstrates the general applicability.

J. Eschenbaecher · S. Wiesner (✉) · K.-D. Thoben
BIBA—Bremer Institut für Produktion und Logistik GmbH, Hochschulring 20,
28359 Bremen, Germany
e-mail: wie@biba.uni-bremen.de

J. Eschenbaecher
e-mail: esc@biba.uni-bremen.de

K.-D. Thoben
e-mail: tho@biba.uni-bremen.de

1 Introduction

Mobility provided through electric vehicles (e-mobility) is one of the big trends in the automotive sector these days. For a more sustainable mobility, electric powered individual transportation is a prevailing subject; not only for automotive players but also for energy and infrastructure providers. The German National Development Plan for e-Mobility aims on pushing research and development, market preparation and market launch for electric mobility. It stipulates to have one million electric vehicles on German streets until 2020 (Die Bundesregierung 2009). The focus of research is on the one hand of technical nature, in means of developing new electric powered cars and infrastructure solutions. On the other hand, the research is focusing on developing and implementing new mobility services, which provide increased value for the customers. Developing new products and implementing new services based on these products is highly interdependent. To successfully provide a new mobility solution, a high degree of collaboration between the different market actors is necessary. Due to this circumstance, the German government started to heavily subsidize collaborative projects promoting e-mobility from 2009.

As an activity out of the many research grants in Germany, 15 model-regions which are investigating the subject e-mobility have been created. One of these regions is the "Modellregion Bremen/Oldenburg" of the project "Personal Mobility Centre" (PMC). In the PMC model region, there is—besides many others—a specific focus on the analysis of current business models, in order to better understand future business models and their probability of success (PMC 2012). Due to present discussions in both industry and general public, many controversial descriptions, opinions and views on e-mobility—especially on potential business models—have been published. For enterprises interested in the new market, such as large automotive companies, it is currently difficult to define a clear strategy and steps to deal with e-mobility (Fraunhofer 2010). Although the German government has just decided to provide another 1 billion Euro subsidy to the community this might not been enough to overcome the technical bottlenecks and challenges. For this reason we have directed our research focus to the discussion of Product lifecycle management (PLM), Extended Products and exemplary business models.

The conception of specific business models for e-mobility is very important for the participating companies, especially for those with a direct customer contact. Markets for e-mobility solutions feature business models that strongly differ from traditional concepts. Car2go in Hamburg—one project out of many from Daimler—is a good example for this wave of new concepts (Car2Go 2012). Such very complex solutions require the interlinking of diverse value chains and business models of the different industry branches, such as automotive, battery industry, energy industry or component and supplier industry. In other words, in order to implement e-mobility, the development of highly innovative and intertwined technical, infrastructural and ecological solutions is necessary. Therefore, if e-mobility is to be introduced, all involved actors need to discuss and coordinate strategic aspects and options to create own strategies (Brüggendick et al. 2008). As a result, a multitude of new business models for e-mobility are currently in development. Such business models for e-mobility include the combination of many components of the value chain, including batteries, materials, cells, infrastructure and their application (Schafhausen 2009). New business models also need to challenge the very difficult situation that is caused by expensive batteries, long charging times and short cruising range (Eßer et al. 2009; Die Bundesregierung 2009).

It has been concluded that e-mobility will create not only new products, such as the electric car, but also a large subset of services. We call the combination of core products, tangible and intangible assets Extended Products, as introduced by Thoben et al. (2001). Although first studies reveal some insights about the design, management and dissolution of Extended Products (Eschenbaecher et al. 2002), there was no uniform approach to identify and depict complex business models, such as those based on Extended Products. The objective of this research is to describe an approach to validate innovative Extended Products for e-mobility with the business model canvas. Its applicability is to be proven by a case study.

First e-mobility as such is presented. Following, approaches and challenges for e-mobility are identified. After this, the concept of Extended Products is introduced in the context of e-mobility. Then the product life cycle in the context of e-mobility is presented, followed by an exemplary analysis of different Extended Product concepts. Finally the business model canvas is presented for e-mobile carsharing. The chapter ends with a conclusion and an outlook.

2 E-Mobility: Approaches and Challenges

The discussion of e-mobility has created many activities in the automotive industry. Generally, all companies agree that the combustion engine will be replaced by another technology sooner or later. Consequently, almost every automotive OEM is investing significantly into developing hybrid cars, electric vehicles and soon first models will be released to the market or are already available. Numerous studies from investment banks, consulting agencies and research institutes try to predict how the automotive market for electric vehicles will develop during the next 10 years. Predictions differ in a wide range from 2 to 25 % (!) market penetration in 2020. In automotive terms, this is only one and a half model cycles from now (ADL 2010).

2.1 Definition of Mobility and Challenges Ahead

The mobility of humans can be seen as starting point for any consideration on e-mobility. Mobility can be divided in two fields, private and professional mobility. Private mobility takes the larger share when comparing between these two. In order to understand the function "to be mobile" (Schneider et al. 2002), three different forms of mobility can be distinguished. These three are mono-, multi- and intermodality which can be selected to fulfill the individual mobility demands.

In which way persons solve their mobility requirements is highly influenced by their individual mobility behavior. The mobility behavior is dependent on socio-cultural and diverse surrounding conditions. Their influence can be very powerful. A decision against a mobility concept is therefore heavily based on social-cultural habits. If somebody aggregates the mobility behavior of larger groups, individual mobility types can be identified and described. With such mobility types, statements about mobility requirements and the openness to mobility concepts can be generated. Mobility providers can more easily design mobility offerings for the various mobility types (see Fig. 1).

Due to the ongoing discussions on reduction of CO_2 emissions, a new debate has started to whether or not rethink public, commercial and private mobility processes. Mobility management stands for the allocation of the necessary information (software) and the fundamental conditions (hardware) in order to fulfill the daily mobility demands (see Fig. 2).

The mobility concepts describe conditions for mobility services and can be considered as part of the hardware. The mobility service provider can indentify mobility types in order to adjust their business models directly to their needs. Investigations around the mobility chain deliver valuable information for the

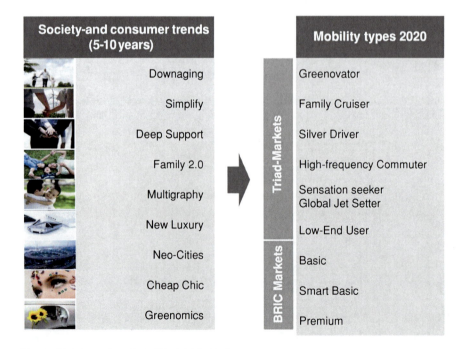

Fig. 1 Global trends and mobility (ADL 2010)

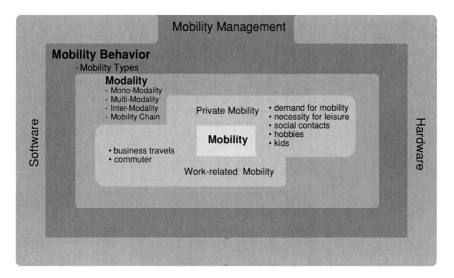

Fig. 2 Definition of mobility

offering of services. In summary, the mobility management can use the knowledge generated and is the last element of the "Schema of mobility" seen in Fig. 2.

2.2 From Selling to Providing

The automotive industry is facing another challenge: The future customers do not necessarily need to own a car. More than 80 % of the 18–29 year old think that city inhabitants do not need to own a car (von Wüst 2011). The urbanite of tomorrow is looking for more flexible models, like leasing or car-sharing, to satisfy his individual mobility demands. In future markets, more and more players like OEMs, leasing or service companies will be involved and fight for the mobility budget of the customers. Consultants from Arthur D. Little identified three so-called "mega trends" (global factors which set up the framework for all areas of business and society for 30–50 years), which will play a role in shifting business models and the uprising of electrified vehicles (ADL 2010).

- Neo-ecology: Initially arisen from the environmental movement of the 80s, society expects a corporate social responsibility. The rise of the oil-price and the CO_2-discussion accelerated this mega trend enormously and products which are not developed considering this trend are almost not marketable nowadays.
- Individualization: This mega-trend describes the release of the consumer from mass movements towards individualized solutions. Traditional lifestyle models are being left throughout all social classes and customers enjoy being not conformed but individuals.

- Mobility: In the 60s and 70s there has been a significant quantitative rise of mobility in triad markets and BRIC markets followed with a little delay. Limitations or harmful impacts of mobility, for example traffic volume and CO_2 emissions, though were reflected in society much later.

Along with the mega trends, social and consumer trends, which affect demand and buying behavior of consumers with a time horizon of 5–10 years, have been identified as well. Downaging, New Luxury, Cheap Chic, Simplify, Deep Support, Family 2.0, Multi-Graphy, Neo-Cities and Greenomics will have an impact on the demand for mobility especially in triad markets (ADL 2010).

These trends are not only leading towards an evolution of automobiles from conventional combustion engines towards electrified vehicles, but they will also lead to new business models, respectively to an extension with intangible product related services.

Today's automotive market is slowly drifting from the conventional model, where a car is bought and owned by the user, towards models where cars are rented, shared or leased. The customer has a requirement for mobility, but that doesn't necessary mean that there is a demand for owning an automobile. Requirements describe the customer's needs, while the demand describes a specific item that satisfies these needs. Walking along with the Extended Product paradigm, which will be detailed in the next chapter, the focus of manufacturers and suppliers is moving from producing and simply selling tangible core products to providing solutions which satisfy the customer's needs.

3 Extended Product Concepts for E-Mobility Life Cycles

After a theoretical application of the concept of Extended Products on e-mobility, the discussion of some examples shall support a better understanding on the variety of options, collaboration opportunities and challenges which need to be addressed along the product life cycle. Finally, the business model canvas is developed for the Extended Product idea "e-mobility approach for car sharing".

3.1 Extended Products and E-Mobility

The basis for the development of the Extended Product concept was a discussion in product engineering which took place in the end of last century (Chao 1993; Aurich et al. 2006; Mont 2002; Bowen et al. 1989). As a conceptual suggestion, the Extended Product concept—introduced in 2001 (Thoben et al. 2001)—offers in our opinion a playground for debating the upcoming issues of e-mobility. One of the key issues regarding e-mobility is the transfer of expectations from "combustion engine" mobility towards electric cars. Criteria such as price, comfort, cruising range and availability of electric cars should be in the same quality as for

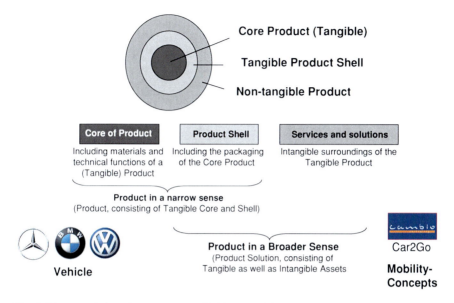

Fig. 3 The extended product concept applied in (e)-mobility

traditional combustion engine cars. New Extended Product concepts will focus on mobility solutions—including several aspects—instead of cars only (see Fig. 3). E-mobility is a good example for a focus shifting from core products towards solutions satisfying the demand for mobility.

The concept of Extended Products is based on this shift of customer expectations when buying new products and services. The *Core Product* is the physical product which is offered on the market, while the *Product Shell* describes the tangible "packaging" of the product (e.g. a **car** including equipment). The *non-tangible products* are intangible additions, which facilitate the use of the product (e.g. maintenance plans or mobility guarantees). **Mobility concepts** go beyond the classic understanding of a product and disconnect the non-tangible (services) from the tangible parts (physical product).

Today, Customers are looking for solutions and benefits (not for products) or even more they are requesting intangibles like fun, success, fame, vanity etc. In this sense, Extend Products aim at the provision of solutions and may intertwine many tangible and intangible product components. In the context of e-mobility, the provision of mobility as such would be the main benefit (see Fig. 4).

Manufacturers need to package their core products with additional services to make them more attractive. This development from products in a narrow sense, like physical products and components systems, as the electric car itself (i.e. the Mitsubishi iMiev) towards products with a broader perspective (e.g. car sharing services) is shown in Fig. 5. One major shift is indicated by the transfer from production of products and components such as cars for a sometimes known and sometimes anonymous market towards mobility solutions.

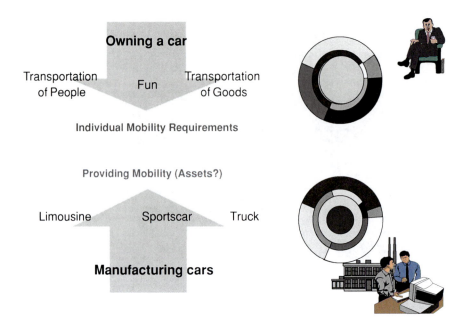

Fig. 4 From "owning a car" to "being mobile"

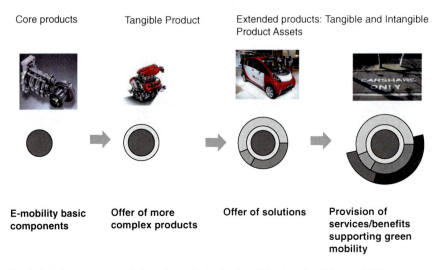

Fig. 5 Moving to an extended product (Eschenbächer 2011, based on Thoben 2001)

The pictured development describes the simultaneous offering of the tangible product (Core Product and Shell) extended with proper tailored services. In this case, both physical products and services contribute to the revenues, their balance needs to be adaptively determined and continuous innovation of services assumes a key competitive advantage (e.g. a car and the associated financial, maintenance, configuration services etc.). Furthermore, decoupling the manufacturing of goods and selling of services, where in most cases physical goods remain the property of the manufacturer and are considered as investment, while revenues come uniquely from the services (e.g. by selling "mobility" instead of the physical car) can be achieved. Several reasons for the described developments can be stated:

- E-mobiles, including batteries, are too expensive,
- Cost needs to be allocated around the product life cycle in order to enable customers to pay for them,
- E-mobility is extremely dependent on the interlocking of the diverse components to achieve good solutions and
- Mobility services, such as car driving, are probably becoming more expensive in contrast to the past.

3.2 Extended Products Life Cycle for E-Mobility

It has been shown, that companies can plan to produce core products on the one hand or to present e-mobility solutions to the customer. Nevertheless the following section shows the implications of offering products or services to the customer when looking into the product life cycle.

The components and the overall product have a life cycle, divided in the Begin, Mid and End of Life. The different phases of the life cycle offer multiple opportunities for Extended Product concepts to the stakeholders in e-mobility. Major players for these new products and services are the car manufacturers (OEMs), battery developers, infrastructure providers, electricity providers and mobility providers. It becomes obvious that these players partly offer competitive, but also complementary products. Moreover, the life cycles of these offers differ massively from each other. In Fig. 6 we have combined the shell concept of Extended Products with the three different lifecycle phases of a product.

3.2.1 Begin of Life

Business models in the Begin of Life (BoL) phase can be the currently known ones for car manufacturers. It starts with the vehicle development, engineering and production (Pahl et al. 2005). To match the future development requirements coming from the market of e-mobility solutions, this phase will be constantly evolving as well as shifting from the involvement of OEMs only towards collaborations between the different players in the e-mobility field. For example, the development of charging stations needs to be standardized and considered already during the vehicle development. Therefore, at least players like battery developers and infrastructure providers need to be involved in the vehicle development. As

Fig. 6 Extended Product oriented product lifecycle

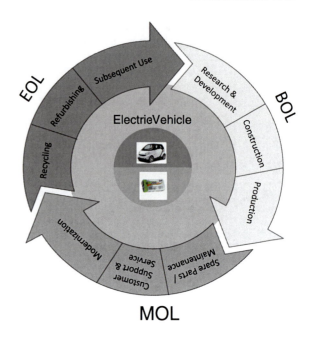

every player wants to get a piece of the cake as big as possible, it seems to be logical that energy providers will not only focus on the revenue of electricity alone, especially due to the fact that the amount of additionally needed electricity for e-mobility will be quite small (Engel 2009).

Charging stations with a monthly paid fee could be a solution for this issue, but the authors subsume that this model will not play any important role in the future. Certain Extended Product concepts presuppose cars which are specially designed for the needs of the specific concepts; e.g. the model introduced by "better place" requires vehicles which are designed with a changeable battery, compatible to their battery changing stations infrastructure (Better Place 2012). Many companies who want to become a player in the e-mobility field, started strategic alliances with big car OEMs. For example Nissan Renault is cooperating with the infrastructure provider "Better Place" (Mullins 2012). There is a strong interdependency between the battery and the vehicle, which necessarily leads to co-creation in the product development phase. Infrastructure providers also need to develop the charging stations in BoL.

3.2.2 Mid of Life

Within the Mid of Life (MoL), car manufacturers traditionally offer spare parts and maintenance services as well as customer support and -service (Klimke 2008). In maintenance services, replacement and modification of vehicle parts could be established as upgrade services. E.g. the replacement of engine or battery for higher performance, respectively range. As technologies will evolve, modernization services might be also feasible for the Mid of Life phase. If, for example, the fuel cell technology reaches its break-through, battery powered vehicles could be equipped with a fuel cell as a so-called range extender. Accompanying further services might be offered like e.g. own car sharing models, and fleet management for business customers.

3.2.3 End of Life

As car producers in the European Union have to take their cars back by law, various recycling related activities can be seen as a source for new services and related businesses models during the End of Life (EoL) phase. Another service could be the replacement of the vehicles battery, as the batteries life might be shorter than the vehicles life. Old batteries need to be either recycled according to national laws or they could be used subsequently for other purposes, e.g. as Uninterruptible Power Supply (UPS) for servers.

A niche in the End of Life phase could be the refurbishing of used vehicles in a way which makes the vehicle more attractive for new potential buyers' e.g. equipping the vehicle with a new battery or new upholstery etc. to bring the vehicle in a "as new" condition to sell them.

3.3 *Product Life Cycle for E-Mobility*

Service and product life cycle concepts have been heavily discussed in the literature (Johnston and Clark 2008; Mont 2002; Bowen et al. 1989). In this chapter we will reference mainly Thoben et al. (2002) with his discussion on the Extended Product life cycle. In order to develop a product life cycle model for e-mobility, several components need to be integrated. The list of stakeholders is long: government, supplier, investors, leasing and banking companies, automotive OEM, infrastructure provider, battery producer, power firms, car sharing provider, recycling actors and charging stations etc.

To develop a life cycle model we also summarize the phases product creation, product development and sales, product usage and product recycling and -disposal (Thoben et al. 2002). Finally, three main components need to be highlighted, which are vehicle, infrastructure and energy. In addition to these groups and elements, several processes must be analyzed. There are maintenance processes regarding battery charging, loading infrastructures or energy management which must be customized to the customers' needs. Both players and processes that are important for the implementation of electric cars were identified (see Fig. 7).

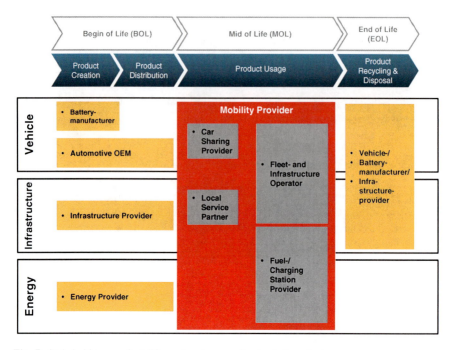

Fig. 7 Stakeholders, product life cycle, player and extended products

The following main elements are included in Fig. 7:

1. The graphic can be understood as business opportunity matrix reflecting the most important components and the product life cycle. In this matrix we have positioned the actors and their products symbolized as Extended Products.
2. The four elements (stakeholders, product life cycle, components and Extended Product concepts) have been brought together to demonstrate the ultimate need for integration in e-mobility.
3. Relevant stakeholders: Battery producer, Automotive OEM, power firms, infrastructure provider and mobility service provider.
4. The product life cycle with 3 phases shown in the top of the diagram. These phases indicate in which phase the one or the other stakeholder as well as Extended Products concepts can be positioned.

On the basis of the Fig. 7, several tasks will be subject of future work:

- How can Extended Products be identified, analyzed and evaluated?
- Which stakeholder will have the highest importance and weight?
- Which areas do include the biggest problems?
- Are there other players?
- Will mobility service providers indeed play an important role in the future?
- How much money are customers finally willing to pay for mobility?

These and other questions need to be discussed in the future—which cannot be done in this chapter. Therefore the focus is given to the analysis of Extended Products in the context of e-mobility to provide first ideas on how to address them.

4 Extended Products in the Context of E-Mobility

The definition of product and services for e-mobility is complex. Many different organizations need to combine resources and competencies to finally enable solutions. In the following, we will provide an example demonstrating how potential collaboration requirements can be identified. To do this, we will highlight three different aspects which subsequently led the analysis:

1. Discussion of the five main players collaborating to provide Extended Product concepts for e-mobility,
2. Examples to explain nodes and edges between e-mobility players and
3. Car sharing to demonstrate the Extended Product concept for e-mobility.

4.1 The Main Players in Extended Product Concepts for E-Mobility

In this section, the five most relevant organizations for creating Extended Product concepts for e-mobility (battery producer, automotive industry, infrastructure provider, energy providing company and mobility service provider) are presented. The different colors in Fig. 8 indicate different potential services (tangible or non-tangible) around the core product.

Due to the end of fossil resources, the automotive manufacturers are looking for new ideas leading to new concepts for mobility (Deloitte 2009). The spare part and maintenance business is very important for automotive OEM and is a part of their general business. In addition to the general services, new services supporting e-mobility solutions are needed.

4.2 Examples for the Collaboration Need Between E-Mobility Players

Very few companies such as VW or BMW (Reithofer 2010) will be able to offer e-mobility solutions within a solo attempt. Consequently, this chapter provides some suggestions and ideas how collaborations could be conceptualized by the Extended Product concept. In the following we will discuss spare part or maintenance for electric vehicles, automotive leasing and e-battery charging service.

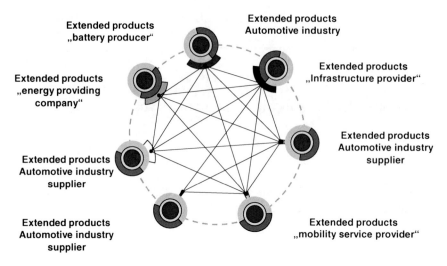

Fig. 8 Extended products for e-mobility—players and potential services

For B2B oriented battery producers, **spare part or maintenance** business is less important, because they are not in direct contact with end customers in contrast to the OEMs and garages. Energy suppliers also enter the market with their core product electricity for driving, but cannot build upon their traditional business concepts, as the margin for simple e-mobile charging is, at least not yet, high enough (Mora 2008). Currently, diverse energy suppliers work on new products and services in the context of e-mobility, such as smart grids, green energy for mobility and car leasing (Wirtschaftswoche 2010).

Automotive leasing can be seen as a further example for collaboration between the five main participants. For automotive OEMs, automotive leasing can be seen as a standardized offer. Likewise, more and more mobility service providing companies offer such services. For energy suppliers, leasing offers are completely new, but gain in importance. New leasing concepts can be seen as a mobility enabling services. Many types can be differentiated such as distance, pool- or full-service leasing packages (Braess and Seifert 2007). So far, leasing offers for electric cars are generally very expensive (MiD 2008). In this respect non-leasing experts such as power firms will have problems to acquire significant market shares, because they have to focus solely on electric cars and cannot revert to conventional power train cars (Strommagazin 2009). OEMs on the other hand can easily provide a combination of both, especially in the early phases of e-mobility.

The last example refers to the definition of an **e-battery charging service**. This is a completely new service for e-mobility. Consequently, many different companies, for example McDonalds, might have an interest to enter this new market (Schwartz 2012). The power firms offer such services as a complementary offer to their usual business. As already mentioned, the additional revenue for charging batteries until 2025 is very low (Mora 2008). The battery charging service is currently not of great interest for the automotive industry, so, no specific offers have been made by those companies. The providers of infrastructure are currently seen as suppliers of the needed equipment. Nevertheless, the petrol stations could extend their business by operating the infrastructure and selling rights of use. Figure 9 shows the many linkages needed to enable an e-battery charging service.

At this point of time, it's not possible to predict which players *will take the lead* in the e-mobility market. Nevertheless, we assume that automotive manufacturers and power firms might play a decisive role due to their ability to provide core products as discussed by Prahalad and Hamel (1990). Additionally, they also interact with the end customer, which is a major competitive advantage. Indeed the automotive industry faces one of the most difficult transitions in history. Fuel prices are going up, the need for mobility is increasing and the competition for e-mobility solutions is just starting. The paradigm shift from fossil fuel combustion engines towards electric cars is tremendous (Rothfuss 2009; WIWO 2010). Most of the business models focus on mobility without ownership, e.g. car sharing. Companies such as BMW experiment with new vehicles in combination with a car sharing solution (Reithofer 2010). Other companies like Toyota promote combined technologies such as Hybrids (combination of combustion and electrical engines), but also invest heavily in electric cars. It is indeed difficult to foresee if the automotive OEM will lose their competitive position, however, mobility service providers and energy suppliers still depend on the vehicle itself. Car sharing as an Extended Product concept for e-mobility.

The explanatory Extended Product summarized in Fig. 10 is based on the offer at German railway stations by the company Flinkster. In the centre are the "non-tangible" core product "mobility by car sharing" and the tangible product

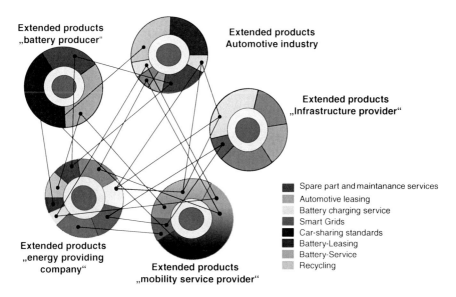

Fig. 9 Nodes and edges between the extended products

Fig. 10 Usage of the extended product concept for car sharing

"electric car". The electric car is continuously occupied by other customers for a limited time period. The intangible assets are mobility extending, ensuring and creating services. This example illustrates the different parts of the extended product concepts and the need to analyze them.

In the above Figure, the different shapes indicate the various services around the core product which is the electric car. It is possible to identify the content of the shapes by discussing them with the diverse players.

5 Application of the Business Model Canvas

Extended Product concepts can be used to identify options for new business models for mobility in general. In general the authors believe that business models will partly change from owning a car to providing individual mobility assets to customers. In this respect in the future being mobile will not automatically mean to own a car. In addition to this we believe that mobility offers will be based on Extended Product driven solutions different levels of product and service assets. This leads to a number of questions which need to be answered in the near future:

- Who will offer mobility services in the future?
- Which role can e-mobility play?
- Will e-mobility become the "next big thing" or simply the next level of individual mobility?

- … or will the concept of mobility change completely?
- Which price are customers willing to pay to fulfill their individual mobility demands?

Consequently we were investigating different models and tools to analyze the value of different business ideas in the area of e-mobility.

Generally the use of business models is highly common to understand whether or not somebody should invest in a business. We will present the business model canvas approach as one way of identifying new business models. Indeed the business model canvas is not made to support collaborations. Nevertheless it strongly helps to illustrate the complexity of Extended Product concepts for e-mobility. The business canvas is a new technique to structure business models and can be very useful for the collaboration planning.

5.1 Business Model Approaches Supporting the Analysis of Extended Product Concepts

Indeed there are many different models published to describe, analyze and evaluate business models. Consequently there was the need to identify an appropriate model fitting to our conceptual debate on Extended Products. After an analysis of different approaches in order to support Extended Product concepts related to business models, the authors have selected the Osterwalder business model canvas. Osterwalder has developed the business model canvas to support the development of business models (Osterwalder 2004). Together with Pigneur, Osterwalder defines the business models as follows (Osterwalder and Pigneur 2010):

> "A business model describes the rationale of how an organization creates, delivers and captures value." Thus, a Value Proposition results from a set of elementary offerings (product, services, their features, etc.) that have value in the eyes of the customers

Altogether nine areas must be analyzed to define a new business model. In the following we will adapt his approach in our example.

5.2 Exemplary Analysis of Flinkster Car-Sharing by Using the Business Model Canvas

People travelling with German railway provider "Deutsche Bahn" always have the option to directly book local public transportation, e-bicycles, cars and—since a few months—electric cars. These mobility options support the customer to reach his final destination from the train station. In this chapter we have analyzed the e-mobility service offered by cars called Flinkster, subsidiary of "Deutsche Bahn". The car rental station is directly at the railway station and the electric cars charging processes are reliable (Teczilla 2010). Based on the identification of different parts of the Extended Product in Fig. 10, a business canvas analysis was created as suggested by Osterwalder. The different colors indicate different levels of activity intensity (see Fig. 11).

In the following, the nine areas in Fig. 11 are briefly presented:

1. Customer segments: The most important customer group is the rail traveler who needs a transport to his final destination. As a second group, ecologically sensitive persons in the neighborhood are addressed.
2. Value Proposition, consisting of three components: The mobility without ownership is easy, because it is a pre-condition of any car sharing. Additionally it must be clarified that the sold product is ecological if the electrical power stems from renewable resources. Finally it has to be ensured that the customer can use the service all over Germany (Schneider et al. 2002).
3. Distribution channels: The electric cars can be ordered by both internet and telephone hotline. Direct links (www.flinkster.de and www.dbcarsharing-buchung.de) are placed strategically.
4. Customer relationships: The offer of Flinkster is predominantly promoted through membership contracts. Furthermore the management of unforeseen events such as the unavailability of certain cars can be prevented. A big deficiency is the non-inclusion of the general "Deutsche Bahn" service centers, because they don't support the booking process. Consequently, there is a medium need for action.

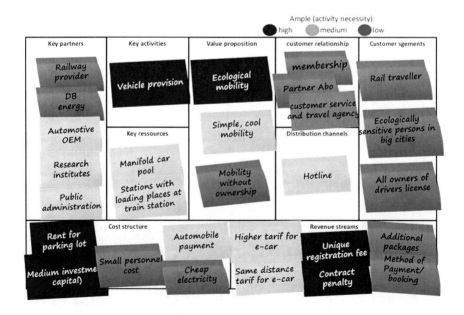

Fig. 11 Business canvas analysis for electric car sharing

Validation of Innovative Extended Product Concepts for E-Mobility   149

5. Revenue streams: The revenues of this offer are realized in different ways. The method of payment is uncritical, because these revenue streams are similar to normal car sharing. To some extent, it will be difficult to charge higher rates for the usage of electric cars. "Deutsche Bahn" (DB) is confident that customers are willing to pay higher prices for e-mobiles (Teczilla 2010).
6. Key resources: Key resources are car pools and stations with charging places. Both criteria have a medium need for action. On the one hand, diverse mobility wishes need to be served by different e-vehicles. On the other hand, batteries must be recharged frequently. For this reason, also a good infrastructure must be available. Generally, the long charging periods substantially lower the value of e-mobility for the customer.
7. Key activities: In our case, the key activity is the provision of enough electric cars which are charged. Every problem can have a large damage on the car-sharing service. Consequently, alternative vehicles should always be available in case a customer returns a vehicle too late.
8. Partner network: "Deutsche Bahn" and "DB Energie" are the key partners.
9. Cost structures: There are many different costs sources, e.g. charging and maintaining the vehicles, based on the complexity of managing electric cars. The big advantage is the closeness to railway stations.

The previous explanations show that the concept of the Extended Product in combination with the business model canvas can be used to analyze business models for e-mobility. Of course this research is in its infancy so more discussion is needed to make the results more solid. As an initial conclusion for the case example the following balance points can be set:

- Applicability to analyze collaboration activities,
- Analysis of Extended Product concepts,
- Qualification of business canvas and
- Inter-linkages between the various players in the market.

6 Conclusions and Future Research

E-mobility and its repercussion for the automotive industry is already getting a lot of attention. At this stage, nobody from research or industry can finally say whether or not electric vehicles will really replace the combustion engine in a large scale. Many uncertainties need to be addressed and hopefully being solved in collaboration of many industry players. One uncertainty—which can be characterized as business model analysis—has been addressed in this chapter. In this respect the chapter has presented the concept of Extended Products in the context of e-mobility. As described, e-mobility will create new products and services supporting mobility. Different players need to combine their resources in order to create the solutions. Core competencies need to be further developed. It has also been discussed that only the collaboration between energy suppliers, automotive OEM, battery producer, infrastructure provider (e.g. for charging stations) and mobility service providers can lead to a successful implementation of e-mobility. The inter-linkages are complex and the collaboration need is high (Eschenbaecher et al. 2009). Generally we believe that a close link between Extended Product concepts and the use of different forms of business models analysis can lead a significant value. Companies need to analytically understand if the one of the other approach in e-mobility can make sense. Consequently the business model canvas has been used to discuss nine business model areas. Finally, the following aspects indicate the most important lessons learnt:

- E-mobility and its repercussion for the automotive industry are already getting a lot of attention.
- Different concepts of Extended Products in the context of e-mobility have been discussed.
- Collaboration between energy suppliers, automotive OEM, battery.

Acknowledgments This work has been partly funded by the PMC and NeMoLand project (PTJ). The authors wish to acknowledge the Bundesministerium für Verkehr, Bau und Stadtentwicklung (BMVBS) for their support. We also wish to acknowledge our gratitude and appreciation to all the PMC project partners for their contribution during the development of various ideas and concepts presented in this chapter.

References

ADL (2010) Zukunft der Mobilität 2020. http://www.adl.com/mobilitaet-2020. Accessed 26 Feb 2010
Aurich JC, Fuchsa C, Wagenknecht C (2006) Life cycle oriented design of technical product-service systems. J Cleaner Prod 14(17):1480–1494
Better Place (2012) http://www.betterplace.com/. Accessed March 22, 2012
Bowen D-E, Siehl C, Schneider B (1989) A framework for analyzing customer service orientations. In: manufacturing. Acad Manage Rev 14(1):75–95
Braess HH, Seifert U (2007) Vieweg Handbuch Kraftfahrzeugtechnik. Vieweg, Wiesbaden
Brüggendick K, Efthimiou V, Woste M (2008) Megatrend Elektromobilität—Strategische Ansätze für Energieversorger. Energiewirtschaftliche Tagesfragen 58. Jg. Heft 12 2008, p 18
Car2Go (2012) http://www.car2go.com/hamburg/de/. Accessed March 20, 2012
Chao P (1993) Partitioning country of origin effects, consumer evaluations of a hybrid product. J Int Bus Stud 24(2):291–306
Die Bundesregierung (2009) Magazin für Wirtschaft und Finanzen. Presse-und Informationsamt der BundesregierungNr:07409/2009. http://www.bundesregierung.de/Content/DE/Magazine/MagazinWirtschaftFinanzen/074/pdf/2009-09-21-magazin-zumausdrucken,property=publicationFile.pdf. Accessed 28 January 2010
Deloitte Consulting GmbH (2009) Konvergenz in der Automobilindustrie—Mit neuen Ideen Vorsprung sichern. Deloitte Consulting GmbH
Eßer A, Fichtner W, Gerbracht H, Kaschub T, Möst D (2009) Elektro-Mobilität, Auf dem Weg in den Wettbewerb. BWK (Das Energie Fachmagazin), Bd. 61 Nr. 11, pp 44–48

Engel T (2009) Vortrag auf Innovationstag e-Mobilität in Bremen im Kongress Center Bremen

Eschenbaecher J, Hirsch B-E, Thoben KD, Schumacher J, Janssson K, Ollus M, Karvonen I (2002) Extended products: results of the first roadmapping study. In: Standforth-Smith B, Chiopazzi E, (eds) Challenges and achievements in e-business and e-work (Venice), pp 677–684

Eschenbaecher J, Seifert M, Thoben K-D (2009) Managing distributed innovation processes in virtual organizations by applying the collaborative network relationship analysis. In: Camarinha-Matos LM, Paraskakis I, Afsamarnesh H (eds) Leveraging knowledge for innovation in collaborative networks IFIP International Federation for Information Processing, Chennai, pp 13–21

Fraunhofer Institut System- und Innovationsforschung (2010) Die Stellschrauben der Elektromobilität. Zur gesellschaftlichen Integration der Elektroautos, Luzern

Schwartz A (2012) Is it green?: McDonald's. http://inhabitat.com/is-it-green-mcdonalds/. Accessed on March 23, 2012

Johnston R, Clark G (2008) Service operations management. Improving service delivery. Financial Times Prentice Hall, Essex

Klimke R (2008) Erfolgreicher Lösungsvertrieb. Gabler, Leipzig

MiD-2008/2 (2010) Ergebnisbericht—Mobilität in Deutschland 2008.MiD—Mobilität in Deutschland, Bonn/Berlin

Mora M (2008) VDE-Studie. Stromverbrauch steigt bis 2025 um 30 Prozent. Pressemitteilung des Verbandes der Elektrotechnik Elektronik Informationstechnik (VDE), Frankfurt, 22 Jan 2008

Mont OK (2002) Clarifying the concept of product–service system. J Cleaner Prod 10(3):237–245

Mullins J (2012). Better Place and Renault launch Fluence Z.E., the first "unlimited mileage" electric car together with innovative eMobility packages, in Europe's first Better Place Center. http://www.betterplace.com/the-company-pressroom-pressreleases-detail/index/id/better-place-and-renault-launch-fluence-z-e-the-first-unlimited-mileage-electric-car-together-with-innovative-emobility-packages-in-europe-s-first-better-place-center. Accessed March 22, 2012

Osterwalder A (2004) The business model ontology. A proposition in a design science approach. Doctoral Dissertation, Université de Lausanne, Switzerland

Osterwalder A, Pigneur Y (2010) Business model generation. Wiley, New Jersey

Pahl G, Beitz W, Feldhusen J, Grote KH (2005) Konstruktionslehre: Grundlagen erfolgreicher Produktentwicklung; Methoden und Anwendung. Springer, Berlin (6 Aufl)

PMC (2012) http://www.personal-mobility-center.de. Accessed March 22, 2012

Prahalad CK, Hamel G (1990) The core competence of the corporation. In: Harvard Business Review, May-June 1990, p 79–91

Reithofer N (2010) Elektroautos. http://www.focus.de/finanzen/news/elektroautos-bmw-erwartet-vorerst-keinen-gewinn_aid_565219.html. Accessed 8 Feb 2011

Rothfuss F (2009) Innovationsnetzwerk Future Car (FuCar). Fraunhofer IAO, IAT Universität Stuttgart. http://www.protoscar.com/media/Innovationsnetzwerk_FuCar.pdf. Accessed 28 Jan 2011

Schafhausen F (2009) Herausforderung Elektromobilität. Lektion 2—Politische Hintergründe, Einordnung in Energieversorgung und Klimapolitik. Euroforum Verlag GmbH, Düsseldorf

Schneider NF, Limmer R, Ruckdeschel K (2002) Mobil, flexibel, gebunden Familie und Beruf in der mobilen Gesellschaft. Campus, Frankfurt am Main

Strommagazin (2009) Energieversorger und Autokonzerne treiben Elektroauto voran. http://www.strom-magazin.de/strommarkt/energieversorger-und-autokonzerne-treiben-elektroauto-voran_25400.html. Accessed 25 May 2010

Teczilla (2010) Deutsche Bahn: Mehr Carsharing mit Elektroautos. http://www.xoomix.de/deutsche-bahn-mehr-carsharing-mit-elektroautos/12660#more-12660. Accessed 10 Feb 2013

Thoben K-D, Eschenbaecher J, Jagdev H (2001) Extended products evolving traditional product concepts. In: Proceedings of the 7th international conference on concurrent enterprising (ICE 2001), Bremen, Germany, pp 429–439

Thoben K-D, Jagdev H, Eschenbaecher J (2002) Emerging concepts in E-business and extended products. In: Gasos J, Thoben K-D (eds) E-business applications—technologies for tomorrow's solutions, advanced information processing series. Springer, New York, pp 17–37

WirtschaftsWoche (2010) Unternehmen und Märkte Galerie "Elektroautos für alle" http://www.wiwo.de. Accessed 26 Sept 2010

Von Wüst C (2011) Waffe der Feinde. http://www.spiegel.de/spiegel/print/d-81562387.html. Accessed May 23, 2012

Author Biographies

Jens Eschenbaecher has graduated as economist in 1996. Between 1997 and 2000 he was employed subsequently at Volkswagen, CSC Ploenzke and Produtec Ingenieurgesellschaft. Between 2000–2007 he was working as both project manager and department leader at Bremer Institut für Produktion und Logistik GmbH (BIBA) in Bremen. He finished his doctoral degree about innovation processes in collaborative networks in 2007 with predicate. Between 2008 and 2012 he worked as Senior Scientist at BIBA. During this timeframe this publication has been prepared. In 2013 he was appointed Professor at the Private Fachhochschule für Wirtschaft und Recht in Vechta.

Stefan Wiesner graduated in Industrial Engineering and Management (focus on Project Management and Development of Production) from the University of Bremen in 2007. Since then he has been employed at the Bremer Institut für Produktion und Logistik (BIBA). Stefan Wiesner is responsible for the BIBA contribution to various national and European research projects that deal with ICT support of Enterprise Networks. In his research he is concentrating on Requirements Engineering for Extended Products in Enterprise Networks.

Klaus-Dieter Thoben studied mechanical engineering at TU Braunschweig. After finishing his doctoral degree with a focus on CAD applications to increase the efficiency of product development at the University of Bremen in 1989, he joined the Bremer Institut für Produktion und Logistik (BIBA) as head of the Department of Computer Aided Design, Planning and Manufacturing. He received the state doctorate (Habilitation) for the domain of Production Systems in 2002. In the same year he was appointed professor at the University of Bremen. In 2003 he became director of the BIBA.

Part II
Challenges for Companies and Politics

Strategic Perspectives for Electric Mobility: Some Considerations About the Automotive Industry

Richard Colmorn and Michael Hülsmann

Abstract Through the competition over technological possibilities, cost-efficient realizations and innovative concepts for the usage of Electric Mobility the automotive industry stands at a strategic crossroad. Hereby, the question is to a lesser extent whether new driving technologies will penetrate the market. Moreover, it is the question of substance, which determinants for the diffusion process of Electric Mobility will become apparent; what are enablers and disablers and what are their resulting strategic implications? For this, the article intends to discuss—inter alia—the following questions: How will the competing forces position themselves? Which potentials for the positioning and standing out on the market will arise? How will the structures and processes of the global value creation networks change? As a result, it will be assumed that an increased complexity will only lead to bounded changes in the strategic behavior in the long-term so that the "rules of the game" will stay mostly the same.

1 Introduction

Currently, a wide spectrum of opinions about the chances of success for Electric Mobility can be observed. Figure 1 represents an exemplary extraction of headlines taken from German newspapers. Hereby, the opinions range from cautious, e.g. "the biggest uncertainty is the end-customer" (Gernot 2011), and negative, e.g. the "Economic development scheme for Electric Mobility crystallizes as a

R. Colmorn (✉) · M. Hülsmann
School of Engineering and Science, Systems Management, International Logistics,
Jacobs University Bremen, Campus Ring 1 28759 Bremen, Germany
e-mail: r.colmorn@jacobs-university.de

M. Hülsmann
e-mail: m.huelsmann@jacobs-university.de

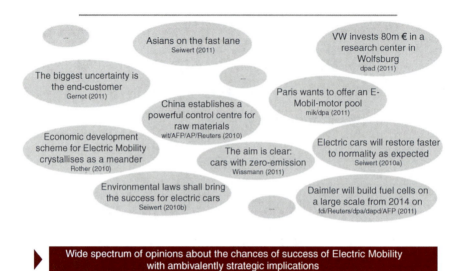

Fig. 1 Wide spectrum of opinions about the chances of success of Electric Mobility (wit, AFP, AP, Reuters 2010; Seiwert 2010b, 2011; Wissmann 2011; fdi, Reuters, dpa, dapd, AFP 2011; mik/dpa 2011)

meander" (Rother et al. 2010), to promising, e.g. "VW invests 80 m & in a research center in Wolfsburg" (dapd 2011a, b) and even optimistic ones, e.g. "Electric cars will restore faster to normality as expected" (Seiwert 2010a).

Depending on which of these perspectives about the chances of success one is willing to take, different and partly conflictive strategic implications can be derived. For example, while a cautious perspective might lead to a reduced investment in the development of such a new technology, a positive opinion can lead to a significant investment as indicated in the example of Volkswagen above.

To put these ambivalently strategic implications straight, the concept of 'path dependencies' as an analytical concept from the discipline of social sciences can be used (Schreyyögg et al. 2003). Hereby, according to Schreyyögg et al. (2003) the timely behavior of processes are uniquely described and resembles a path in a structural manner so that the progress is determined by the previous states and tends to reach one of several states (Schreyyögg et al. 2003; Ackermann 2001; Schäcke 2006; Dievernich 2007). Moreover, each path can be classified within four typical stages: the generating momentum, the path shaping, the lock-in and the de-locking situation (Fig. 2). Therefore, due to the wide spectrum of opinions about the chances of success the research question about the impact of electric mobility on a constricting versus widening corridor of strategic paths of development arises. On the one side, a constricting corridor of strategic paths can exist because of an institutional lock-in situation so that on the basis of the inferior driving technologies the development of electric mobility might be constrained, because the full potentials with respect to the efficiency can not be unfolded due to

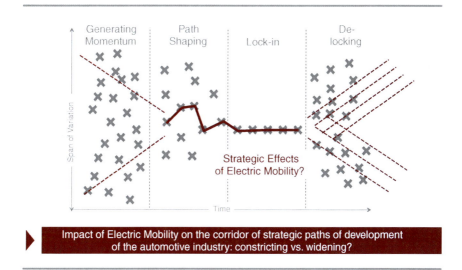

Fig. 2 The corridor of strategic paths of development (Schreyyögg et al. 2003)

e.g. certain standards in the production processes. On the other side, environmental conditions that have to be considered from the outside, such as for example customer expectations or political requirements, might lead to different possible solutions such as for example battery electric vehicles or fuel cell vehicles, which chances of success cannot be adequately predicted. As a result, a widening corridor of strategic paths of development can occur, because assumptions about the events that are expected to arise—and therefore which alternative to select—cannot be predicted. Against this background, it is prudent to derive more information about the current situation of automotive industry so that lock-in situations can be avoided that describe an inflexibility to leave an inefficient situation. For this reason, it is the objective of the paper to analyze the current situation of the automotive industry so that implications e.g. for the development stable and successful business models can be derived.

The above-outlined research question refers to the scientific discipline of strategic management in which a "strategic fit"—i.e. a concerted configuration between the company and the environment—plays a guiding principle (Bea and Haas 2005; Welge and Al-laham 2001). Therefore, it is necessary to determine the requirements of a company's environment and internal potentials of a company as strategic success factors, before implications of the management for a strategic fit can be derived. For this reason, Sect. 2 will use Porter's Five Forces as a methodological approach for determining the market attractiveness through the depiction of the barriers for a market entry. Section 3 intends to use the concepts of strategic success factors and positioning models, because the approaches can be used to exemplary analyze the internal situation. Section 4 will outline the traditional architecture of the value creation networks in the automotive industry, because based on this it can be derived that the strategic network management will have to change the processes and structures in the global value creation networks due to an increased complexity.

2 How Will the Competing Forces Position Themselves?

Porter's Five Forces is a common used methodological approach to analyze the external competing situation of an industry (Porter 1998). Its central assumption consist of the idea that the market attractiveness is determined through the intensity of competitive rivalry—i.e. the barriers to market entry (Welge and Al-laham 2001). Against this background, the section will exemplary discuss Porter's five competing forces—i.e. the bargaining power of suppliers and customers, the threat of potential competitors and the threat of substitute products and the resulting competitive rivalry within the automotive industry—with respect to the question about a widening or restricting corridor of strategic paths of development (Fig. 3).

With regard to the threat of potential competitors—that deals with the possibility of a market entry of competing companies—our research implies that both traditional car manufacturers will extend their portfolio to the production of electric cars as well as new car manufacturers are expected to make a step into the market. For example, the industry journal of the Bundesverband Elektromobilität BEM e.V. shows the announcement of the market introduction of several electric vehicles for the years 2012 and 2013, e.g. the R8 e-tron of Audi, the Megacity

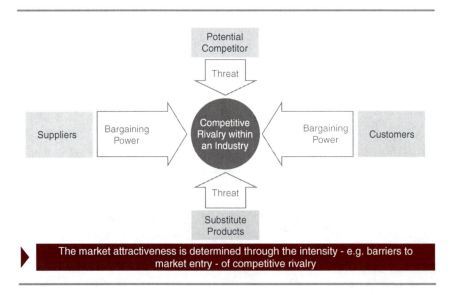

Fig. 3 Porter's five forces for the analysis of the competing situation in the automotive industry (with reference to Porter 1998)

Strategic Perspectives for Electric Mobility 159

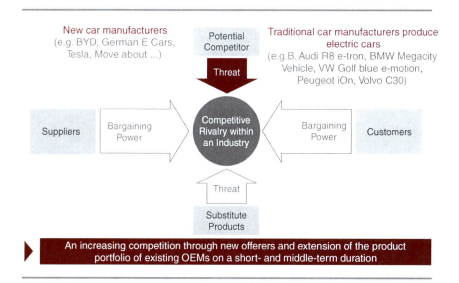

Fig. 4 Threat of potential competitors

Vehicle of BMW, the Golf blue e-motion and e-Up of Volkswagen, the iOn of Peugeot, the C30 of Volvo as well as the Roadster of Tesla, the E6 of BYD, the Stromos of German E Cars or the Think of Move About (cf. e.g. BEM 2011; Ford Motor Company 2012; Fairley 2010; Dutta 2011; Neue Mobilität 2011). Against this background it is assumed that the competition will increase on a short- and middle-term due to new offerers and an extension of the product portfolio of traditional car manufacturers (cf. Frenken et al. 2004; Hildermeier and Villareal 2011; Bento 2010), (Fig. 4). For example, along the supply chain new companies such as electrical suppliers have to be integrated, while new start ups on the outside of the supply chain can become a potential threat in terms of competitors.

For studying the bargaining power of the customers—that deals with the possibility of customers to negotiate terms of conditions—our research could find indicators that the price is still "the top priority for both conventional and the electric vehicles with range ranked second." (Lieven et al. 2011) Hence, some authors argue that the customer satisfaction and customer retention is essential for gaining a competitive advantage (Kley et al. 2011), while, for example, the scientist about future trends Peter Wippermann explains that the meaning to save time and to contribute to the environment increases significantly for certain customer segments (cf. e.g. Wippermann 2010). In this conjunction it can be assumed that a changed structure of customer preferences will increase the bargaining power of the customer because of the need to adapt to these emancipated customer requirements (Fig. 5) (cf. the discussion about the change of customer requirements in: Dutta 2011).

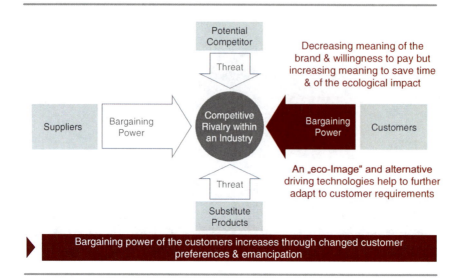

Fig. 5 Bargaining power of customers

Against this background, the threat of substitute products—that deals with the possibility of alternative products, services or technologies to replace the considered product, service or technology of the company—can also be derived because the need to develop alternative driving technologies (e.g. fuel cell vehicles or hybrid cars) (Chan 2007) can be traced back to the above-mentioned increased ecological customer requirements, binding legal regulations and a decreasing availability of fossil fuels from few and often politically instable regions (e.g. (Wallentowitz et al. 2011). Furthermore, the development of new and extended business models (e.g. concepts of Car Sharing) is expected to increase the amount of alternative products and services (Fig. 5) (cf. e.g. Schmidt 2010; Cepolina and Farina 2012; Loosea et al. 2007) (Fig. 6).

For the study of the bargaining power of the suppliers—that deals with the possibility of suppliers to negotiate the terms of conditions—it can be argued that due to an abolition of existing and the adding of new—especially electrical—components the trend of a power shift towards the suppliers will continue (cf. e.g. Hülsmann and Colmorn 2010a, b; Dombrowski et al. 2011). Finally, it is expected that due to the increasing power shift of the value creation networks towards the suppliers will be amplified through the increasing modularisation of the vehicle construction (cf. e.g. Colmorn and Hülsmann 2012; Wondrak et al. 2010). Therefore, it can be assumed that the architectures for the value creation and the configurations of power will become more complex and result in a further power shift towards the suppliers (Fig. 7).

Consequently, based on the above-mentioned assumptions it can be derived that the fifth force of the competitive rivalry within an industry—that integrates the

Strategic Perspectives for Electric Mobility

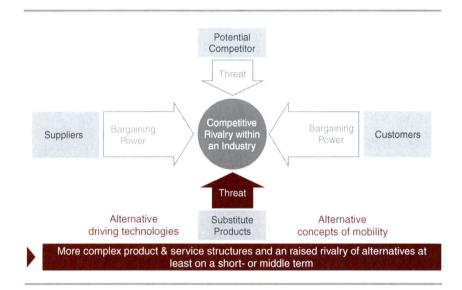

Fig. 6 Bargaining power of customers

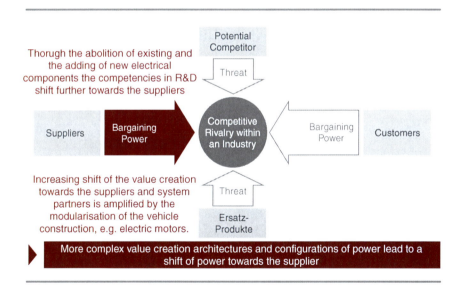

Fig. 7 Bargaining power of suppliers

above-mentioned threats and bargaining power so that implications about the intensity of competition can be derived—will increase and the forces will position towards customers and suppliers, while the diffusion processes of innovations will decide over diversity and standards of the suppliers and products structures.

Therewith, the existing competition will shift, because projects of traditional car manufacturers targeted towards to raise their prestige can be expected. But what does that mean for the company's to successfully position on the market?

3 Which Potentials for the Positioning and Standing Out on the Market Will Arise?

According to Grant (2005) the positioning against the environment and standing out on the market is determined through the selection of strategies (cf. Grant 2005; Zott and Amit 2008) The selection of strategies is determined through capabilities of a company, while these capabilities are the result of the internal resources (Barney 1991). Therefore, by taking an external perspective and considering the competing forces in the automotive industry in the previous section, this section intends to focus the attention on internal strategic success factors. According to Bea and Hass (2005) success factors are the strengths and weaknesses of a company that are critically responsible for its sustainable success and can be distinguished into achievement potentials or management potentials with different measurable features (cf. Bea and Haas 2005). For example, in Fig. 8 some strategic success factors for gaining a competitive advantage are exemplary introduced.

For example, the differentiation as one strategic approach to positively differentiate from a competitor depends on e.g. the value of the brand or the marketing mix, because these are aspects that can be recognized by the end-customer and evaluated by them as a positive (or negative) incentive. Therefore, car manufacturers try to promote e.g. their image as one factor to influence their competitive advantage.

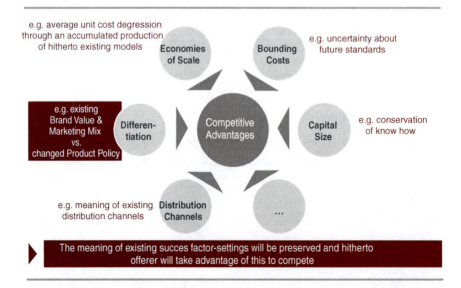

Fig. 8 Exemplary illustration of the success factors for gaining a competitive advantage

Strategic Perspectives for Electric Mobility

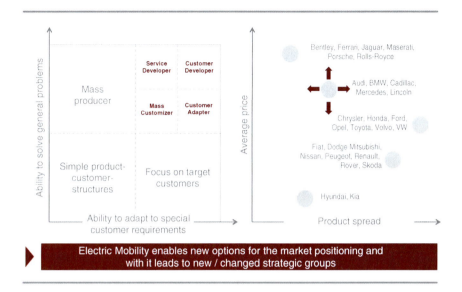

Fig. 9 Exemplary illustration of position models (with reference to Hülsmann et al. 2011)

Our research focused—inter alia—on the capital size responsible e.g. for securing know how, on bounding costs with respect e.g. about the uncertainty of future standards, on economies of scale with regard to e.g. a average unit cost digression and on distribution channels because of the meaning of already existing channels. Hence, it can be assumed that the existing settings of success factors will stay mostly the same, because the discussions in the scientific literature about strategic success factors (Lieven et al. 2011) and their logic can be totally adapted to electric mobility. Moreover, we expect that traditional car manufacturers will use their existing capabilities, e.g. the firm's capital, to further successfully compete.

Against this background using different positioning models can derive the potentials for a market positioning and standing out on the market. For example, Fig. 9 illustrates how the ability to solve general problems was plotted against the ability to adapt to special customer requirements or the average price was plotted against product spread of a company (cf. Hülsmann et al. 2011). Hereby, it can be derived that new options for a market positioning and as result new strategic groups will arise.

4 How Will the Structures and Processes of the Global Value Creation Networks Change?

To answer the question about the change of the structures and processes of the global value creation networks of the automotive industry, our research started with an inventory taking of the traditional architecture of the value creation

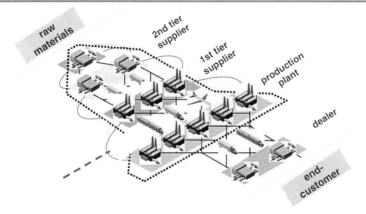

Fig. 10 Exemplary illustration of the traditional architecture of value creation in the automotive industry (from Hülsmann and Colmorn 2010a, b; with reference to Gehr and Hellingrath 2009)

(Fig. 10). So far, the architecture of the value creation in the automotive industry is often still described through the hierarchy of the car manufacturer and its suppliers (first tier supplier for complex systems, second tier supplier for subsystems and modules and third tier supplier for components) implying a linear flow of material from the raw materials to the end-customer (Oh et al. 2010).

Hence, our research approaches to consider the international value creation system of the automotive industry as a network of networks. Reasons for this can be traced back to the findings of the several linkages between the car manufacturers, several linkages along the supply chain between different suppliers so that the supply chains are highly interconnected and several linkages between suppliers and car manufacturers so that a supplier delivers the same components to different car manufacturers. The result of this interconnectivity is an increased complexity, because more and more companies and their behavior have to be taken into account so that slightly changes on "one side of the network" might have dramatic impacts on the "other side of the network". For example, a production stop of a highly specialized supplier can lead to a total production stop of whole supply chain (Colmorn et al. 2012; Colmorn and Hülsmann 2012a, b).

Therefore, in combination with the results of the increased competitive situation in the automotive industry the implications of the investigation of the—structural and dynamical—complexity within the automotive industry leads to the conclusion that it will become more and more harder to plan and control. This is the case because the timely behavior on the level of a system is the result of non-linear feedback loops of a manifold set of cause-and-effect-structures (cf. Mainzer 2008). For example, due to the power shift from the car manufacturers to the suppliers

Interaction	Local interdependencies of system components always have global effects
Nonlinearity	Small disturbances might lead to totally different results through nonlinear interdependencies (cf. butterfly effect)
Path dependency	The timely behavior depends not only on the current but also on the previous system states
Attractors	Complex system can to aspire to certain states - independently from their initial conditions
Emergence	The behavior on the level of the system can not be explained by an isolated analysis of the elements of the system
Self-organisation	Information procession enables the creation of stable structues to build an inner balance

▶ Value creation networks containing the characteristics of complex systems more and more harder to plan and to control

Fig. 11 Complexity-based implications for the network management

and new players as outlined-above the typology of the strategic network of the automotive industry will be macerated towards a more decentralized distribution of power.

In conclusion, for studying the challenges of the automotive industry due to the development of electric mobility, our research derived reasons that support the idea to understand the development of electric mobility as a paradigm shift (Colmorn and Hülsmann 2012a, b). Hence, the "rules of the game" within the automotive industry will stay mostly the same, but the management has to face an increased complexity that makes it harder to manage and control the value creation system.

5 Concluding Remarks

The paper contributed to the question about the—constricting versus widening—corridor of strategic paths of development for the automotive industry due to electric mobility by using a complexity-based perspective.

With respect to the question about the competing forces will position themselves, it was argued that the competition will increase and the forces will position towards customers and supplier. Moreover, diffusion processes of innovations will decide over the diversity and standards of the supplier and product structures. With respect to the question which potentials for the positioning and standing out on the market will arise, it was argued, that shifts in the setting of success factors are still unclear. Hence, the importance of existing resources and competencies remains crucial, but new options are thinkable. Finally, with respect to the question which structures and processes of the global value creation networks will change, it was argued that due to an increasing network complexity the ability to plan and control will decline.

As a result, an increase of the strategic complexity in the short-term and an increase of the logistic complexity in the long-term could be concluded. Hence, it is assumed that this increased complexity will only lead to bounded changes in the strategic behavior in the long-term so that the "rules of the game" will stay mostly the same (Fig. 11). Due to the allowance of both external competing forces and internal potentials of success different strategies for a successful diffusion of technologies for electric mobility can be evaluated. Future research is targeted towards the investigation of the network typology for the development of electric mobility in order to enable an economic evaluation of the cooperative strategic partnerships against the background of such a complexity-based perspective.

Acknowledgments This research was supported by the Federal Ministry of Transport, Building and Urban Development in the context of the "Model Regions Electric Mobility".

References

Ackermann R (2001) Pfadabhängigkeit. Institutionen und Regelreform, Tübingen
Barney J (1991) Firm resources and sustained competitive advantage. J Manage 17(1):99–120
Bea FX, Haas J (2005) Strategisches management, 4. neu bearb. Aufl. edn. Lucius and Lucius, Stuttgart
BEM (2011) Elektroautos in Deutschland. Neue Mobilität—Das Magazin vom Bundesverband für eMobilität 2:24
Bento N (2010) Dynamic competition between plug-in hybrid and hydrogen fuel cell vehicles for personal transportation. Int J Hydrogen Energy 35(20):11271–11283
Cepolina EM, Farina A (2012) A new shared vehicle system for urban areas. Transp Res C: Emerg Technol 21(1):230–243
Chan CC (2007) The state of the art of electric, hybrid, and fuel cell vehicles. Proc IEEE 95(4):704–718
Colmorn R, Hülsmann M (2012a) Industriedynamiken und das Management von Ambidextrien— Eine Komplexitätstheoretische Analyse der System Dynamics im Paradigmenwechsel vom Verbrennungs- zum Elektromotor. In: Proff. Schönharting H, Schramm J, Ziegler J (Hrsg.): Zukünftige Entwicklungen in der Mobilität—Betriebswirtschaftliche und technische Aspekte, 2012, Springer Gabler, Wiesbaden, pp 245–272
Colmorn R, Hülsmann M (2012b) Complexity and company performance—Four assumptions about the dependency of the focal firm from the network. In: 29 Deutscher Logistik-Kongress, 17–19 Oktober 2012, Konferenzband, 16 pages (in print)

Colmorn R, Hülsmann M, Brintrup A (2012) The structure of the value creation network fort he production of electric vehicles—Lessons to be learned from complex network science. In: Conference proceedings international conference robust manufacturing control (RoMaC)—Innovative and Interdisciplinary approaches for global networks, Bremen, 12 pages (In print)

Dapd (2011) VW investiert 80 Millionen Euro in Elektro-Zentrum in Wolfsburg. Available at http://www.ad-hoc-news.de/vw-investiert-80-millionen-euro-in-elektro-zentrum-in–/de/News/21982622. Last accessed 18 Aug 2012

Fdi, Reuters, dpa, dapd, AFP (2011) Wasserstoffantrieb: Daimler baut Brennstoffzellen-Großserie ab 2014. Available at http://www.spiegel.de/auto/aktuell/wasserstoffantrieb-daimler-baut-brennstoffzellen-grossserie-ab-2014-a-766256.html. Last accessed 18 Aug 2012

Dievernich FEP (2007) Pfadabhängigkeit im Management : wie Führungsinstrumente zur Entscheidungs- und Innovationsunfähigkeit des Managements beitragen. Kohlhammer, Stuttgart

Dombrowski U, Engel C, Schulze S (2011) Changes and challenges in the after sales service due to the electric mobility. Service operations, logistics, and informatics (SOLI), 2011 IEEE international conference on, pp 77

Dutta S (2011) Green marketing : a strategic initiative. Available: http://www.smbs.in/adminpanel/document/1332594339IJMCS%20(RJ).pdf#page=40. Last accessed 19 Aug 2012

Fairley P (2010) Speed bumps ahead for electric-vehicle charging. Spectrum, IEEE 47(1):13–14

Ford Motor Company (2012) The 2013 ford focus electric. Available at http://www.ford.com/electric/focuselectric/2012/. Last accessed 19 Aug 2012

Frenken K, Hekkert M, Godfroij P (2004) R&D portfolios in environmentally friendly automotive propulsion: Variety, competition and policy implications. Technol Forecast Soc Chang 71(5):485–507

Gehr F, Hellingrath B (2009) Logistik in der Automobilindustrie: Innovatives Supply Chain Management für wettbewerbsfähige Zulieferstrukturen, 1 Aufl. Springer, Berlin

Gernot R (2011) Die größte Unsicherheit ist der Kunde. Neue Mobilität—Das Magazin vom Bundesverband für eMoblität, no 03, pp 116–117

Grant RM (2005) Contemporary strategy analysis, 5th edn. Blackwell, Malden

Hildermeier J, Villareal A (2011) Shaping an emerging market for electric cars: How politics in France and Germany transform the European automotive industry. Available at http://revel.unice.fr/eriep/index.html?id=3329. Last accessed 19 Aug 2012

Hülsmann M, Colmorn R (2010b) Strategische Perspektiven für Logistikdienstleister—Komplexitätstheoretische Überlegungen zur Elektromobilität. In: 27. Deutscher Logistik-Kongress, 20–22 Oktober 2010, Konferenzband, pp 128–160

Hülsmann M, Colmorn R (2010b) Strategische Perspektiven für Logistikdienstleister—Komplexitätstheoretische Überlegungen zur Elektromobilität, In: BVL-Kongressband 2010, DVV, Hamburg

Hülsmann M, Korsmeier B, Brenner V, Decker J, Krähe M (2011) (Re-) Positioning of logistics service providers in a globalised world—an empirical survey on positioning models and strategic capabilities, In: BVL-Kongressband 2011, DVV, Hamburg

Kley F, Lerch C, Dallinger D (2011) New business models for electric cars—A holistic approach. Energy Policy 39(6):3392–3403

Lieven T, Mühlmeier S, Henkel S, Waller JF (2011) Who will buy electric cars? An empirical study in Germany. Transp Res D Trans Environ 16(3):236–243

Loosea W, Mohra M, Nobisb C (2007) Assessment of the future development of car sharing in germany and related opportunities, Routledge

Mainzer K (2008) Komplexität, 1. Aufl., UTB/Fink-Verlag, Paderborn

Mik, dpa (2011) Umwelt-Initiative: Paris will E-Mobil-Flotte anbieten. Available at http://www.spiegel.de/auto/aktuell/umwelt-initiative-paris-will-e-mobil-flotte-anbieten-a-753636.html. Last accessed 18 Aug 2012

Neue Mobilität (2011) Elektroautos in Deutschland, Januar 2011, Neue Mobilität—Das Magazin vom Bundesverband für eMoblität, no 02

Oh S, Ryu K, Moon I, Cho H, Jung M (2010) Collaborative fractal-based supply chain management based on a trust model for the automotive industry. Flex Serv Manuf J 22(3):183–213

Porter ME (1998) Competitive strategy : techniques for analyzing industries and competitors; with a new introduction, Free Press, New York

Rother FW, Seiwert M, Wildhagen A, Rees J, Klesse HJ, Haerder M (2010) Förderkonzept für Elektroautos erweist sich als Irrweg. Available at http://www.wiwo.de/unternehmen/nationale-plattform-elektromobilitaet-foerderkonzept-fuer-elektroautos-erweist-sich-als-irrweg-/5229310.html. Last accessed 18 Aug 2012

Schäcke M (2006) Pfadabhängigkeit in Organisationen—Ursache für Widerstände bei Reorganisationsprojekten. Duncker and Humblot, Berlin

Schmid SA, Mock P, Friedrich H (2010) Welche Antriebstechnologien prägen die Mobilität in 25 Jahren? emw—Zeitschrift für Energie, Markt, Wettbewerb, no 4, pp 6–10

Schreyyögg G, Sydow J, Koch J (2003) Organisatorische Pfade—Von der Pfadabhängigkeit zur Pfadkreation? In: Schreyyögg G, Sydow J (eds) Strategische Prozesse und Pfade. Gabler Verlag, Wiesbaden, pp 257–294

Seiwert M (2010a) Elektroauto wird schneller Normalität, als wir dachten. Available at http://www.wiwo.de/unternehmen/bmw-elektroauto-wird-schneller-normalitaet-als-wir-dachten/5208248.html. Last accessed 18 Aug 2012

Seiwert M (2010b) Umweltgesetze sollen für Erfolg der E-Autos sorgen. Available at http://www.wiwo.de/unternehmen/la-auto-show-umweltgesetze-sollen-fuer-erfolg-der-e-autos-sorgen/5698448.html. Last accessed 18 Aug 2012

Seiwert M (2011) Unternehmen and Märkte—Überfälliger Schwenk. Wirtschaftswoche 4:42–46

Wallentowitz H, Freialdenhoven A, Olschewski I (2011) Strategien zur Elektrifizierung des Antriebsstranges: Technologien, Märkte und Implikationen. Vieweg + Teubner Verlag—GWV Fachverlage GmbH, Wiesbaden

Welge MK, Al-Laham A (2001) Strategisches Management: Grundlagen—Prozesse—Implementierung, Gabler, Wiesbaden

Wippermann P (2010) Die Mobilität von morgen, In: Das neue Automobil: Konzepte—Technologien—Visionen, Automobilwoche, edn

Wissmann M (2011) Das ist ein Marathon, Handelsblatt GmbH, no Nr 4, pp 48–49

wit/AFP/AP/Reuters (2010) Export Seltener Erden—China gründet mächtige Rohstoffzentrale. Available at http://www.spiegel.de/wirtschaft/unternehmen/export-seltener-erden-china-gruendet-maechtige-rohstoffzentrale-a-736841.html. Last accessed 18 Aug 2012

Wondrak W, Dehbi A, Willikens A (2010) Modular concept for power electronics in electric cars. In: 6th international conference on Integrated power electronics systems (CIPS), p 1

Zott C, Amit R (2008) The fit between product market strategy and business model: implications for firm performance. Strateg Manag J 29(1):1–26

Author Biographies

Richard Colmorn studied Physics at the University of Dortmund and graduated from the university of Bremen with his diploma thesis about the self-organization of multi-agent systems in a market similar model. Since may 2009 Richard Colmorn is a PhD-Student and research associate in the working group "Systems Management" of Prof. Hülsmann. He focuses on research in the field of strategic management, logistics networks and complexity theory. Furthermore, he takes the responsibility for the project on E-Mobility in the model region of Bremen/Oldenburg.

Michael Huelsmann holds the chair of "System Management" at the School of Engineering and Science at Jacobs University Bremen. He focuses on Strategic Management of Logistics Systems. Additionally he leads the sub-projects "Business model concept/product idea" and "Sustainable Innovation and technology strategies" in the framework of the electric mobility model region Bremen/Oldenburg.

How Knowledge-Based Dynamic Capabilities Help to Avoid and Cope with Path Dependencies in the Electric Mobility Sector

Philip Cordes and Michael Hülsmann

Abstract Companies involved in the electric mobility sector such as car manufacturers, battery producers or infrastructure providers are confronted with risks of technological and institutional path dependencies and resulting lock-in situations. One approach that might enable these companies to avoid and cope with such lock-ins is the idea of knowledge-based dynamic capabilities. Accordingly, companies need to be able to replicate their organizational resources through knowledge codification and transfer, and to reconfigure their resources through knowledge abstraction and absorption. This chapter reveals exemplary effects that result from such knowledge-based activities on the abilities to maintain or increase their technological and strategic flexibility. Accordingly, companies in the electric mobility sector have e.g. the chance to reduce risks of path dependent developments through transferring their knowledge internally and between each other in order to trigger a combination of knowledge resources, which in turn increase the varieties of their decision alternatives.

1 Introduction

The worldwide increasing pollution of the environment with CO_2 and the entirety of its consequences is a main driver for an ongoing development in direction of less internal combustion engines and more electric mobility (Brauner 2008; Pehnt et al. 2011; Sammer et al. 2011). A large amount of stakeholders (Freeman 1984)

P. Cordes (✉) · M. Hülsmann
School of Engineering and Science, Systems Management, International Logistics,
Jacobs University Bremen, Campus Ring 1, 28759 Bremen, Germany
e-mail: p.cordes@jacobs-university.de

M. Hülsmann
e-mail: m.huelsmann@jacobs-university.de

(e.g. car manufacturers, consumers of mobility, politics, electricity provider, etc.) have high expectations that in the not all that far future both the environment as well as the economy can profit from this development (Pehnt et al. 2011). However, since the shifts from traditional car manufacturing and traditional mobility concepts to electric mobility implies drastic changes for these stakeholders, especially for car manufacturers (Wallentowitz and Freialdenhoven 2011), customers as well as other companies that provide the necessary infrastructure (Brauner 2008), the risk of path dependencies arise. In result, two main risks occur: First, an institutional lock-in situation (see e.g. Schreyögg et al. 2003) could retard the development from traditional combustion engines to more environmental friendly electric vehicles. Second, technological path dependencies (see e.g. David 1985) could force the development of electric engines and vehicles to develop on the basis of inferior technologies that might hinder the electric mobility to unfold its full potential in efficiency and environmental friendliness. The latter in turn could affect again the solidification of the institutional paths that have developed over the years. Both have the three main principles of path dependencies in common: First, history matters, which means that the reason for the restricted amount of alternatives lies somewhere in the past. Former decisions shape the path on which the future develops (David 1994; Liebowitz and Margolis 1995). Second, increasing returns have occurred, which refers to an increase of the probability that a certain alternative is selected, because the same alternative has been selected before (Dievernich 2007; Schreyögg et al. 2003). Third, a lock-in situation finally occurs, which describes an inflexibility to leave an inefficient system state, i.e. to render other decisions than the ones that have been solidified on the previous path (Dievernich 2007; Schreyögg et al. 2003).

Both path dependency risks reveal problems for those companies that have or plan to develop business models that are dependent on the diffusion of electric mobility. Hence, the question arises, how companies can take countermeasures in order to enable and maintain a large amount of managerial options. One concept that is discussed as a potential enabler for avoiding path dependencies is the dynamic capabilities-approach (O'Reilly and Tushman 2008; Schreyögg and Kliesch-Eberl 2007).

Consequently, the following research question arises: How do dynamic capabilities of companies within the electric mobility sector affect the risk of institutional and technological path dependencies?

In order to answer this question, the chapter aims on a descriptive level to present the theory of path dependency, the dynamic capabilities approach in general and the knowledge-based conceptualization of dynamic capabilities following Burmann (2002). These theoretical frameworks will serve as the basis for the analysis of path dependencies and possible ways to avoid or cope with them in the electric mobility sector.

On an analytical level, the chapter aims first to reveal current and prospective risks of path dependencies in the electric mobility sector, different companies as well as customers are confronted with. Second, it shall be analyzed how the ability to replicate and reconfigure organizational resources contribute or limit these risks.

On a pragmatic level, the chapter intends finally to deduce implications for the electric mobility sector for avoiding and coping with the identified risks of path dependencies.

The chapter will proceed as it follows: After an introduction in Sects. 1, and 2 introduces the theory of path dependency in general as well as its implications for the electric mobility sector in order to reveal associate risks. Section 3 presents the dynamic capabilities in general as well as its knowledge-based conceptualization following Burmann (2002) in order to lay the analytical framework for Sect. 4, in which an analysis of the dynamic capabilities' effects on the risk of path dependencies in the electric mobility sector will be conducted. Section 5 finally subsumes the results and deduces implications for the management of companies within the electric mobility sector as well as further research requirements.

2 Risks of Path Dependencies in the Electric Mobility Sector

2.1 General Characteristics of Path Dependencies and Lock-In Situations

Path dependency theory revealed the insight that decisions rendered in the past influence the scope of managerial decision alternatives available today and imply the risk that their amount is reduced over time. Consequently, decisions have a formative character for subsequent decisions (David 1994; Liebowitz and Margolis 1995), wherefore Schreyögg et al. (2003) called the underlying principle '**history matters**' (Schreyögg et al. 2003)—the first main characteristic of path dependencies. The development of companies and entire industries are mainly based on decisions that are rendered within them. Thereby, decisions on the management level of companies play the most important role and are at least partly irreversible, and so are parts of the developments within organizations and industries. Historical events that determine the future development of a company or an entire industry in a negative way are called by David (1985) 'historical accidents' that shape the broadness of the path on which the company/industry can develop and hence, restrains the remaining managerial options in the future (David 1994). According to Van Driel and Dolfsma (2009), organizations are generally sensitive to such bygone decisions or events that function as initial conditions and hence as a starting points for a path dependency (Van Driel and Dolfsma 2009). Hence, business processes and developments are characterized by an essentially historic character (Liebowitz and Margolis 1995).

The second main characteristic of path dependencies is the occurrence of '**increasing returns**' (Schreyögg et al. 2003). Following the above-mentioned assumptions, it can be assumed that economic actors have never a totally free choice. Instead, decisions that they or others to which they are related in any way rendered in the past restrict their scope of available decision alternatives today. Self-reinforcing effects intensify this effect by increasing the probability that a certain decision is rendered at least similarly again in the next time step when it has been rendered before (Arthur 1989, 1990).

The evolvement of a '**lock-in situation**' is the third main characteristic of path dependencies and constitutes the main critical result of their developments (Schreyögg et al. 2003). The first identified lock-in situation was the QWERTY-typewriter keyboard as the quasi-standard among all industries and private households that use computers, although superior typewriter configuration exist (David 1985). A lock-in situation in such a technological sense refers to a situation in which users adopt one particular technology due to historical events and increasing returns, although superior technologies are available.

Transferred to an organizational context, lock-in situations reflect basically an inflexibility to change decision-making behavior which equals a reduction of available managerial options (Schreyögg et al. 2003). Combinations of such managerial options are regarded as corporate strategies when they aim to create a certain industry position (Porter 1999). Hence, such behavioral lock-ins that result from self-reinforcing effects that emanate from certain strategy choices can reduce the amount of strategies companies can apply in order to generate and maintain competitive advantages (Schreyögg et al. 2003). In consequence, such companies are becoming inflexible to react to market changes or to main developments in their organizational environments.

Consequently, the literature on path dependencies distinguishes between technological and institutional path dependencies (Schreyögg et al. 2003). The former refers to a situation in which only one or a few technologies remain being selectable by customers although superior alternatives might be available (e.g. the QWERTY-keyboard) (David 1985). The latter describes situations in which the scope of alternatives of decision makers is decreased in a way that only a few or in an extreme situation only one alternative is left being selectable (Ackermann 2001; Dievernich 2007; Schäcke 2006; Schreyögg et al. 2003). Hence, the question arises, what implications occur for the electric mobility sector?

2.2 Implications of Path Dependencies for the Electric Mobility Sector

The described development in direction of more electric mobility combined with the general risk of institutional and technological lock-ins reveals the need to investigate possible risks that might result from path dependent developments in the electric mobility sector. Therefore, the above-mentioned distinction between managerial and technological path dependencies will be picked up in order to deduce associate implications.

On an **institutional level**, the electric mobility sector consists of a large amount of differing economic actors—from car manufacturers over rental car companies to industrial customers with large transport fleets and the end-consumer who rents or buys an electric vehicle. Many of them have experienced great success with the traditional internal-combustion engine in the past. Especially car manufacturers (such as BMW or VW) are still very successful with the further development of the traditional car types. But also other actors such as rental car companies (e.g. Sixt or Avis) earned most of their money up to now with standard traditional cars with internal combustion engines. The car pools of the greatest share of companies consist also only of non-electric vehicles. And finally, end consumers are very satisfied with the cars that were and are still available.

Path dependency theory suggests that decisions that have been made in the past and that led to successful results are likely to be rendered at least similarly again. A similar argumentation line is deduced in the dominant logic approach. Accordingly, managers build schemes that reflect their point of view and that solidify more and more with the success of these decisions in the past. They are one main reason for the development of path dependencies. Hence, all the economic actors that experienced success with the internal combustion engines are at risk of evaluating decisions that conform to this success in the past as better compared to decisions that in a way compete with the solidified ones.

This theoretically derived assumption can be underlined by observations in practice: According to a capgemini study from 2012, car manufacturers overestimated their innovational strength and did not invest enough money into the development of new technologies in niche markets, such as electric engines (Dammenhain 2012). This indicates that the industry faces the risk of institutional path dependencies based on increasing returns that have led to a solidification of the "old" internal combustion technology for mobility (Göllinger 1997). Hence, it will be even harder for the electric mobility to break this path that has developed over almost a centenary.

Regarding the perspective of **technological path dependencies (prospective)**, there is currently a large amount of companies that are engaged the development of electric mobility and the associate infrastructure. Hence, there is also a large amount of different approaches and technologies that are tested at the moment (Brauner 2008). However, at the time one company starts to be more successful than another, the infrastructure will develop with respect to the certain approach or technology favored by this special company. Hence, the risk occurs, that such a path might hinder the development of other technologies that might be superior since they are simply not compatible with the developed infrastructure. One recent example is the question of battery supplies. Once, an infrastructure is build for the supply of power, a path emerges that forces the car manufacturers to build compatible cars. Otherwise the car manufacturers' customers cannot use the infrastructure.

Therewith, companies in the electric mobility sector might choose inferior technologies in order to either ensure compatibility with existing infrastructure or in order utilize existing knowledge and capabilities e.g. for avoiding sunk-costs.

Hence, the electric mobility sector faces also the risk of a technological lock-in situation in which an inferior technology becomes prevalent although superior alternatives might be available in the near future.

Consequently, path dependent developments endanger the diffusion of superior technologies, infrastructures and associated business models of the companies involved in the electric mobility sector through decreasing the remaining scope of available decision options. In other words, such a lock-in situation is characterized by inflexibility to change the already developed path. Therefore, the question arises, how to avoid and to cope with such lock-in situations through increasing and maintaining the involved companies' decision flexibilities.

3 Dynamic Capabilities: A Knowledge-Based Approach

3.1 A General Understanding of Knowledge-Based Dynamic Capabilities

The management literature is in relative agreement on the question how to ensure strategic flexibility. Accordingly, companies need dynamic capabilities, which are seen by O'Reilly and Tushman (2008) as the pre-requirement to overcome inertia and path dependencies (O'Reilly and Tushman 2008). This approach was introduced into the management literature by Teece et al. (1997) who define dynamic capabilities as "[…] the capacity to renew competences so as to achieve congruence with the changing business environment […]" through "[…] appropriately adapting, integrating, and reconfiguring internal and external organizational skills, resources and function-al competences to match the requirements of a changing environment" (Teece et al. 1997). Therewith, the underlying idea picks up the main assumptions of the resource-based view (Barney 1991; Penrose 1959; Selznick 1957; Wernerfelt 1984) as well as the competence-based view (Prahalad and Hamel 1990; Sanchez 2004; Teece et al. 1997). Both argue basically that only those companies gain and sustain competitive advantages that possess resources and competences that fulfill the so-called VRIN- respectively the VRIO- criteria— i.e. when they are valuable, rare, inimitable and not substitutable respectively organizationally embedded (Barney 1991, 2002).

However, both approaches fail to explain, why there are substantial differences in the performance of organizations that are equipped with the same or at least similar resources and competences (Freiling 2004). Therefore, Teece and Pisano (1994) argue that the essential capability is to be able to change resources and competences over time and under consideration of environmental changes (Teece and Pisano 1994).

According to Burmann (2002), the strategic flexibility that results from such dynamic capabilities can be operationalized through a focus on knowledge-based activities in organizations (Burmann 2002). Thereby, Burmann (2002) follows the assumption of a multitude of different authors who state that the resource knowledge plays an especially vital role in the creation of success of organizations (Drucker 1993; Grant 2002; Kusunoki et al. 1998; Quinn 1992; Spender and Grant 1996). Therefore, this assumption can be adopted for the electric mobility sector.

Burmann (2002) operationalized his understanding of dynamic capabilities and resulting strategic flexibility by stating that companies need to be able to replicate their organizational resources through knowledge codification and knowledge transfer, and to reconfigure their resources through knowledge abstraction and knowledge absorption (Burmann 2002). The underlying argumentation is that organizational flexibilities occur only if a company is able to…

- … identify and externalize organizational knowledge through its **codification**
- … make the organizational knowledge available to the entire organization through internal its **transfer**
- … devolve organizational knowledge to new fields of appliances respectively markets through its **abstraction**
- … combine the organizational knowledge with new organization-external knowledge through knowledge **absorption** (Burmann 2002).

Consequently, the question arises what these general dimensions of dynamic capabilities could mean in concrete.

3.2 Knowledge Codification, Transfer, Abstraction and Absorption

3.2.1 Knowledge Codification

Following the assumption that knowledge is the most important resource for any company—and therewith also for companies involved in the electric mobility sector—it is necessary to be able to treat knowledge as an economic good. That, in turn, is only possible according to Ancori et al. (2000), if a company is able to "[…] put [it] in a form that allows it to circulate and be exchanged" (Ancori et al. 2000). In order to do so, it is necessary to transform implicit knowledge (also called as tacit) within the organization that evolved over the years through the absorption of data, information and the accumulation of practical experiences (Burmann 2002) into explicit knowledge (Nonaka et al. 1997). A prerequisite for doing that is the transformation of experience and information into symbolic forms (Teece 1981), which in turn, have to be understandable and interpretable by the organization members that are supposed to decode the codified knowledge. Therefore, knowledge codification is understood in the following as the transformation of organizational knowledge into understandable and interpretable symbolic forms.

3.2.2 Knowledge Transfer

In order to be able to diffuse relevant knowledge within an organization or even within an industry like the electric mobility sector, it is furthermore necessary to transfer the codified knowledge from different application fields to others. Thereby, Jane Zhao and Anand (2009) note that "[...] when firms transfer organizational capabilities from one unit to another, they transfer not only individually held skills, but also organizationally embedded knowledge or collective knowledge" (Jane Zhao and Anand 2009). Szulanski (2000) describes such a transfer as "[...] a process in which an organization recreates and maintains a complex, causally ambiguous set of routines in a new setting" (Szulanski 2000). Since relevant knowledge cannot be narrowed down to routines, it might be reasonable to apply a wider perspective: According to Bou-Llusar and Segarra-Ciprés (2006), knowledge transfer is "[...] the exchange of knowledge between units within a firm (internal transfer) or between different firms (external transfer)" (Bou-Llusar and Segarra-Ciprés 2006). Additionally, a complete knowledge transfer passes different stages: (1) The initiation stage in which a managerial decision triggers to diffusion of codified knowledge. (2) The implementation stage in which the source and the recipient of the knowledge actually exchange resources. (3) The ramp-up stage, in which the received knowledge is utilized by the recipient. (4) The integration stage in which the received knowledge leads to first positive effects in its new application fields (Szulanski 1996). Consequently, knowledge transfer is understood in the following as a process in which individual or collective knowledge is devolved from one application place to another (Burmann 2002).

3.2.3 Knowledge Abstraction

In order to use organizational knowledge not only in different application fields, for which a simple transfer of knowledge would be sufficient, but in totally different areas of application, the knowledge has to be taken out of its context. Such a decontextualization of knowledge is called knowledge abstraction, when its essentials—i.e. its underlying structure, which reveals certain cause-and-effect relationships (Burmann 2002) that are not directly observable in the knowledge's codification (Boisot 1998)—are applied elsewhere. That in turn enables to generalize knowledge for making it usable in a wider range of applications. Since abstraction requires reducing the knowledge to its essentials, it can be seen as the key process for the transformation of experience-based knowledge into theoretical knowledge on causal interrelations (Nahapiet and Ghoshal 1998). Consequently, knowledge abstraction will be understood in the following as the ability of disengaging knowledge from its context and engaging it in other contexts.

3.2.4 Knowledge Absorption

In order to be able to reconfigure organizational resources under consideration of not only internal but also environmental changes, it is necessary to internalize external knowledge and combine it with the already existing knowledge. The underlying idea can be found in the concept on absorptive capacity. Accordingly, firms need to be able "[...] to recognize the value of new external information, assimilate it, and apply it to commercial ends" (Cohen and Levinthal 1990). In order to do so, knowledge has to be acquired, assimilated, transformed and finally exploited (Zahra and George 2002). Thereby, acquiring knowledge means to identify and obtain externally generated knowledge that is critical to a firm's operations and functionality. Assimilating knowledge refers to an organization's ability to analyze, process, interpret and understand such information. Transforming knowledge means to "[...] develop and refine the routines that facilitate combining existing knowledge and the newly acquired and assimilated knowledge" (Zahra and George 2002). Exploiting knowledge finally means to "[...] refine, extend, and leverage existing competencies or to create new ones" (Zahra and George 2002) through the incorporation of acquired or transformed knowledge into the operations of an organization. Consequently, knowledge absorption is understood in the following as an organizational process that internalizes external knowledge through its acquisition, assimilation, transformation and exploitation (Boisot 1995; Zahra and George 2002).

Figure 1 subsumes these four described knowledge-based activities as dimensions of dynamic capabilities.

Having defined these four dimensions of dynamic capabilities, the question arises, what companies in the electric mobility sector can learn from them. In other words: What are effects of knowledge-based dynamic capabilities on the risks of institutional and technological path dependencies for companies involved in the electric mobility sector?

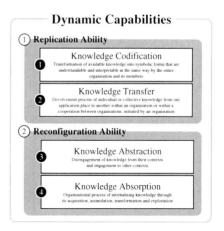

Fig. 1 Dimensions and sub-dimensions of dynamic capabilities (based on Burmann 2002)

4 Effects of Knowledge-Based Dynamic Capabilities on the Risks of Path Dependencies in the Electric Mobility Sector

4.1 Multitude of Combinatorial Interrelations Between Dynamic Capabilities and Path Dependencies

In order to reveal the effects that arise when firms within the electric mobility sector increase their abilities to codify, transfer, abstract and absorb knowledge, it is first necessary to span the field of possible interrelations. Figure 2 illustrates the resulting combinatorial possibilities between the dimensions of dynamic capabilities and the risks of institutional as well as technological path dependencies and resulting lock-ins.

Consequently, each field in the figure illustrates a room in which a multitude of causal interrelations can occur. Additionally, there are causal interrelations between the dimensions of dynamic capabilities as well as between the technological and institutional path dependency risks, which are ignored for the moment for a reasonable complexity reduction. Therefore, the following discussion of contributions and limitations entails exemplary effects for each field, which can be seen as a starting point for further research.

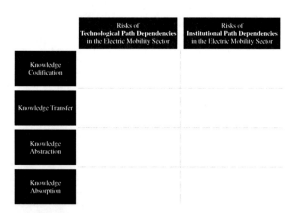

Fig. 2 Fields of combinatorial possibilities of interrelations between dynamic capabilities in risks of path dependencies in the electric mobility sector

4.2 Exemplary Effects of Knowledge Codification on Path Dependencies in the Electric Mobility Sector

An increase of the ability of companies in the electric mobility sector to codify their knowledge leads to both contributions as well as limitations for reducing risks of technological and institutional path dependencies.

A **contribution** ensues from an essential necessity for a successful codification of knowledge: According to Zollo and Winter (2002), one has to form a mental model out of knowledge in order to be able to transform it into symbolic forms (Zollo and Winter 2002). Therewith, the economic actors within the electric mobility sector have to bring their knowledge (e.g. on battery technologies or infrastructures) to their minds so as to be able to codify it. That, in turn, facilitates the generation of new ideas how to change technologies if necessary or reasonable or how to change the corporate strategies of their respective companies (e.g. car manufacturers). Therewith, knowledge codification does not only reduce the tacitness of knowledge on electric mobility markets and technologies, but also does it increase the deepness with which this knowledge is anchored within the economic actors in the electric mobility sector. That, in turn increases the variety of possible decision alternatives on both the institutional level for selecting from a wider range of strategies in the electric mobility sector as well as the technological level for selecting from a wider range of possible technology decisions such as batteries produced or infrastructures created. Consequently, knowledge codification can reduce the risk of institutional and technological risks of path dependencies and resulting lock-in situations in the electric mobility sector.

A **limitation** resulting from knowledge codification to the ability to avoid and cope with risks of path dependencies is based on a fundamental principle of such codification: Best practices are written down in forms of guidelines. Hence, these guidelines can include knowledge, which is based on assumptions that result from historic success—such as the superiority of the internal combustion engine. Therewith, the risk occurs that already locked-in knowledge is embedded within the transformed symbolic forms, which can lead to a solidification of learned routines. That, in turn, leads to the opposite of technological and strategic flexibility and triggers increasing returns that might lead to path dependencies. This assumption is ratified by an empirical study by García-Muiña et al. (2009). They showed that "[…] an excessive presence of codified knowledge, strongly institutionalized in the heart of the company, can put a serious brake on the creativity, intuition and employees' radical improvisation skills that major innovative activity requires" (García-Muiña et al. 2009) (see also Casper and Whitley 2004; Crossan and Bedrow 2003; Hedlund 1994; Vera and Crossan 2005).

Furthermore, knowledge codification is an expensive undertaking since the employees of the respective companies need time for forming their mental models out of their knowledge and writing it down in a way that other organization members can understand it and interpret it within the meaning of the codifying employee (Burmann 2002; Zollo and Winter 2002). Hence, since it is not assured that the codification of certain knowledge leads to concrete benefits it is also not assured that the respective costs are overcompensated by subsequent economic effects. The bounded financial resources cannot be invested into other projects that might create or at least maintain technological and strategic flexibility for companies in the electric mobility sector. Hence, codification could also trigger path dependent effects and hence the risk of associated lock-in situations.

Nevertheless, despite the presented limitations, knowledge codification can contribute fundamentally to avoiding risks of path dependencies and resulting lock-ins. However, a pre-requisition is that the codification processes are constantly monitored regarding the occurrence of possibilities negative side effects, as it is proposed by Schreyögg and Kliesch-Eberl (2007), in order to enable a constant dynamization of the stock of existing knowledge and competences (Schreyögg and Kliesch-Eberl 2007).

4.3 Exemplary Effects of Knowledge Transfer on Path Dependencies in the Electric Mobility Sector

One contribution of knowledge transfer to avoiding and coping with risks of path dependency in the electric mobility sector results from the possibility to reveal new combinations of already existing knowledge, that in turn leads to the creation of new knowledge. For instance, the question regarding the batteries that are used by the car manufacturers or the infrastructure that is necessary in order to build a profitable business model with fleets of electric cars for rent can be confronted with new solutions. Kogut and Zander (1992) call this combinative capabilities (Kogut and Zander 1992). In the electric mobility sector such capabilities are especially important for cooperations—e.g. between car manufacturers, battery producers and infrastructure providers. Only through the transfer of knowledge between such companies it is possible to create value that incorporates the knowledge of all relevant participants in the creation of the entire "good" electric mobility. Hence, lock-ins through technological and institutional path dependencies can be reduced through both internal knowledge transfer as well as knowledge transfer between cooperating firms.

One **limitation** results from the above mentioned risk of codification of 'locked-in' knowledge. The transfer of such knowledge would in turn lead to a solidification of knowledge that has been proven successful in the past, but might not be appropriate anymore. E.g. if car manufacturers transfer their 'locked-in' knowledge on internal combustion engines, which entails a one-sided evaluation of the advantageousness in comparison to electric vehicles to firms that are willing to buy vehicle fleets, they might be already biased when they render their decision. Hence, the institutional path that already started to develop in the car manufacturers' employees' minds could perpetuate in the respective firms that buy vehicle fleets. Once they invested in traditional cars with combustion engines and resigned to build the infrastructure for electric vehicles, increasing returns occur and the probability that they repeat their decision for internal combustion engines increases. Hence, such a knowledge transfer between companies within a cooperation can trigger path dependencies and hence increases the risk of associated lock-ins.

In sum, knowledge transfer can enable organizations in the electric mobility sector to replicate their existing resources, which however entails the mentioned associated risks. Consequently, a constant monitoring of the transferred knowledge is necessary as well as a constant evaluation of the contextual appropriateness, in order to avoid running the risk of lock-ins through transfer of locked-in knowledge.

4.4 Exemplary Effects of Knowledge Abstraction on Path Dependencies in the Electric Mobility Sector

A **contribution** that results from the ability to abstract knowledge from its original context and apply it in a different context to the risks of path dependencies is an increased scope of application areas of knowledge within the electric mobility sector. For example, car manufacturers who are able to disengage their knowledge on how to optimize the efficiency of internal combustion engines and engage it to the development of different battery types can increase their variety of decision alternatives. That in turn leads to a better ability to adapt to a changing infrastructure that supports e.g. only certain types of batteries. The underlying assumption is underlined by Burmann (2002) who states that knowledge abstraction enables the companies and their management to identify new application fields of information of causal relations that underlie their organizational knowledge (Burmann 2002). Consequently, companies in the electric mobility sector—especially those that have a successful history with non-electric mobility products such as car manufacturers—that are able to decontextualize the knowledge that they already posses and apply this in the context of electric mobility products or services gain flexibility potentials through this dimension of dynamic capabilities. Hence, on both levels the technological as well as the institutional level it is possible to contribute to a wider range of possible decision alternatives through knowledge abstraction, which in turn reduces the risks of technological and institutional path dependencies.

A **limitation** stems from the fact that knowledge abstraction requires the employment of valuable and therewith expensive resources—especially employees with expertise in different fields. Only employees that have strong expertise in different application areas are able to do both decontextualize knowledge and identify new application fields for it. In the example, car manufacturers are required to use their employees with the most knowledge on internal combustion engines and battery technologies for the abstraction of knowledge. Expertise in turn is expensive, which results in a bounding of personal resources without a coercive and direct monetary effect. These bounded resources cannot be invested at the same time e.g. in market research in order to monitor the electric mobility sector and its developments for being able to react appropriately to environmental changes. Hence, this effect could lead to not being able to recognize when a certain path begins to shape and increasing returns develop, which can finally lead to lock-in situations on both the technological as well as the institutional level.

Subsuming, abstraction of relevant knowledge helps to avoid and cope with risks of path dependencies in the electric mobility sector. However, a constant monitoring of the associated costs is necessary.

4.5 Exemplary Effects of Knowledge Absorption on Path Dependencies in the Electric Mobility Sector

One **contribution** of knowledge absorption to the abilities to avoid and cope with developing path dependencies in the electric mobility sector results from the ability to internalize relevant knowledge from the companies' relevant environments. Through monitoring the electric mobility sector the companies can be enabled to acquire the information that seems to be relevant for the respective economic actors (e.g. battery technology developments for infrastructure provider), assimilate and transform it (e.g. converting the information on battery technology developments into requirements for the infrastructure) and finally exploit it (e.g. adapting the provided infrastructure in order to make it compatible to new battery technologies). The underlying assumption by Cohen and Levinthal (1990) has been validated by authors like Fosfuri and Tribó (2008) who showed in an empirical study that those companies that practice an active knowledge absorption can gain in innovativeness through being ahead of their competitors (Fosfuri and Tribó 2008). This general mechanism can be assumed to be valid also for the companies involved in the electric mobility sector, which in turn would mean that the risk of technological and institutional path dependencies is decreasing when the respective firms know constantly about their organizational environment and are able to internalize the relevant aspects.

A **limitation** results from the risk of an information overload, which can occur when the knowledge absorbing companies are not able to filter the incoming data into relevant and irrelevant knowledge. If a company internalizes too much data, which it is not able to process, the risk occurs that it becomes paralyzed with the work associated with acquiring, assimilating and transforming knowledge without any additional benefits such as an increased innovativeness. One example is an infrastructure provider that is overloaded with information on battery technologies and is therefore not able anymore to improve the business model itself. This in turn, would even solidify certain paths that have already proven successful in the past, because the companies are not able anymore to evaluate other decision alternatives. In the example, the infrastructure provider runs the risk of just sticking to the old business model because it does not have any resources left to identify optimization possibilities. Consequently, the risk occurs that excessive knowledge absorption would even lead to both technological as well as institutional lock-in situations.

In sum, knowledge absorption can contribute fundamentally to avoiding and coping with lock-ins in the electric mobility sector. However, with recourse to Schreyögg and Kliesch-Eberl (2007) again, it is necessary to monitor these processes in order to avoid information overload through the absorption of a too extensive amount of external knowledge (Schreyögg and Kliesch-Eberl 2007).

5 Conclusion

The research question that was addressed in this chapter was: How do dynamic capabilities of companies within the electric mobility sector affect the risk of institutional and technological path dependencies? The key findings of the qualitative and exemplary-based analysis of knowledge-based dynamic capabilities' effects on the risks of path dependent developments in the electric mobility sector are the following:

First, it can be stated that the companies involved in the electric mobility sector face two different risks of path dependencies: On a technological level, the risk arises that a certain electric mobility technology (e.g. a certain battery type) becomes the quasi standard although superior ones are available due to sunk costs and necessities for complementarity between vehicles and infrastructure. On an institutional level, the risk arises that the respective companies are not able to leave their strategic paths, which have been proven successful over the years but are not appropriate anymore in the future considering the evolvement of new technologies and business models.

Second, an economically reasonable approach to avoid and to cope with such path dependencies is the activation of knowledge-based dynamic capabilities through knowledge codification, transfer, abstraction and absorption. Each of these four dimensions contributes in a certain way to the risks of technological and institutional lock-in situations in the electric mobility sector. E.g. knowledge transfer between car manufacturers, battery producers and infrastructure providers enables the combination of the knowledge of each participating actor, which increases the variety of decision alternatives on both levels: On the technological level, car manufacturers, infrastructure providers and battery producers become aware of all the technological possibilities that are determined through the interplay between these companies. On the institutional level, the respective managers become aware of all the resulting strategic opportunities, which they can use in order to create and maintain competitive advantages.

Third, there are also strong limitations that have to be considered when knowledge codification, transfer, abstraction and absorption is actively triggered by the involved companies. For example, the transfer of knowledge on batteries or internal combustion engines implies the risk that it includes a bias, which stems from an already developed path dependency, in the course of which certain alternatives have already been excluded.

Although the shown effects have an exemplary character and do not provide a holistic picture, they still give insights into possibilities of avoiding and coping with such path dependencies in an evolving sector such as the electric mobility. Hence, companies such as car manufacturers or infrastructure providers are well-advised when they start codifying and transferring their knowledge for an active replicate of their organizational resources and abstracting and absorbing knowledge for a reconfiguration of their resources, while at the same time monitoring all possible negative effects that might occur in order to be able to countersteer when necessary.

Consequently, further research requirements result mainly from the qualitative character of this study: It is first necessary to identify more possible effects for getting nearer to a holistic perspective. Second, the dimensions of knowledge-based dynamic capabilities in the electric mobility sector have to be operationalized in order to enable a measurement for an empirical study. Third, the empirical study needs to be conducted in order to get insights about a net effect in order to see if the contributions or the limitations to the risks of path dependencies in the electric mobility sector overweigh and which countersteering actions might be advisable.

Acknowledgments This research was supported by the Federal Ministry of Transport, Building and Urban Development in the context of the 'Model Regions Electric Mobility'.

References

Ackermann R (2001) Pfadabhängigkeit, Institutionen und Regelreform, Tübingen
Ancori B, Bureth A, Cohendet P (2000) The economics of knowledge: the debate about codification and tacit knowledge. Ind Corp Change 9(2):255–287
Arthur WB (1989) Competing technologies, increasing returns, and lock-in by historical events. Econ J 99(394):116–131
Arthur WB (1990) Positive feedbacks in the economy. Sci Am 262(2):92–99
Barney JB (1991) Firm resources and sustained competitive advantage. J Manage 17(1):99–120
Barney JB (2002) Gaining and sustaining competitive advantage. Prentice Hall, Upper Saddle River, NJ
Boisot M (1995) Information space: a framework for learning in organizations, institutions and culture. Thomson Learning Emea, London
Boisot M (1998) Knowledge assets: securing competitive advantage in the information economy. Oxford University Press, Oxford
Bou-Llusar JC, Segarra-Ciprés M (2006) Strategic knowledge transfer and its implications for competitive advantage: an integrative conceptual framework. J Knowl Manage 10(4):100–112
Brauner G (2008) Infrastrukturen der Elektromobilität. e & i Elektrotechnik und Informationstechnik 125(11):382–386
Burmann C (2002) Strategische Flexibilität und Strategiewechsel als Determinanten des Unternehmenswertes. Deutscher Universit? ts Verlag, Wiesbaden
Casper S, Whitley R (2004) Managing competences in entrepreneurial technology firms: a comparative institutional analysis of Germany, Sweden and the UK* 1. Res Policy 33(1):89–106
Cohen WM, Levinthal DA (1990) Absorptive capacity: a new perspective on learning and innovation. Adm Sci Quart 35:128–152

Crossan MM, Bedrow I (2003) Organizational learning and strategic renewal. Strateg Manag J 24(11):1087–1105
Dammenhain K (2012) Chancen der Elektromobilität [Homepage of Automobil Produktion], [Online]. Available: http://www.automobil-produktion.de/2012/02/chancen-der-elektromobilitaet/. Acessed 06 2012
David PA (1985) Clio and the economics of QWERTY. Am Econ Rev 75:332–337
David PA (1994) Why are institutions the 'carriers of history'?: path dependence and the evolution of conventions, organizations and institutions. Struct Change Econ Dynam 5:205–220
Dievernich FEP (2007) Pfadabhängigkeit im management: wie Führungsinstrumente zur Entscheidungs- und Innovationsunfähigkeit des managements beitragen, Kohlhammer, Stuttgart
Drucker PF (1993) Post-capitalist society. Harper, New York
Fosfuri A, Tribó JA (2008) Exploring the antecedents of potential absorptive capacity and its impact on innovation performance. Omega 36(2):173–187
Freeman RE (1984) Strategic management: a stakeholder approach. Pitman, Boston
Freiling J (2004) A competence-based theory of the firm. Int Rev Manage Stud 15:27–52
García-Muiña FE, Pelechano-Barahona E, Navas-López JE (2009) Knowledge codification and technological innovation success: empirical evidence from Spanish biotech companies. Technol Forecast Soc Chang 76(1):141
Göllinger T (1997) Das Innovationspotential der E-Mobility. In: Schriftenreihe: Arbeitspapiere des Instituts für ökologische Betriebswirtschaft Institut für ökologische Betriebswirtschaft e.V., Siegen
Grant RM (2002) The knowledge-based view of the firm. In: Choo CW, Bontis N (eds) The strategic management of intellectual capital and organizational knowledge. Oxford University Press, New York, pp 133–148
Hedlund G (1994) A model of knowledge management and the N-form corporation. Strateg Manag J 15:73–90
Jane Zhao Z, Anand J (2009) A multilevel perspective on knowledge transfer: evidence from the Chinese automotive industry. Strateg Manag J 30(9):959–983
Kogut B, Zander U (1992) Knowledge of the firm, combinative capabilities, and the replication of technology. Organ Sci 3(3):383–397
Kusunoki K, Nonaka I, Nagata A (1998) Organizational capabilities in product development of Japanese firms: a conceptual framework and empirical findings. Organ Sci 9(6):699–718
Liebowitz SJ, Margolis SE (1995) Path dependence, lock-in, and history. J Law Econ Organ 11(1):205–226
Nahapiet J, Ghoshal S (1998) Social capital, intellectual capital, and the organizational advantage. Acad Manage Rev 23(2):242–266
Nonaka I, Mader F, Takeuchi H (1997) Die organisation des Wissens : wie japanische Unternehmen eine brachliegende Ressource nutzbar machen, Campus-Verl., Frankfurt/Main [u.a.]
O'Reilly CA, Tushman ML (2008) Ambidexterity as a dynamic capability: resolving the innovator's dilemma. Stanford University Graduate School Of Business research paper no. 1963, vol 28. Available at SSRN: http://ssrn.com/abstract=978493. pp 185–206
Pehnt M, Helms H, Lambrecht U, Dallinger D, Wietschel M, Heinrichs H, Kohrs R, Link J, Trommer S, Pollok T (2011) Elektroautos in einer von erneuerbaren Energien geprägten Energiewirtschaft. Zeitschrift für Energiewirtschaft, pp 1–14
Penrose ET (1959) The theory of the growth of the firm. Blackwell, Oxford
Porter ME (1999) Wettbewerbsstrategie : Methoden zur Analyse von Branchen und Konkurrenten = (Competitive strategy), 10., durchges. und erw. Aufl edn, Campus-Verl., Frankfurt/Main [u.a]
Prahalad CK, Hamel G (1990) The core competence of the corporation. Harvard Bus Rev 68(3):79–91
Quinn JB (1992) The intelligent enterprise a new paradigm. The Executive 6(4):48–63

Sammer G, Stark J, Link C (2011) Einflussfaktoren auf die Nachfrage nach Elektroautos. e & i Elektrotechnik und Informationstechnik 128(1):22–27
Sanchez R (2004) Understanding competence-based management—identifying and managing five modes of competence. J Bus Res 57(5):518–532
Schäcke M (2006) Pfadabhängigkeit in Organisationen—Ursache für Widerstände bei Reorganisationsprojekten, Duncker & Humblot, Berlin
Schreyögg G, Kliesch-Eberl M (2007) How dynamic can organizational capabilities be? Towards a dual-process model of capability dynamization. Strateg Manag J 28(9):913
Schreyögg G, Sydow J, Koch J (2003) Organisatorische Pfade—Von der Pfadabhängigkeit zur Pfadkreation. In: Schreyögg G, Sydow J (eds) Strategische Prozesse und Pfade. Gabler Verlag, Wiesbaden, pp 257–294
Selznick P (1957) Leadership in administration: a sociological interpretation. Harper, Row, New York
Spender JC, Grant RM (1996) Knowledge and the firm: overview. Strateg Manag J 17:5–9
Szulanski G (1996) Exploring internal stickiness: impediments to the transfer of best practice within the firm. Strateg Manag J 17:27–43
Szulanski G (2000) The process of knowledge transfer: a diachronic analysis of stickiness. Organ Behav Hum Decis Process 82(1):9–27
Teece DJ (1981) The market for know-how and the efficient international transfer of technology. Ann Am Acad Polit Soc Sci 458:81–96
Teece DJ, Pisano G (1994) The dynamic capabilities of firms: an introduction. Ind Corp Change 3(3):537–556
Teece DJ, Pisano G, Shuen A (1997) Dynamic capabilities and strategic management. Strateg Manag J 18(7):509–533
Van Driel H, Dolfsma W (2009) Path dependence, initial conditions, and routines in organizations: the Toyota production system re-examined. J Organ Change Manage 22(1):49–72
Vera D, Crossan M (2005) Improvisation and innovative performance in teams. Organ Sci 16(3):203–224
Wallentowitz H, Freialdenhoven A (2011) Treiber für Veränderungen. Strategien zur Elektrifizierung des Antriebsstranges, pp 3–37
Wernerfelt B (1984) A resource based view of the firm. Strateg Manag J 5(5):171–180
Zahra SA, George G (2002) Absorptive capacity: a review, reconceptualization, and extension. Acad Manage Rev 27(2):185–203
Zollo M, Winter SG (2002) Deliberate learning and the evolution of dynamic capabilities. Organ Sci 13(3):339–351

Author Biographies

Philip Cordes works as a post-doctoral fellow in the department «Systems Management» at Jacobs University Bremen. He wrote his Ph.D. on effects of organizational design options on the evolvement of knowledge-based dynamic capabilities. His further exploratory focuses are autonomously co-operating logistic systems and strategic management in the field of the music industry. He deals with complexity-based theories of complex adaptive systems, with contingency-based descriptions of organizational structures as well as dominant management logics and path dependencies.

Michael Huelsmann holds the chair of "System Management" at the School of Engineering and Science at Jacobs University Bremen. He focuses on Strategic Management of Logistics Systems. Additionally he leads the sub-projects "Business model concept/product idea" and "Sustainable Innovation and technology strategies" in the framework of the electric mobility model region Bremen/Oldenburg.

Safety Aspects of Electric Vehicles: Acoustic Measures, Experimental Analysis, and Group Discussions

Kathrin Dudenhöffer and Leonie Hause

Abstract The first prototypes and serial models of electric cars already drive almost soundless, especially at lower speed. While beneficial for residents noiseless traffic could be a danger for blind and visually impaired persons. This study analyzes people's sound perceptions and sense of safety in traffic situations with electric cars and aims to discover the risk potential of electrified road traffic. Integrated into the study were five pairs of vehicles, each consisting of a BEV (battery electric vehicle) and one or two identical vehicles with ICE (internal combustion engine). The study covers acoustic measures of the vehicles, which were conducted on a vehicle measurement site on the one hand and at the test location on the other hand. Sound perceptions and safety estimations were measured via experimental design and questionnaires (n = 260). Moreover, the researchers conducted deepening group discussions with blind and visually impaired people (n = 28). Main result of this study is that there exist problems of perception for quiet electric vehicles even for higher speeds of 30 km/h. Additionally, the analysis shows that modern petrol cars are almost as quiet as electric cars. This indicates that a broad solution for quiet cars in general is necessary.

1 Introduction

One argument has drawn criticism for a while: Electric cars are quiet and therewith a potential danger especially for blind and visually impaired persons (Oppegaard 2007; Rice 2007). These persons rely on vehicle sounds for warning and as help

---

K. Dudenhöffer (✉) · L. Hause
Universität Duisburg-Essen, Bismarckstraße 90 47057 Duisburg, Germany
e-mail: kathrin.dudenhoeffer@uni-due.de

L. Hause
e-mail: leonie.hause@volkswagen.de

for orientation on the roads. In addition, people whose perception is restricted due to other reasons like senior citizens are affected (NFB). Also for children cars driving up silently could be dangerous (ANEC 2010). U.S. a2011nd Japanese Experts develop a guideline, which recommends fitting electric and hybrid vehicles driven in electric mode with an artificial sound (UNECE 2011). The Nissan Leaf for example provides a turbine sound in lower speeds (Leaf Owners Group 2010). For reversing the technicians integrated beeping sounds as already known by large trucks. Now the European Union is planning a similar guideline (DELTA SenseLab 2011). With these developments all attempts by technicians and communes to reduce vehicle noise and offer a quiet living atmosphere are thwarted.

The World Health Organization (2010) estimates the global number of people visually impaired to be 285 million (4.2 % of the total population), of whom 39 million are blind (0.6 % of the total population). Raising numbers are anticipated, because of a shifting age pyramid in the society of developed countries and a higher risk potential for older people to lose visual quality (people 50 years and older are 82 % of all blind, WHO 2010).

Nevertheless, numerous studies meanwhile approved that traffic noise can have heavy consequences (for a review cf. Babisch 2006). Stansfeld and Matheson (2003) found strong evidence for effects of environmental noise on health, especially annoyance, sleep and cognitive performance of adults and children. So, relationships between environmental noise and specific health effects like cardiovascular disease, cognitive impairment, sleep disturbance and tinnitus are shown (WHO 2011). Unfortunately, noise abatement (silent asphalt, quiet goods trains, and noise barriers) is highly expensive and leads to a poor benefit-cost ratio (Klaeboe et al. 2011). The diffusion of electric cars would implicate noise reduction as a free add-on.

1.1 Research Background

At the time of designing this study only two research projects existed, which focused on acoustic perceptions of electric vehicles (JASIC 2009; NHTSA 2010). Both studies used hybrid cars driving in electric mode for the tests. Vehicle sound levels were measured at test tracks. The results were that sound levels of the identical vehicles with internal combustion engine (ICE) were above the sound levels of the hybrid cars in electric mode. The vehicle sounds approached in an interval between 15 and 30 km/h. This is caused by the increasing tire noise. Standing with running engine the hybrid cars were too quiet to be detected. These studies indicate a perception problem of quiet electric cars below 30 km/h and while standing.

Additionally in both studies tests of sound perception were conducted via laboratory experiments. Blind and visually impaired persons had to listen to recorded vehicle sounds and should state when they heard the approaching vehicle. Thus, time-to-vehicle-arrivals were computed for different background noise levels. The average time-to-vehicle-arrival of hybrid cars was significantly shorter than of ICE vehicles. The critical speed, below which hybrids were not well detectable, has been located between 20 and 30 km/h. The proposed solution was an artificial sound for vehicles in electric mode at lower speeds. According to the NHTSA a characteristic engine noise has been preferred by the blinds.

Based on these studies the U.S. Congress passed a guideline known as the 'Pedestrian Safety Enhancement Act of 2010'. It notes that pedestrians should be able to detect a nearby vehicle in "critical operating scenarios" including constant speed, accelerating or decelerating.

One disadvantage of these studies was that only hybrid cars were included. In contrast, our study focuses on pure battery electric vehicles (BEV). The second disadvantage is seen in the methodology. Both studies use laboratory experimental designs in that the participants listen to vehicle noises at a computer. Our approach is more oriented on realistic situations to enhance the validity of the study.

This study was part of the project 'ColognE-mobil' within the model region for electric mobility Rhine-Ruhr. The following questions were focused: Are battery electric vehicles (BEV) quieter than vehicles with internal combustion engines (ICE)? How do affected persons perceive sound differences and how do these people evaluate their personal safety in such situations? Is the installation of an artificial sound the optimal solution? And: What could be an optimal solution, which satisfies both interests (warning of affected groups and quiet living atmosphere)?

The remainder of the manuscript is organized as follows. Section 2 demonstrates the methodology of the study. In Sect. 3 the results are presented, which are discussed in Sect. 4. Section 5 gives a conclusion.

2 Methodology

To test how dangerous quiet electric cars really are and if artificial sounds are the optimal solution, an experimental study with three runs in July and November 2010 as well as April 2011 has been conducted. Each run included an experimental part and sound level measures.

2.1 Test Vehicles

In total, eleven different cars were tested (Table 1). For comparison reasons available battery electric vehicles and one or two identical models with ICE were chosen (cf. JASIC 2009; NHTSA 2010). In total, five battery electric vehicles (Fig. 1) and six vehicles with ICE were included. In the first run two pairs of transporters (small and medium) were tested, while in the second and third run passenger cars were focused.

Table 1 Test vehicles (in total)

Producer	Model	Engine	Run
German E-Cars	Stromos (basis: Suzuki Splash)	56 kW E-Engine, 19.2 kWh battery	3
Opel	Agila (identical to Suzuki Splash)	1.2 l Petrol, 69 kW @ 6,000 rpm	3
Mega	E-City	1.25 l Petrol, 60 kW @ 5,800 rpm	3
Ford	Fiesta (similar to Mega e-City)	4 kW E-Engine	3
Smart	Fortwo	1.0 l Petrol, 62 kW @ 5,800 rpm	2
Smart	Fortwo	0.8 l Diesel, 40 kW @ 3,800 rpm	2
Smart	Fortwo	20/30 kW E-Engine, 16.5 kWh Battery	2
Ford	Transit	2.4 l Diesel, 103 kW @ 3,500 rpm	1
Ford	Transit	3-Phases 90 kW	1
Peugeot	Partner	1.6 l Petrol, 88 kW @ 6,000 rpm	1
Peugeot	Partner	Not known	1

German-E-Cars Stromos

Mega e-City

Smart Fortwo e.d.

Ford Transit electric

Peugeot Partner electric

Fig. 1 Test vehicles (BEV)

2.2 Sound Level Measures

Since the acoustic perception of electric vehicles is critical at approximately 30 km/h or 20 mph (at higher speeds tire noise prevails, cf. JASIC 2009; NHTSA 2010), passing was conducted at 30 km/h. This is a realistic driving condition, because in 30 km/h or 20 mph zones streets are regularly supposed to crosssing without traffic lights. Additionally, at this speed the stopping distance could be problematic. While this is 18 m on dry asphalt with a speed of 30 km/h or 20 mph, it is only 4 m at 10 km/h or 6 mph on dry streets (according to the rule of thumb).

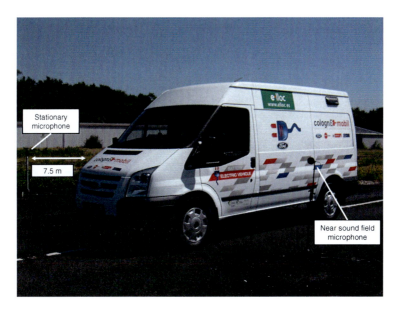

Fig. 2 Sound level measures at the test track

Hence, the danger of collision is small at lower speeds. The second condition was constant driving, since extremely high-speed drives are seldom in residential areas' city traffic. This driving condition resembles common driving in moving traffic. Tachometer variations were considered.

The sound level measures were conducted at the test track in Cologne-Merkenich by the project partner Ford Motor Company Germany (Fig. 2). The measure chain according to ISO 362 comprised mobile microphones B&K 4188 in 7.5 m distance. The measures contain the mean of maximum sound levels LHS and RHS. For all drives the speed was measured at the microphone line. This corresponds to the average speed during the test drive. Caused by these special measure conditions only speeds of 30 km/h or 20 mph and higher could be measured.

Additional sound level measures were conducted at the experiment location.[1] For the experiments a one-way street with little traffic in a residential area near the University of Duisburg has been chosen as test location. At both sides cars were parked and the road surface was new and without repairs (important for reliability). Since standards for such acoustic measures did not exist so far and the main focus was to control measure conditions and to capture the probands' perceptions most realistically, the researchers developed an own standard. The binaural head simulator was positioned at the roadside similar to a test person with the ear height

---

[1] These acoustic measures were conducted in cooperation with the Fraunhofer Institute for Building Physics. The measure equipment comprised two binaural head simulators (HEAD HMS III digital), SQLab (Heim, DATaRec3-Serie), a HEAD Recorder and Artemis V11 (recording and analysis software, HEAD Acoustics).

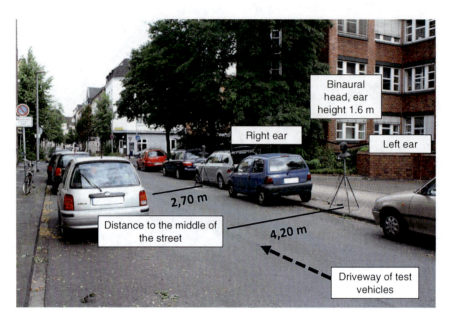

Fig. 3 Sound level measures at the experiment location

in 1.6 m (Fig. 3). In contrast to the measures at the test track, lower speed of 10 km/h or 6 mph on could be measured here. Walking speed is at 7 km/h or 4 mph and is measured with 10 km/h or 6 mph, which is the lowest speed to be measured reliably. Peak values (A valued, FFT 4,096 lines, 50 %, Hanning) per total passing (7.2 s/60 m) were considered. Caused by the different conditions the measures at the test track cannot be compared to the measures at the experiment location. Nevertheless the sound levels of different speeds and vehicles can be compared for each test location separately.

2.3 Design of the Experiments

The test persons got the task to wait at the roadside and to cross the street after the passing of each test vehicle. To prevent sequence effects the vehicle order was defined in a plan of measuring repetitions (balancing). The first passing was used as baseline to adjust knowledge and expectation levels and was not included into analysis. The probands did neither know that electric cars were included into the study nor the vehicle order. The type of the vehicle (BEV or ICE vehicle) was garbled by masking labels. After each passing and crossing the test persons had to answer a questionnaire concerning acoustic perception and sense of safety in this situation. The research team helped the blind and visually impaired persons with the questionnaires (Fig. 4). For each passing the street was closed temporarily to

Fig. 4 Researcher interviewing blind participant

prevent the passing of other non-test cars. In addition to the questionnaires a buzzer system (Activote) was used to conduct objective time-to-vehicle-arrival measures (cf. NHTSA 2010). Each participant received an electrical buzzer with the instruction to press the button as soon as he or she hears the converging vehicle. Time and distance of the vehicle were calculated by comparing the individual times of perception and the arrival time of the vehicles at the waiting position of the test persons.

The evaluation of vehicle sounds was measured via semantic differentials. The pairs of adjectives describing the vehicle sounds were developed on the basis of established semantic differentials for resident surveys concerning vehicle noise perception. In cooperation with experts from the Fraunhofer Institute for Building Physics the adjectives were adjusted to the focal topic. Studies concerning subjective perception of vehicle sounds have not been conducted so far.

For data analysis multivariate analyses of variance (MANOVAs) were used. They have been controlled for effects of visual level (visual impaired or blind vs. visual not impaired) and aural level (deaf with using a hearing aid vs. not deaf). In addition, means have been compared via t-tests. The program used for analysis was IBM SPSS Statistics 19.

2.4 Sample Description

The study participants had been recruited via flyers, organizations of blind and visual impaired persons, get-togethers of senior citizens, schools and the University of Duisburg-Essen. In total, 240 persons took part at the experimental study. The broad sample comprised visually impaired and not impaired persons, children, pupils (up from the 9th class), students, and senior citizens:

- 40 % female, 58 % male, 3 % not specified
- Average age: 35 years, range: 5–94 years
- Visual impaired or blind: 15 % (36 persons)
- Deaf: 14 %
- Driver's license: 51 %
- Only 6 % had already driven an electric vehicle, 26 % did not know this kind of vehicle

2.5 Group Discussions

Since the problem of perception is specific for blind and visual impaired persons, these people were especially integrated into the study through group discussions. For the researchers it was important to discuss the findings with them and to think together about possible solutions. One discussion was conducted before the second run in October 2010 with members of the 'Blinden- und Sehbehindertenvereine in Nordrhein-Westfalen'. The second discussion was conducted subsequent to the third run in April 2011 with participants of the study.

3 Results

3.1 Results of Sound Level Measures

Figure 5 shows that the tested battery electric cars [57–58 dB(A)] were quieter at the measured speeds than the comparable cars with ICE [59–62.5 dB(A)] indeed. But the findings also show that the sound levels between BEV and ICE vehicles differ less than expected. The lowest sound level of all tested vehicles with ICE showed the Opel Agila with 59 dB(A). This car is only 1 dB louder than the Smart BEV. The difference between Opel Agila (petrol) and its comparable BEV Stromos is only 2 dB. Such small differences are almost not perceivable. In contrast, sound level differences of 4.5 dB as measured between the Smart BEV and Smart Diesel are easy to perceive. This is caused by the logarithmic character of decibel, which can be explained as follows: A level change of 3 dB is perceived 1.2-times louder than the former sound and a level change of 6 dB is perceived 1.5-times

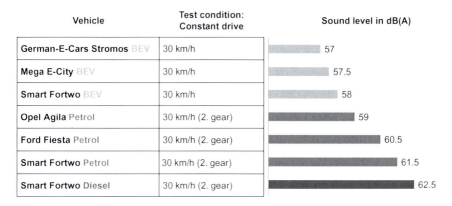

Fig. 5 Sound level measures at test track (only passenger cars)

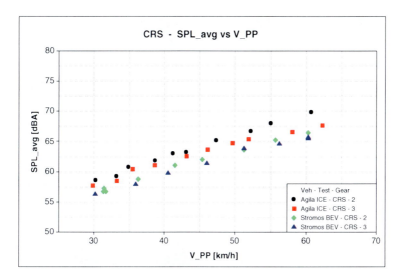

Fig. 6 Sound level measures at test track (Agila/Stromos) due to remodeling the Stromos possesses two gears, which is not common for electric cars

louder than the former sound. For comparisons: 50 dB corresponds to quiet radio music, 60 dB to a normal conversation at low volume, and 70 dB to a lawn mower.

At higher speeds the sound levels of BEV and ICE vehicles should converge further because at approximately 30 km/h or 20 mph tire noise prevails. But for the vehicle pair of Stromos/Agila (BEV/petrol) the curves are so similar that a further approach is barely visible (Fig. 6).

At the experiment location the sounds were measured at 10, 20, 30, and 40 km/h. Since these measures were conducted in realistic situations with background noise, the measures are not as reliable as the measures at the test track.

Table 2 Average level of sound level measures at the experiment location (Stromos/Agila)

Speed at constant drive (km/h)	Stromos [BEV dB(A)]	Agila (Petrol)	Difference (dB)
10	49	52.5 dB(A) (1. gear)	3.5
20	53.5	54.5 dB(A) (2. gear)	1
30	58.5	57 dB(A) (2. gear)	1.5
40	60.5	62.5 dB(A) (3. gear)	2

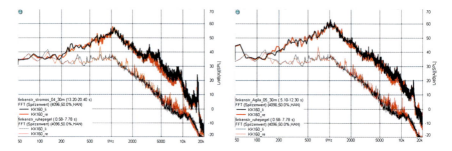

Fig. 7 Acoustic measures at experiment location (Stromos *left*/Agila *right*)

Therefore these measures have to be interpreted cautiously. As assumed the measures indicate larger differences between BEV and petrol vehicles at lower speeds (Table 2). At 10 km/h (which corresponds to walking speed) the difference is the biggest with 3.5 dB. Nevertheless, both vehicles are extremely quiet at this speed. In a quiet environment they are only marginally louder than the background noise. In louder environments it is assumed that both vehicles will be masked completely by background noise.[2] As well-known in the meantime the BEV was acoustically not perceivable while standing with running engine, because of missing engine noises (cf. NHTSA 2010; JASIC 2009).

The binaural head measures at constant speed of 30 km/h of the vehicles Stromos and Agila at the experiment location showed very similar curves (Fig. 7). This confirms again that both vehicles are barely different in the sound levels and for that in the objective acoustic perception.

3.2 Results of the Experiments

For vehicles with similar sound characteristics sound perception profiles result as presented in Fig. 8. The semantic differential for Agila and Stromos shows clearly that the test persons, including the blinds, did not perceive differences in the subjective perception between BEV and modern petrol vehicle. Concerning the

---

[2] Environmental background noise of approximately 30 dB is referred to be quiet, of 50 dB is referred to be moderate loud (cf. NHTSA 2010).

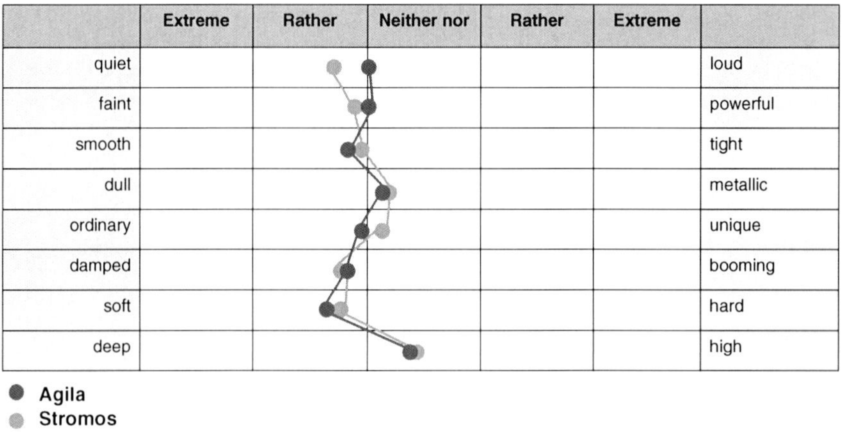

Safety Aspects of Electric Vehicles

	Extreme	Rather	Neither nor	Rather	Extreme	
quiet						loud
faint						powerful
smooth						tight
dull						metallic
ordinary						unique
damped						booming
soft						hard
deep						high

● Agila
◐ Stromos

Fig. 8 Sound perceptions in the third run (Agila/Stromos)

requested items the BEV Stromos and the petrol vehicle Agila almost conform congruently. All differences have not been significant in the MANOVAs. In contrast, the expected differences between BEV and ICE vehicle became visible in the comparison of Smart BEV and Smart Diesel. Here the BEV was significantly more quiet, faint, damped and unique compared to its ICE twin (Fig. 9).

Like the semantic differential of Stromos and Agila, the buzzer measures did not show significant differences between BEV and petrol vehicle (Fig. 10). For both vehicle types the percentage of people who heard the approaching vehicle too late or not at all were almost equal.

Additional questions to test perception differences between BEV and ICE vehicle showed a very similar picture. Differences concerning the question 'Did

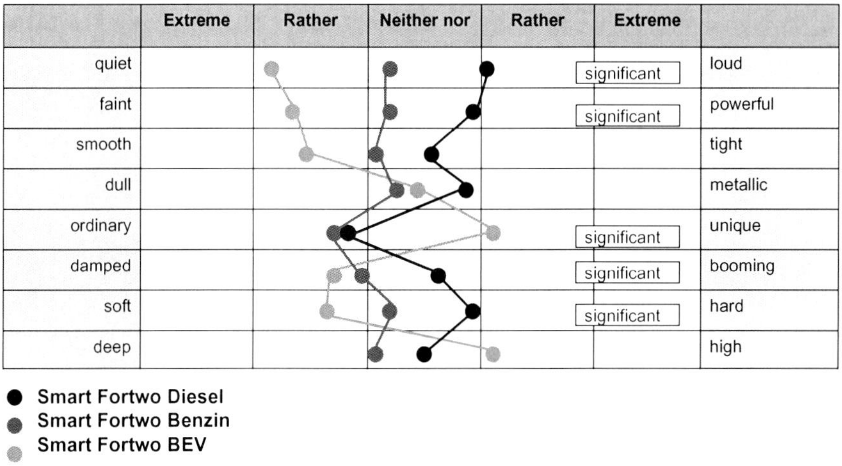

	Extreme	Rather	Neither nor	Rather	Extreme	
quiet					significant	loud
faint					significant	powerful
smooth						tight
dull						metallic
ordinary					significant	unique
damped					significant	booming
soft					significant	hard
deep						high

● Smart Fortwo Diesel
◐ Smart Fortwo Benzin
◌ Smart Fortwo BEV

Fig. 9 Sound perceptions second run (Smart Fleet)

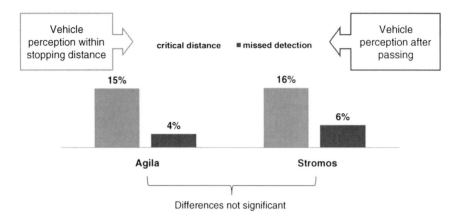

Fig. 10 Results of buzzer measures

the vehicle emerge unexpected?' were not significant for the cars Stromos/Agila, but significant for Smart BEV/Diesel (BEV emerged more unexpected). The sense of safety was equal for Stromos/Agila, but significantly different for Smart BEV/Diesel (questions: 'How safe was this traffic situation for you/for others (e.g. children)?'). The same results were obtained by the questions 'Did you fear an accident in this situation?' and 'Please guess: How many accidents will happen to pedestrians in such situations?'

In general the situation of pedestrians in road traffic is evaluated as 'rather safe'. The analysis did not show any influences of gender, age or visual impairment. But under certain circumstances like rain, heavy wind or snow blind and visual impaired persons feel significantly more unsafe. In addition, the sound of a standing or slowly driving vehicle is more important to blind and visual impaired than to other people (Fig. 11). For blind people vehicle sounds are essential to orientate in road traffic and to detect risks.

	Extremly unimportant	Rather unimportant	Neither nor	Rather important	Extremely important
	-2	-1	0	1	2
Sound of a standing vehicle		-0.4		0.8	
Sound of a slowly driving vehicle			0.4		1.3
Sound of a fast driving vehcile!				1.0	1.5

● Not visually impaired persons
● Blind and visually impaired persons

Fig. 11 Importance of vehicle sounds in different situations

3.3 Results of Group Discussions

The first group discussion showed that blind and visual impaired persons fear the diffusion of electric cars, because they worry about being limited in their safe mobility and to be excluded from social life. For this reason organizations of blinds and visual impaired demand fitting quiet electric cars with artificial sounds to make them perceivable. For blind and visual impaired persons it is important to be able to follow the drive of a vehicle via its soundscape. According to the study participants a common, traditional engine sound is better predictable than uncommon, new sounds (like Nissan Leaf or Chevrolet Volt). Also tram sounds are harder to assess than ICE vehicle sounds. In addition, blind people are geared to standing cars via engine sounds, for example at crossings without signals for blinds. By engine sounds they realize where a crossing is possible. Thus vehicles with automatic start/stop are extremely dangerous for blinds, because they cannot be heard while standing.

In the second group discussion the participants had to evaluate the vehicle pair Stromos/Agila in different driving modes at lower speeds, because of the surprising results of the sound level measures and experimental study. According to the probands the BEV was quiet, but perceivable at 20 km/h constant drive. Nevertheless the situation was evaluated as dangerous. At 10 km/h constant drive they described the vehicle as quieter and the situation as more dangerous, because of missing tire noise. Also parallel parking at the roadside was hardly perceivable and therefore dangerous. Also tested was the situation "stop at cross-walk". Here the test persons evaluated decelerating as quiet, but perceivable, while accelerating was too quiet to be heard. As mentioned above, the standing vehicle was not detectable. According to the test persons the perception of the identical ICE vehicle Agila was similar to the BEV. Again, parallel parking was almost unperceivable.

4 Discussion

The sound level measures and survey results of this study show that not only electric vehicles are quiet, but also modern petrol cars. Perception and sense of safety were not significantly different for these vehicles at a speed of 30 km/h. Therefore it is not the optimal solution to provide battery electric vehicles with artificial sounds, but not modern petrol cars. Consistently, all quiet vehicles independent of their driving mode should be changed by artificial sounds, if they lie below a certain decibel limit. So far, a higher crashworthiness of modern petrol cars has not occurred. Thus, it could be reasoned that it would be the wrong way to destroy one of the main advantages of electric vehicles, which is soundless driving.

Nevertheless, the concerns and fears of blind and visual impaired people should not be ignored. Since not only electric vehicles, but also modern petrol vehicles and especially vehicles with automatic start/stop are hard to perceive or are not perceivable at all, the problem must not be underestimated. At crossings without signals for blinds these people are increasingly endangered. One solution would be a nation-wide configuration of traffic lights and sideways for blind people, which seems to be a long-lasting and expensive undertaking.

Another solution could be an electrical warning system for potential risk groups like blind and visual impaired persons, children, seniors and deaf people. Modern vehicles already possess assistant systems like park distance control systems, automatic low beam, brake assistant, spacer, and night vision devise. Via cooperation and communication between certain sensors, an assistant system could be developed, which warns the affected persons and signals the car driver in the same time that endangered persons are nearby. This system would endure the safety of affected groups and prevent a sound competition on the roads. Since there exists a relationship between traffic noise and illness, the side effect of electric cars could be used to support a healthy noise atmosphere.

5 Conclusion

Due to the plan of the European Union to pass a guideline for fitting electric cars with artificial sounds, a large-scale study was conducted comprising sound level measures, experiments, and group discussions in a period of 18 month. The aim was to show, if battery electric cars are quieter than vehicles with internal combustion engine, how the sounds are perceived by pedestrians and how they evaluate the safety of these situations. A total of 240 persons took part in the experimental studies. 36 participants were blind or visual impaired.

While initial results of the studies indicated that BEV are quieter than ICE vehicles indeed, the comparison of a modern petrol car with a BEV in the third run of the study showed that those differences decrease. With the development of more insulated ICE vehicles and the spread of automatic start/stop the soundscapes of vehicles with different drives will approach increasingly.

Therefore it is not the optimal solution to fit BEV with artificial sounds. A solution for all interest groups could be an electronic assistant system, which is able to warn the affected person and the driver of the vehicle at the same time. Thus, the safety of blind and visual impaired as well as seniors, children and deaf people will be guaranteed in electrified road traffic. Additionally the spread of electric cars would help to reduce vehicle noise and a healthy living atmosphere in cities.

Acknowledgments The researchers thank Dr. Armin Bruno (Ford Motor Company Germany) for the supportive sound level measures at the Ford test track in Cologne-Merkenich as well as the Fraunhofer Institute for Building Physics for the accompanying measures at the test location.

References

ANEC (2010) Silent but dangerous: when absence of noise of cars is a factor of risk for pedestrians. http://www.anec.eu/attachments/ANEC-DFA-2010-G-043final.pdf. Accessed 13 March 2012

Babisch W (2006) Transportation noise and cardiovascular risk: updated review and synthesis of epidemiological studies indicate that the evidence has increased. Noise Health 8:1–29

Delta SenseLab (2011) White paper on external warning sounds for electric cars, 28 March 2011. http://live.unece.org/fileadmin/DAM/trans/doc/2011/wp29grb/QRTV-06-04e.pdf. Accessed 12 March 2012

JASIC (2009). A study on approach warning systems for hybrid vehicle in motor mode. Presented at the 49th World Forum for Harmonization of Vehicle Regulation (WP.29) GRB Working Group on Noise, 16–18 Feb 2009, Document Number: GRB-49-10

Klaeboe R, Veisten K, Amundsen A, Akhtar J (2011) Selecting road-noise abatement measures: economic analysis of different policy objectives. Open Transp J 5:1–8

Leaf Owners Group (2010) Nissan leaf sounds. http://leafownersgroup.com/leaf-media-documents/91-nissan-leaf-sounds-new-short-version-video.html. Accessed 12 March 2012

NFB (2011) National Federation of the Blind. Committee on automobile and pedestrian safety/quiet cars. http://quietcars.nfb.org/. Accessed 13 March 2012

NHTSA (2010) Quieter cars and the safety of blind pedestrians. National Highway Traffic Safety Administration, Washington, DC, 2010

Oppegaard B (2007) Silent Threat, Columbian, 25 May 2007

Rice P (2007) Hybrid Cars silent, advocates for blind warn, Albuquerque Tribune, 23 May 2007

Stansfeld SA, Matheson MP (2003) Noise pollution: non-auditory effects on health. Br Med Bull 68(1):243–257

UNECE (2011) Proposal to develop a global technical regulation concerning quiet vehicles. http://www.unece.org/fileadmin/DAM/trans/main/wp29/WP29-155-42e.pdf. Accessed 12 March 2012

World Health Organization (2010) New estimates of visual impairment and blindness: 2010. http://www.who.int/blindness/en/. Accessed 12 March 2012

World Health Organization (2011) Burden of disease from environmental noise. http://www.euro.who.int/__data/assets/pdf_file/0008/136466/e94888.pdf. Accessed 12 March 2012

Author Biographies

Kathrin Dudenhoeffer is mentoring coordinator of the faculty of engineering and a researcher at the Center Automotive Research of the University Duisburg-Essen. She is studying for a doctorate at the Mercator School of Management. Her current research interests include: Acceptance of electric vehicles, safety aspects and new markets (India, China) as well as fleet management and flexibility. She has studied business studies and communications science in Munich.

Leonie Hause is employed at the Volkswagen AG in the field of Marketing Product and Communication. Prior to that, she was a researcher at the Center Automotive Research of the University Duisburg-Essen. She has studied industrial engineering and management in Recklinghausen.

Transition Management Towards Urban Electro Mobility in the Stuttgart Region

Alanus von Radecki

Abstract This chapter analyzes the introduction of electro-mobility in the Stuttgart Region. It studies the actors that are relevant for the creation of the new mobility system and the management structures that govern and accompany the implementation process of electro mobility. The guiding question is whether the governance and management structures in the Stuttgart Region meet key criteria of an ideal Transition Management system as defined by Transition Management theory. It aims at discovering whether the current structure provides a good basis for the change of the urban mobility system towards electro mobility, or if certain preconditions for the success of an electro-mobile system are still lacking. I argue that the Transition Management framework not only serves as guiding framework for policy design, but also works well as an analytical tool for the analysis of concrete transition processes. Applying it to the introduction of electro mobility in the Stuttgart Region clearly shows that the development of a 'guided vision' and the creation of informational transparency currently are the biggest challenges for electro mobility in the Stuttgart Region.

1 Introduction

Wherever we look in the last 2 decades we can see society, politicians and economists struggle with the challenges of unsustainable ways of production, consumption or lifestyle leading to complex problems on a global scale which seem to be almost impossible to dominate by mankind. Rising energy consumption and lack of efficiency as well as energy production, housing, city planning or industrial agriculture are examples of these kinds of problems. They always

---

A. von Radecki (✉)
Fraunhofer-Institut für Arbeitswirtschaft und Organisation IAO, Nobelstraße 12,
70569 Stuttgart, Germany
e-mail: Alanus.Radecki@iao.fraunhofer.de

involve a complex mélange of technical artifacts, social, economic and political structures, but also habits, culture and personal preferences; a mixture that renders the search for adequate actions and strategies extremely difficult. There has been intensive debate on how to approach these problems (Bateman et al. 2011; Bradshaw 2000; Bartelmus 2008; Zhao 2011; French 2007; Meza and Meza 2008; Cooke 2010; Paredis 2011; Boersema and Reijnders 2009).

> This debate stems from three kinds of conviction: that current patterns of economic development are environmentally unsustainable; that these patterns of development are nevertheless deeply entrenched by technological, economic, institutional and cultural commitments; and that alternative technological and institutional configurations can be designed that will deliver both environmental and economic benefits over the longer term (Berkhout 2005: 57)

This chapter uses a set of theoretical concepts from various disciplines—referred to as 'Transition of socio-technical regimes'—in order to analyze the case of electro mobility in the Stuttgart Region. Therefore the system of urban mobility is perceived and described as a 'socio technical regime,' exposed to similar problems as mentioned above, prone to several external and internal pressures and equipped with adaptive capacity for a more or less successful transition towards a future system of electro mobility. The purpose of this research is to find out, whether the steps already taken towards an electro mobile system can be perceived as successful and promising and whether we can find leverage points for a more successful and sustainable transition. Therefore the study draws on the Transition Management approach—a concept developed by Dutch scientists in order to improve long-term governance. The concept of Transition Management is used here as an analytical tool for allowing a comparison of an encountered real situation with an ideal transition.

The chapter is structured as follows. Section 2 sets out the theoretical framework and discusses its main components. Section 3 describes the socio technical regime of urban mobility in the Stuttgart Region with an actor centered approach. Section 4 relates the consequences of socio-technical regimes—the need for governance—to the Transition Management approach and Sect. 5 then analyzes the Transition Management in the Stuttgart Region. A conclusion shows possible leverage points for future transitions towards electro mobility in the Stuttgart Region as generated from the analysis in Sect. 6.

2 Socio Technical Regimes: The Theoretical Background

Since the mid-1980s many researchers, from a wide range of disciplines, started to reorient their analysis in search of answers to pressing problems that had a direct and causal relationship with unsustainable processes of socio-economic development. The main question that stood behind many research projects and publications was: how to conceive and steer transitions of highly complex systems that depend on and reach into multiple spheres of our lives towards a more sustainable equilibrium?

The field of science and technology studies assessed technology networks and diffusion processes in search for adequate answers (Summerton 1994). Economic path-dependency theories approached the question by analyzing and systemizing increasing returns to adoption of new technologies, resulting in new insights about diffusion mechanisms (Arthur 1990). Evolutionary economists had an interest in long-term changes that impacted the world economy (Barnett 1998), or addressed the persistence of existing technologies by introducing the concept of 'technical regimes' (Nelson and Winter 1982). These efforts provide the theoretical background for the study at hand and I shall make use of them to construct a theoretical framework that allows me to explain and analyze the shift towards electro mobility.

2.1 Socio-Technical Regimes

From my point of view, the most promising theoretical approach for conceptualizing change of complex and interwoven systems like urban mobility comes from a group of scholars that combine the concepts of Transition Management (Voß et al. 2009), System Innovations (Schnabel et al. 2010) and Transformation of socio-technical regimes (Smith et al. 2005) to create a set of loosely linked theories on how to change our hyper complex world towards a more sustainable way of living and acting. I refer to this set of concepts as the 'Transition of socio-technical regimes' (Berkhout et al. 2004).

The underlying idea of socio-technical regimes (STR) emphasizes that actors, technologies and social structures within a complex system, like urban mobility, are embedded in broad social, environmental and economic systems that are mutually interwoven (Rip and Kemp 1998). Analyzing artificially separated systems like 'the economy of electric cars' (Biere et al. 2009) or 'the range of Lithium-ion batteries' (Seiffert and Walzer 1989) will necessarily fail to encompass the complexity of the existing structures and processes and is therefore not capable of assessing the multiple implications of a changing system. A better—and perhaps more realistic—way of grasping socio-technical change towards more sustainable patterns can be found in looking at transitions of such regimes. Socio-technical regimes are defined here as *"relatively stable configurations of institutions, techniques and artifacts, as well as rules, practices and* networks that determine the 'normal' development and use of technologies (Smith et al. 2005: 1493). This definition includes all actors, processes and social structures that are related to the use and reproduction of a regime[1] into the scope of analysis.

---

[1] By referring to the term 'regime' I imply the defined "socio-technological regime".

We can think of the regime of urban mobility as consisting of the municipality, important industrial actors like car producers and their sub-contractors, providers of public transport systems, political bodies, users of cars, bicycles, trains, and their convictions and psycho-physical predispositions. Personal habits and preferences belong to the regime, but also rules and regulations on the national, the international and the local level. Further interest groups like automotive clubs or associations for business development, and industries that are indirectly linked to urban mobility like oil companies or construction enterprises. These actors and institutions have to be understood as standing in close interdependent relations with technological structures like mobility artifacts and infrastructure.

The most valuable notion of this concept for the study at hand is its central focus on agency. Functionalistic avowals of structures and systems are avoided. A socio-technical regime consists of a complex mélange of conscious actors that have intentions, interests, habits and convictions. Analyses of its transition pathways therefore have to focus on agency and discourses.

In a static world any existing socio-technical regime would infinitively keep on reproducing itself. Since there is no such thing, regimes are always in movement, sometimes changing slowly and sometimes very suddenly. Smith et al. point out two basic elements that are responsible for regime transitions: "The articulation of selection pressures and the adaptive capacity available to facilitate regime transformation" (Smith et al. 2005: 1492).

2.1.1 Selection Pressures on STRs

We can understand 'selection pressures' as every process, discourse or event that puts pressure on an existing socio-technical regime to change. In the case of the regime of urban mobility it is distinguished here between environmental pressures, public pressures, economic pressures and political pressures. This practice might raise objections since there seems to be a contradiction with the concept of socio-technical regimes: clearly defined classes of pressures clash with a fluid and rhizomic concept that emphasizes agency and interests. However, it is important to cluster selection pressures in order to operationalize research and to categorize responses. Which role do environmental pressures play for economic actors? And what does it mean, when political actors mainly bring forth economic pressures as arguments? By categorizing selection pressures we obtain a powerful instrument for assessing interests and discourses.

It is important to note here, that these selection pressures are defined as concrete pressures on actors of the socio-technical regime at stake. This concept is derived from the empirical analysis of a concrete socio-technical regime and no general validity for these categories is claimed her. By environmental pressures it is referred to measurable processes from the natural environment that have an impact on the STR like global warming, air pollution, the finite nature of fossil fuels expressed in rising oil prices, battery life cycles, etc. Since an absolute proof of environmental pressures is difficult for factors like climate change, a large scientific consensus on an environmental problem is counted here as environmental pressure.[2] Public pressures are all sorts of arguments brought forth publicly by interest groups, private individuals, NGO's, enterprises etc. that affect the actors of the regime at stake. Economic pressures include all processes that put economic or financial constraints on the actors of the regime—these can be competitors, new innovations, market forces or small administrative budgets. Political pressures, finally, are defined here as interventions from political bodies on regional, national or international levels that aim at altering the current regime.

It is important to notice that:

> [t]here is typically no shortage of pressures acting on any given regime, often pushing in opposing directions. In practice, it is therefore not simply the existence of such pressures that is decisive. Instead, it is what we term the articulation of pressures for any given regime transformation" (Smith et al. 2005: 1495).

This is the reason why the main focus of this study is not on key figures and collected parameters of e-Mobility in Stuttgart. Rather, the analysis concentrates on actors, relationships and discourses in the regime of urban mobility in the Stuttgart region, for it is in the realm of interests, institutions, communication and power, where decisions about regime transitions are being made.

2.1.2 Adaptive Capacity

The second basic element of regime transition—adaptive capacity—is defined as "capacity and resources to respond to the selection pressures bearing on [the regimes]" (Smith et al. 2005: 1495). It can best be assessed by uncovering and analyzing the functions that are essential to the continuance and reproduction of the socio-technical regime. "The better that regime members are able to fulfil these functions, the better the regime as a whole will be able to respond to selection pressures" (Smith et al. 2005: 1495). Jacobsson and Johnson defined five functions that socio-technical systems[3] perform and that characterize their adaptive capacity: "1. Creation of new knowledge. 2. Influence over direction of search processes among users and suppliers of technology. 3. Supply of resources (capital, competences, input materials, political resources) 4. Creation of positive external economies. 5. Formation of markets" (Jacobsson and Johnson 2000).

There is, however, more to the adaptive capacity than just fulfilling the functions of an existing socio-technical regime, since functions can change with innovations, new perceptions and new technologies. It becomes clear, when looking at the level of regime actors: other factors will decide about their ability to adapt to and influence regime change. Since the regime is constituted by

---

[2] A good example is the 4th report on climate change of the IPCC: (Pachauri and Reisinger 2007).

[3] The term "socio-technical system" is used here interchangeably with "socio-technical regime" for the regime-concept was created out of the system approach.

interactions, technologies, discourses, governance structures and resource flows between its members, factors like agency and power will determine how regime members adapt to regime transitions. Giddens defines Agency as "the ability to take actions and make a difference over a course of events" (Smith et al. 2005: 1503). I use the term agency to describe the ability to contest to selection pressures, based on resources, political influence, innovativeness and discourse dominance. The specific distribution and interaction of these factors determine the agency of regime actors.

The characteristic interaction of regime actors, defined by their agency, forms the adaptive capacity of the socio-technical regime. Since there are many different actors with diverging visions and interests, adaptive capacity is mainly a social construct that depends on negotiations and cooperation. The interaction of regime actors

> …takes place in a social setting (the regime) and must act through networks of actors and institutions, which are not necessarily cooperative or pliant. The ability to make a difference thus requires the exercise of political, economic and institutional power (Smith et al. 2005: 1503).

I follow Smith et al. in their use of the term 'power' as the ability to exert multiple ways of influence upon other regime members and actors external to the socio-technical regime (Smith et al. 2005: 1503). This can occur by applying the resources strategically, but also by keeping issues off the political agenda or impacting discourses and visions in a dominant way. It is essential, however, to conceive power in socio-technical regimes as bound into existing structures.

In this configuration, the adaptive capacity of a socio-technical regime depends on its ability to align the resources, interests, power, discourse dominance and innovativeness of its members by successful governance in order to fulfill the functions defined above.

3 The STR of Urban Mobility in the Stuttgart Region

A key question is to define, which actors belong to the STR, which actors are key actors, and which ones are secondary actors. As Smith et al. put it:

> One way to delineate boundaries for regime membership is to analyse the degree to which different actors participate in carrying out functions reproducing the regime. […] Those actors who contribute intensively to reproducing the regime will be core members of the regime. Actors whose involvement is less intensive will be peripheral members of the regime (Smith et al. 2005: 1505)

A profound description of a socio technical regime would have to include not only the actors but also culture, habits, geographical, economic and political structures and power relations. This unfortunately goes beyond the scope of this chapter. For the purpose of this study—to analyze structures of transition management in the Stuttgart Region—it is sufficient to concentrate on the actor analysis.

Transition Management Towards Urban Electro Mobility

The assessment of the changing regime of urban mobility in the Stuttgart region has provided a list of 42 actors that are directly or indirectly involved in the reproduction of the regime of urban mobility in the Stuttgart Region. Most of them are represented in the guide to the model region electro mobility, provided by the Economic Development Corporation "Wirtschaftsförderung Region Stuttgart" (Rogg 2011), and we find them scattered across the four sectors that are important for regime transitions: the economy, politics, civil society and science.

For the identification and clustering of key actors, a criteria based approach by Zimmermann was used (Zimmermann 2006). The criteria for identification were:

(a) A key actor scores high on at least three of the four criteria 'legitimacy', 'resources', 'networks' and 'relevance for electro mobility'.
(b) A core actor scores high on all four of criteria.
(c) At least one representative of each relevant sector (economy, politics, research, civil society) is included in the actor analysis.
(d) If there are sections with differing roles within one sector, they shall be included, if possible and reasonable.

Of the 42 actors, four were identified as 'core actors', scoring high on all criteria. These are:

- The State parliament of Baden Württemberg (governing coalition).
- The City of Stuttgart (Administration).
- The economic development corporation 'Wirtschaftsförderung Region Stuttgart.'
- The regional Parliament 'Verband Region Stuttgart'.

Another 10 actors were identified as key actors scoring high on at least 3 of the 4 criteria. To these 'Elmoto' was added as key niche actor for electro mobility, following criteria (c) of the identification method. The final list of key actors for transitioning the regime of urban mobility in the region Stuttgart towards electro mobility is provided in Table 1.

Semi standardized Interviews were conducted with representatives of each of the identified key actors in order to find out about their agency, their formulation of selection pressures and the governance of change towards electro mobility in the Stuttgart Region. Since this chapter concentrates on the latter, let us now turn towards Transition Management.

4 Governance of Socio-Technical Regimes

It was shown how an interdependent network of regime actors creates and reproduces the dominant STR, in this case the regime of urban mobility. Since there is no central coordination unit within the regime, regime transitions cannot be initiated and steered hierarchically. They have to be managed and negotiated by

Table 1 Key actors of the urban mobility regime in the Stuttgart region

Sector	Sector roles	Actors	Interviewees
Economy	Automotive	Daimler AG	Sebastian Hoffmann
	Automotive supplier	Robert Bosch GmbH	Kerstin Mayr
	Energy company	EnBW	Dipl. Ing. Lars Walch
	E-Mobility Startup	Elmoto	Stefan Lippert
	IT company	SAP	Mafred J. Pauli
	Standardisation	DEKRA	Andreas Richter
Politics	State level	Landesagentur eMobil BW	Dr. F. Loogen
		Ministry of finance and economy	Dr. Decker
		State parliament	Hr. Haller (SPD)
			Hr. Schwarz (Grüne)
	Local level	Landeshauptstadt Stuttgart	Günter Stürmer
		Verband Region Stuttgart	Dr. Bopp
Society	Associations	Wirtschaftsförderung Region Stuttgart	Dr. Rogg
		IHK Region Stuttgart	Manfred Müller
Science	R&D	Fraunhofer IAO	Florian Rothfuss

a multitude of regime actors, none of which carries absolute power over decisions, and most of which have a different perception of the problem and the transition pathways.

The question guiding the study at hand is how the current regime of urban-mobility in the Stuttgart Region can be transitioned towards a system of electric mobility. Asking about governance of socio-technical regimes therefore points us to the question, whether the existing governance structures are likely to support a transition towards a system of electric mobility and in what way?

A suitable framework for answering this question can be found in the realm of policy analysis or—in our case even more specifically—urban governance. The concept of governance describes a transformation in the mode of political coordination that is mainly characterized by a reduction of hierarchies between state and society (Einig et al. 2005: 2). Governmental actors withdraw from the direct regulation and exertion of control but provide incentives for private actors to take measures in the intended way. The landscape of actors that become involved in questions relevant for society broadens and new actors enter the field. The beginnings of this transformation of public policies reach back to the 1980s, when state-actors felt an increasing burden of high debt and had to search for new ways of financing policies. Regional governments soon started to follow the idea of Public Private Partnership (PPP) (Häussermann et al. 2008: 267). Governance, however, is in two ways more than a classical PPP: first, the actors involved in governance do not only belong to the state and the private sector but also to civil society; and second, the structures normally do not consist of dyadic—but rather of multi-actor constellations (Einig et al. 2005: 2).

For the study at hand I refer to Häussermann et al. for the definition of governance:

Governance encompasses all interacting and intervening forces in the interplay of politics, economy and society. Thus, governance describes a network of different actors from the public and private domain. (Häussermann et al. 2008: 349)[4]

There are various special challenges that transitions of socio-technical regimes pose for governance: (a) the long term perspective of the intended change on the social, economic, cultural and technical level; (b) the involvement of a multitude of different actors that goes along with it; (c) the interconnectedness of social, economic and environmental systems, (d) path dependency and (e) uncertainty of future developments. Researchers interested in system innovations and transitions of socio-technical regimes have come up with a governance model for long-term policy design, which takes into consideration these special challenges. Stemming from systems theory, evolutionary economics and integrated assessment, Transition Management combines important concepts from long-term policy design with a down-to-earth approach to governing transitions of socio-technical regimes (Meadowcroft 2009: 324).

The concept of Transition Management is widely acknowledged to represent an applicable and practical form of governing regime transformations. Since the Netherlands adopted a Transition Management approach in the Fourth Dutch National Environmental Policy Plan in 2001 (VROM 2001), many European scholars—mostly from the Netherlands—began to use this concept for the analysis of policy design and policy implementation (Avelino 2009; Kern and Howlett 2009; Meadowcroft 2009; Smith et al. 2005; Voß et al. 2009; Berkhout et al. 2004; Loorbach 2007).

A good description of Transition Management is given by Meadowcroft:

At the core of 'Transition Management' is the challenge of orienting long-term change in large socio-technical systems. 'Transitions' are understood as processes of structural change in major societal subsystems. They involve a shift in the dominant 'rules of the game', a transformation of established technologies and societal practices, movement from one dynamic equilibrium to another—typically stretching over several generations (25–50 years). 'Management' refers to a conscious effort to guide such transitions along desirable pathways. (Meadowcroft 2009: 324)

Based on the theoretical principles of Transition Management and its linkages to complexity theory and governance models, Grin, Rotmans and Schot developed a practical management framework for the analysis and the design of governance structures that deal with transitions of socio-technical regimes (Grin et al. 2010). The authors define four main activities that are essential for the successful management of regime transitions. By analyzing, whether a certain governance structure provides for the implementation of these activities, one will be able to draw conclusions about this governance structure and about its possible success or failure. The four key elements of Transition Management—as defined by Loorbach are:

---

[4] Translation by author.

4.1 Problem Structuring and Development of a Transition Arena[5]

The Transition Arena can be thought of as a protected network of 'frontrunners', normally consisting of regime members, in which "long-term reflection and prolonged experimentation" (Grin et al. 2010: 157) is supported and sustained. The Transition Arena serves for establishing a common perception of the necessity for regime change and of the challenges that are to be met.

4.2 Development of 'Guiding Visions' and a Transition Agenda

Besides the inspiring impact of guiding visions on the involved actors, the most important function of guiding sustainability visions is seen in the establishment of a consensus among the stakeholders on "what sustainability means for a specific transition theme" (Grin et al. 2010: 158) and on how the future regime is likely to look. This will allow regime actors to backcast transition pathways from a commonly envisioned future to the current state of the regime and anticipate the best steps for transitions. Based on these guiding visions the transition agenda uses concrete objectives, instruments and interim objectives for approaching the regime transition on the ground.

4.3 The Use of Transition Experiments and Programs for System Innovations

'System improvements' and 'system innovations' are the central elements of Transition Management. System improvement is defined as "incremental adjustments to existing practices to address perceived problems" (Meadowcroft 2009: 329). The term 'system innovation' is used by a multitude of scholars in different ways. In this chapter the definition of F. Geels is used who states that "system innovations are defined as large-scale transformations in the way societal functions such as transportation, communication, housing, feeding, are fulfilled." (Geels 2004: 19). Radical innovations with the ability of triggering system innovation are usually not generated within a socio-technical regime; they are generated in technological niches.[6] One important task of Transition Management therefore is

---

[5] The full framework can be found in Loorbach (2007).

[6] "Technological niches form the micro-level where radical novelties emerge. These novelties are initially unstable sociotechnical configurations with low performance. Hence, niches act as 'incubation rooms' protecting novelties against mainstream market selection. Niche-innovations

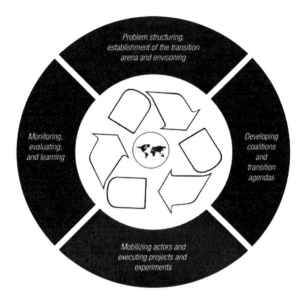

Fig. 1 The transition management cycle (Loorbach 2007: 115)

to protect these niches from market forces and help them grow to a state in which autopoietic reproduction becomes possible. Electro mobility can be seen as an excellent example of a technological niche and the struggle about its role within the future regime of urban mobility is the corresponding intent to govern systems innovation and manage the transition towards a more sustainable socio-technical regime.

4.4 Monitoring and Evaluation of the Transition Process

Continuous monitoring and evaluation of the transition process and of Transition Management processes assures that policy learning can take place and governance mechanisms can be adopted.

Figure 1 shows these four key elements of Transition Management in a recursive cycle, indicating the iterative nature of the approach.

Although the concept was originally designed with the intention of guiding policy design, the study at hand refers to the Transition Management framework in an analytical way. It is used here as an ideal for governance of regime transitions, against which actual governance structures and mechanisms can be compared. The key elements of Transition Management serve as key references for the analysis of governance structures towards electric mobility in the Stuttgart region.

(Footnote 6 continued)
are carried and developed by small networks of dedicated actors, often outsiders or fringe actors" (Geels and Schot 2007: 400).

5 Transition Management in the Stuttgart Region

Based on the theoretical framework as outlined above, the key actors of the urban mobility regime in the Stuttgart Region were analyzed with regard to their involvement and perception of the transition towards electro mobility.

5.1 Problem Structuring and the Transition Arena

The first step towards a successful transition in the direction of a new socio-technical regime is the problem-structuring activity. "By organizing the problem-structuring as a participatory process, it might lead to a shared conceptualization of the system at hand and the problems it is confronted with, and thereby also create a stronger sense of urgency to act" (Loorbach 2007: 116). In the Stuttgart Region, a systematic problem structuring of urban mobility and e-mobility in a narrow sense has not taken place. This is due to the fact that the involved actors have failed to consider themselves as part of a special regional actor group that co-operates towards a new regime of urban mobility. Companies like Daimler or Bosch orient their activities rather on a national or international scale and perceive e-mobility as one future field of industrial activity. Until now, efforts for integrating e-mobility in innovative regional mobility projects are limited to pilot projects with different Partners like "MeRegio Mobil,"[7] "Vito E-Cell," (Rogg 2011: 42) or the "F-Cell symposium."[8] On a decentralized basis, however, a process of problem structuring has taken place. Players like the Wirtschaftsförderung Region Stuttgart or the e-Mobil BW agency[9] have successfully used their potential for integration and have brought actors together for the application to national funding on e-mobility. Examples are the successful application of the Stuttgart Region as a model region for electro mobility (Bundesministerium für Verkehr 2010), the project "MeRegio mobil" or the "Elektronauten" project by the regional energy provider: EnBW (2011). Problem-structuring was therefore bound to concrete project proposals and has been rather short-term oriented. This is reflected in the varying formulations of perceived selection pressures: although most actors share the opinion that questions of environmental sustainability urge the regime of urban mobility to change, the perceptions of the key pressures on the regime of urban mobility differ profoundly. Some actors—like the Wirtschaftsförderung Region Stuttgart or the energy provider EnBW—see the biggest pressure coming from the economic realm. A growing world market of individual mobility that will depend less and less on fossil fuels puts economic pressure on a region that structurally depends on

---

[7] http://www.meregiomobil.de (29.07.2011)

[8] http://www.f-cell.de

[9] This agency has been created in 2010 with the purpose to link actors and projects engaged with e-mobility in the federal state of Baden-Württemberg.

car production. Future competitiveness of the Stuttgart Region is therefore bound to its capacity to respond to these pressures. This is the main argument of the Wirtschaftsförderung but also of political actors like Mr. Haller (spokesman for transport policy of the SPD in Baden-Württemberg). Others point towards the finite nature of fossil resources (Florian Rothfuss, Fraunhofer IAO and Manfred Müller, chamber of industry and commerce) or underline the flexibility of electricity as an energy carrier (EnBW). Actors also differ in their perception of environmental pressures: political actors like Mr. Bopp from the regional parliament or Dr. Decker from the ministry of economy tend to emphasize the local environmental pressures like air pollution and noise pollution whereas actors like Stefan Hoffmann (Daimler AG) or Manfred Pauli (SAP) see global environmental pressures like climate change as most important drivers towards e-mobility.

None of these arguments is wrong and they are certainly all valid. By emphasizing a certain argument, however, an actor also predefines its adaptation strategy to a limited extent. A transition pathway towards electro mobility will be different if the actor's main aim is to create economic development within the region than if he mainly strives for environmentally sustainable mobility in urban structures. With regard to this, Berkhout et al. state:

> we argue that specific configurations of selection pressures on the socio-technical regime will account for specific, historically-situated transformations processes. Relating the context of transformation to transformation processes must become a starting point for analysis, particularly for Transition Management advocates seeking the purposive strength of regime change (Berkhout et al. 2004: 66).

A second important precondition for a successful Transition Management is the establishment of a Transition Arena. With the establishment of the state agency e-Mobil BW in 2010 the federal state of Baden-Württemberg—and with it the Stuttgart Region—has made a big step towards a conscious management of regime change towards electro mobility. Its main tasks are cross linking of actors and creation of networks in order to support the inevitable structural change towards electro mobility (e-mobil BW 2011). In this way it provides the basis for the creation of a regional Transition Arena as "a group of people that reach consensus with each other about the need and opportunity for systemic change and coordinate amongst themselves to promote and develop an alternative" (Van der Brugge and van Raak 2007: 10). Statements of many of the involved actors show that the role of e-mobil BW and its activities are perceived in exactly this way. Mr. Pauli from SAP, for example, emphasized that the agency has found a good balance between bringing actors and information together and leaving space for entrepreneurial competition, and Stefan Hoffmann (Daimler) pointed out that a main contribution of the agency is to successfully connect the various pilot projects in the region and provide for information exchange. Critical voices call for more visibility of the agency (Mr. Walch, EnBW) or more political autonomy that would allow for the implementation of pilot policies and for testing their suitability for electro mobility (Stefan Lippert, Elmoto). Apart from the e-mobil BW agency, there are several initiatives and events that foster debate and exchange of ideas and information between regime actors involved in the reshaping of the urban mobility system. The Wirtschaftsförderung Region Stuttgart initiated the project "Elmo's—Electromobility for Cities and Regions"(Verband Region Stuttgart 2011) for accelerating the introduction of electro mobility. The city of Stuttgart established a centre for electro mobility where companies are given the possibility to present their products, and society and involved actors can communicate and interchange ideas and concepts (Matthieß 2010). This centre is used by e-mobil BW and Incovis—a local consultancy for the automotive sector—for a series of presentations, talks and discussions around pressing problems and innovative solutions in the field of electro mobility (Schwing 2011). Taking these examples together, a network between important actors has been created in the Stuttgart Region that comes pretty close to Loorbach's definition of an ideal transition arena:

"A transition arena is not an administrative platform, new institution or consultative body, but a societal network of innovation and innovators; an experimental playground" (Loorbach 2007: 118). However, an important quality of transition arenas is the ability to select its members.[10] In the case of the Stuttgart Region, the transition arena does not provide for this. It is an open network without restrictions on access or membership. Virtually all interviewed actors pointed out: anybody who wants to, and feels capable of contributing, can participate in projects initiated and coordinated by the e-mobil BW agency. The discussions and conferences are even more open and capability is no self-selection criteria. Therefore one important criteria of a transition arena seems unmet. A critical analysis would be needed here to find out, whether this type of openness of transition arenas could actually result in a higher variety of pilot projects and—ultimately—in higher resilience of transition pathways.

5.2 Development of 'Guiding Visions' and a Transition Agenda

Asked about their visions for electro mobility, regime actors in the Stuttgart Region show high diversity in their responses. Fourteen interviewees told me fourteen different visions. This indicates that a common definition of possible scenarios or transition pathways has not taken place in the Stuttgart Region and shared visions about e-mobility do not exist. Loorbach sees several important functions of visions

---

[10] Van der Brugge writes about the Transition Arena: It is not a typical democratic stakeholder process, but a participatory network of innovators, and selection is based on capabilities and knowledge rather than on power or authority. Initially, only a relatively small number of forerunners from various fields are involved. They are expected to have capabilities such as: (1) being able to reflect on a high level of abstraction; (2) being able to look beyond the limits of their own working field; (3) being able to propagate ideas in their home network; (4) being visionary; and (5) being able to work creatively in a team (Van der Brugge and van Raak 2007: 10).

[…] but for transition visions the over-arching goal is to stimulate a sense of shared direction and ambition amongst a variety of actors. The objective obviously is to create consensus upon a long-term orientation and convergence in terms of action (Loorbach 2007: 118).

This underlines the importance of 'guiding visions' in the process of Transition Management, which has been pointed out by other authors too: "The development of a systemic vision, or a new Leitbild of techno-economic and social development, can be seen as a key element of the network-enabling innovation policy" (Schienstock 2005: 107).

In the Stuttgart Region all actors share the notion that a transition towards electro mobility is important—and eventually without alternative; but there is no consensus, not even a discussion, about the details of the imagined future. These differ widely among the actors: some interviewees like Dr. Rogg (Wirtschaftsförderung Region Stuttgart) or Mr. Walch (EnBW) emphasize that fuel cell technology will play a role of increasing importance within the future system of electro mobility, others, like Dr. Loogen (e-mobil BW) perceive future electro mobility as highly integrated in multimodal mobility structures that link mobility demand with a sustainable and highly efficient mobility supply. There is no agreement on the main scope of electro mobility: many actors believe that it will stay an urban topic, others, however, see the necessity of addressing inter-urban mobility also with e-mobility solutions. The time span for the development of e-mobility is a third issue of disagreement: while some interviewees see a high momentum in the current development of e-mobility with the chance to reach the declared German goal of 1 million electric cars in 2020 much earlier than prospected (Mr. Hoffmann, Daimler AG), others are pretty skeptical about the time frame and do not see a significant market infiltration until 2040 or even later (Hr. Müller, IHK). Actors also express controversial visions and positions when asked about the role of public infrastructure. For some actors (e.g. EnBW, SAP) charging stations and e-mobility infrastructure pose the most important field of development. Ideas and projects range from the integration of charging infrastructure into urban structures and city planning to the development of innovative billing models. According to Mrs. Mayr (BOSCH) future e-mobility will solve the charging problem by integrating inductive charging devices into special highway lanes that allow for charging while being in motion. Other actors like Daimler do not see a special need for public infrastructure in the beginning and rather count on customers charging their electric cars at home.

These differences in the visions on future e-mobility lead to disparities in the perceived value of possible transition pathways. At the moment, there are several interesting ideas in the Stuttgart Region fighting for predominance. The problem is that the context is complex and highly contested and nobody is able to tell at the moment, which future will prevail. The comment I heard from virtually every actor I interviewed was: 'it remains to be seen how e-mobility will develop' and 'the field is completely open now and we cannot know what will happen.' These comments are certainly right. Transition Management acknowledges that

> [...] because of the innovative and complex nature of the Transition Management process, it is impossible to predict outcomes. However, by following the different steps in Transition Management (constantly adapted to the specific circumstances and context), the chances that shared problem definitions and visions and ultimately changed behaviour and new forms of cooperation emerge are greatly enhanced" (Loorbach 2007: 223).

The magic word here is 'dialogue.' If there is a common vision, actors can search for the best solutions to reach it, but if visions differ greatly and coordination does not take place, it becomes hard to realize an envisioned future. The lack of 'guided visions' therefore poses the biggest challenge to the transition of the urban regime of mobility in the Stuttgart Region. Chances are high that multiple pathways develop parallel to each other and in the end solutions by the actors with the highest relative power prevail. Programs for improving the conditions for electro mobility in the Stuttgart Region should therefore start by using the already established transition arena for the development of 'guided visions.' Socio-technical scenarios—as described by Elzen et al. (2004)—could then pose a promising starting point for the shared development of 'guiding visions.'

Transition agendas are a more controversial issue than 'guided visions.' Most interviewees express retentions towards a transition agenda, fearing that an agenda would restrict the organic and evolutionary development of the electro mobile regime. As M. Pauli (SAP) puts it:

> You could say for example: today we don't have any real infrastructure, we have almost no e-cars on the street, why should anyone develop a navigation system that is especially adapted to electric cars? But it is worked upon. An agenda, a plan would define to first concentrate on the important issues, but thereby it would stop this branch of development, which in turn is benefiting others. (Manfred Pauli (2011), SAP, interviewed by author)

However, an agenda in terms of Transition Management is somewhat different from a traditional management schedule:

> A transition agenda [...] does not need to be fully consistent or based on consensus at the level of ambitions, goals, beliefs and expectations. In a sense, a transition agenda more or less needs a certain element of dissent, conflict and difference of opinion so that it facilitates innovation, competition and learning (Loorbach 2007: 121).

The particular strength of a transition agenda is its openness, and accessibility to dialogue and critical reflection. The content itself can be disparate and contested. Unfortunately, this concept poses a big challenge to companies that perceive e-mobility as a possible field of future profits. Apart from joint activities in pilot projects, companies fail to give insight into their programs and strategies, mainly for fear of competition and for protection of innovations. This strategy results in an external orientation of the actors in the Stuttgart regime of urban mobility when searching for an agenda. The only reference point is a rough schedule given by the German "Nationale Plattform Elektromobilität," which focuses on the creation of a lead market for electro mobility and structures the adoption of electro mobility mainly into three phases: (1) Preparation of market (until 2014), (2) Market ramp-up (until 2017), (3) Mass market (by 2020) (Nationale Plattform Elektromobilität (NPE) 2011: 5). Many important issues in the field of urban mobility, like cross linkages with civil society, renewable energies or public transport stay totally untouched by this focus on markets. A real transition agenda developed by regime actors could help include these topics into the pathway towards electro mobility and thereby facilitate future innovations.

5.3 Transition Experiments and Programs for System Innovations

The pilot projects that are being carried out in the Stuttgart Region pose the actual backbone of the regional approach to electric mobility. They contain the seeds for the transition of the incumbent regime. However, they are not the focus of this analysis. The report of the model region for electro mobility gives a good oversight about all projects (Haas et al. 2011). The main intention of most of the projects matches well with the requirements of Transition Management: they experiment within niches of the mobility system with the intention to find out about practical challenges and opportunities of electro mobility. The most important insight gained from the analysis of joined projects towards development and implementation of electro-mobile solutions, however, is that without state funding incentives for sustainable transitions would be far too low and none of the above mentioned projects would come into place. State funding serves as catalyst for the creation of transition arenas around development projects.

5.4 Monitoring and Evaluation

Evaluation and adaptive learning are the central components that ensure flexibility and systems improvement within the Transition Management approach. In the Stuttgart Region interviewees had the most controversial positions when asked about evaluation and adaptive learning. Mr. Haas (Wirtschaftsförderung Region Stuttgart) and Mr. Walch (EnBW), for example, stated that there is an abundance of evaluation and sufficient accompanying research to the pilot projects in the region. Mr. Müller (IHK) did not know of one single evaluation and Mr. Haller (SPD member of the state parliament) did not think that evaluation is really necessary, for after all it is just a trend and a fashionable topic. The common denominator in the case of evaluation is a missing information management.

Although most pilot projects in the region have programs for monitoring and evaluation, results are not systematically distributed within the local urban mobility community. Many interviewees found that a coordination of evaluation is not existent in the Stuttgart Region and that this would be highly desirable for an effective Transition Management. This raises the question: which institution would be suitable for a coordinated evaluation and information transfer?

Since the project landscape in the Stuttgart region is highly fragmented, one possible option could be assigning this task to the e-mobil BW agency. There are, however, different opinions about this solution: Dr. Decker (state ministry of economy) sees e-mobil BW as the institution responsible for this task, since it is neutral and already poses the main transition arena for electro mobility in Baden-Württemberg. Mr. Pauli (SAP) also sees the theoretical duty of e-mobil BW to provide for evaluation and information sharing; however, he argues that this task would be too complex for the agency since it would have to dive into all projects and topics in detail, which he perceives as not feasible. Mr. Rothfuss (Fraunhofer IAO) disagrees, arguing that the state agency should not be endowed with evaluation activity in order to maintain its neutrality and not enter competition with scientific research institutes. This would inevitably lead to a loss of acceptance of e-mobil BW and would therefore be rather counterproductive to the whole process of Transition Management. As we can see—just like the lack of a shared vision—the topic of monitoring and evaluation is an unanswered question within the transition regime in the Stuttgart Region. Since most projects have their own evaluation, a possible solution could be to develop an information platform for reports and evaluations of different e-mobility projects in order to ensure transparency and exchange of results and ideas. The provision and maintenance of this information platform then could be a task for the state agency e-mobil BW.

6 Conclusion

The introduction of e-mobility in the Stuttgart Region is not coordinated with a Transition Management approach so far. In fact there is little coordination at all, if we leave aside the actions undertaken by the agency e-mobil BW. However, using the framework of Transition Management for the analysis of the current situation can show us two things:

First the framework that was originally created as a guiding tool for long-term governance of transitions of socio-technical regimes also serves well as an analytical tool. A comparison of a real transition situation to the ideal framework is able to detect chinks and problems within the process of transition and to generate hints and strategies for an improved and more sustainable Transition Management. A critique of this deployment of Transition Management framework argues that it analyzes an actor constellation that does not intrinsically perceive itself as a group and conclusions drawn out of it are therefore flawed and unrealistic. It is doubted here that actors directly involved in systems innovation and regime transitions have to perceive themselves as a fixed group in order to successfully manage transitions. The contrary issue is the case: the openness of a transition arena also bears the possibility for more participation and therefore for a greater innovativeness. Analyzing such a network with the Transition Management framework only has to cope with difference that actors do not know in detail, who else is part of the group, but most interviewees showed a pretty good intuitive knowledge about the actors and individuals involved in the transition management towards electro mobility.

Second: conclusions drawn from the analysis of the introduction of electro mobility in the Stuttgart Region show two possible leverage points for the improvement of the regime transition: 'Guiding visions' and transparency. If actors together develop a "Leitbild" of a future mobility system, transition pathways can be oriented at a commonly envisioned future state. A higher coherency of projects and policies and fewer conflicts will be the expected result. An information platform for the central publication and distribution of project evaluations, best practices and state of the art innovations could then boost networks of innovation and thereby accelerate regime transition.

References

Arthur B (1990, c1984) Competing technologies: an overview. In: Dosi G (ed) Technical change and industrial transformation. St. Martin's Press, New York, pp 590–607
Avelino F (2009) Empowerment and the challenge of applying transition management to ongoing projects. Policy Sci 42(4):369–390
Barnett V (1998) Kondratiev and the dynamics of economic development. Long cycles and industrial growth in historical context. St. Martin's Press, New York
Bartelmus P (ed) (2008) SEEA—the system for integrated environmental and economic accounting. Springer, Dordrecht
Bateman IJ, Mace GM, Fezzi C, Atkinson G, Turner K (2011) Economic analysis for ecosystem service assessments. Environ Resource Econ 48(2):177–218
Berkhout F, Smith A, Stirling A (2004) Socio-technical regimes and transition contexts. In Elzen B, Geels FW, Green K (eds) System innovation and the transition to sustainability. Theory, evidence and policy. Edward Elgar, Cheltenham, UK, pp 48–75
Berkhout F (2005) Technological regimes, environmental performance and innovation systems: tracing the links. In: Weber KM (ed) Towards environmental innovation systems. Springer, Berlin [u.a.], pp 57–80
Biere D, Dallinger D, Wietschel M (2009) Ökonomische Analyse der Erstnutzer von Elektrofahrzeugen. ZS Energ. Wirtsch 33(2):173–181
Boersema JJ, Reijnders L (eds) (2009) Transitions to sustainability as societal innovations. Springer, Dordrecht
Bradshaw B (2000) Environmentally sustainable economic development (book review): ByAsayehgn Desta. Praeger, Westport, CT, 1999. Human Ecology 282:327–329
Bundesministerium für Verkehr, Bau und Stadtentwicklung (BMVBS) (2010) *Umsetzungsbericht zum Förderprogramm "Elektromobilität in Modellregionen" des BMVBS*. Stand August 2010. Edited by NOW GmbH Nationale Organisation Wasserstoff und Brennstoffzellentechnologie. Berlin. http://www.now-gmbh.de/fileadmin/user_upload/Publikationen_Downloads/Infomappe_Elektromobilitaet_2010_2011/NOW-Umsetzungsbericht_2010.pdf. Accessed 29 July 2011
Cooke P (2010) Socio-technical transitions and varieties of capitalism: green regional innovation and distinctive market niches. J Knowl Econ 1(4):239–267
Einig K, Grabher G, Ibert O, Strubelt W (2005) Urban Governance. Bonn (Informationen zur Raumentwicklung, 9). Edited by Stadt-und Raumforschung (BBSR) Bundesinstitut für Bau. Bundesinstitut für Bau-, Stadt- und Raumforschung (BBSR). http://www.bbsr.bund.de/nn_22710/BBSR/DE/Veroeffentlichungen/IzR/2005/

Heft0910UrbanGovernanceEinfuehrung,templateId=raw,property=publicationFile.pdf/ Heft0910UrbanGovernanceEinfuehrung.pdf. Accessed 14 Apr 2011

EnBW (2011) Tanken à la carte. Immer e-mobil mit der Elektronauten-Ladekarte. http:// www.enbw.com/content/de/privatkunden/innovative_tech/e_mobility/elektronauten-ladekarte/120118_Fly_Ladekarte_Online_mSt.pdf. Accessed 15 Mar 2012

Elzen B, Geels FW, Hofman PS, Green K (2004) Socio-technical scenarios as a tool for transition policy: an example from the traffic and transport domain. In: Elzen B, Geels FW, Green K (eds) System innovation and the transition to sustainability. Theory, evidence and policy. Edward Elgar, Cheltenham, UK, pp 251–281

e-mobil BW (2011) *Wir über uns*. Stuttgart: e-mobil BW. http://www.e-mobilbw.de/Pages/wir-ueber-uns.php. Accessed 31 July 2011

French DA (2007) Managing global change for sustainable development: technology, community and multilateral environmental agreements. Int Environ Agreements: Politics, Law Econ 7(3):209–235

Geels F, Schot J (2007) Typology of sociotechnical transition pathways. Res Policy 36(3):399–417

Geels FW (2004) Understanding system innovations: a critical literature review and a conceptual synthesis. In: Elzen B, Geels FW, Green K (eds) System innovation and the transition to sustainability. Theory, evidence and policy. Edward Elgar, Cheltenham, UK, pp 19–47

Grin J, Rotmans J, Schot JW (2010) Transitions to sustainable development. New directions in the study of long term transformative change. Routledge, New York

Haas H, Reiner R, Gregori E (2011) Modellregion Elektromobilität Region Stuttgart. Wirtschaftsförderung Region Stuttgart GmbH (WRS). http://cars.region-stuttgart.de/sixcms/media.php/1455/modellregion-elektromobilitaet-region-stuttgart.pdf. Accessed 29 Mar 2011

Häussermann H, Läpple D, Siebel W (2008) Stadtpolitik. 1. Aufl., Originalausg. Frankfurt am Main: Suhrkamp

IPCC (2007) Climate change 2007. Synthesis report. An assessment of the intergovernmental panel on climate change. Edited by IPCC. Geneva, Switzerland. http://www.ipcc.ch/pdf/assessment-report/ar4/syr/ar4_syr.pdf. Assessed 6 May 2011

Jacobsson S, Johnson A (2000) The diffusion of renewable energy technology: an analytical framework and key issues for research. Energy Policy 28(9):625–640

Kern F, Howlett M (2009) Implementing transition management as policy reforms: a case study of the Dutch energy sector. Policy Sci 42(4):391–408

Loorbach DA (2007) Transition Management. New mode of governance for sustainable development. Internat. Books, Utrecht

Matthieß A (2010) Stuttgart: Mein Motor. Zentrum E-Mobilität. Landeshauptstadt Stuttgart. Stabstelle des Oberbürgermeisters. http://www.stuttgart.de/img/mdb/item/403986/56877.pdf. Accessed on 1 Aug 2011

Meadowcroft J (2009) What about the politics? Sustainable development, transition management, and long term energy transitions. Policy Sci 42(4):323–340

Nationale Plattform Elektromobilität (NPE) (2011) Zweiter Bericht der Nationalen Plattform Elektromobilität. Gemeinsame Geschäftsstelle Elektromobilität der Bundesregierung. Berlin. http://www.bmu.de/files/pdfs/allgemein/application/pdf/bericht_emob_2.pdf. Accessed 17 May 2011

Nelson RR, Winter SG (1982) An evolutionary theory of economic change. Belknap Press of Harvard University Press, Cambridge

Paredis E (2011) Sustainability transitions and the nature of technology. Found Sci 16(2–3):195–225

Pauli M (2011) Elektromobilität in der Region Stuttgart. Interview with Alanus von Radecki on 15 June 2011

Rip A, Kemp R (1998) Technological change. In: Rayner S, Malone EL (eds) Human choice and climate change. Battelle Press, Columbus, Ohio, pp 327–399

Rogg W (2011) Kompetenzatlas Elektromobilität Region Stuttgart. Wirtschaftsförderung Region Stuttgart GmbH (WRS). http://ecars.region-stuttgart.de/wp-content/uploads/2011/05/Kompetenzatlas_Elektromobilit %C3%A4t_RegionStuttgart.pdf. Accessed on 30 May 2011

Schienstock G (2005) Sustainable development and the regional dimension of the innovation system. In: Weber KM (ed) Towards environmental innovation systems. Springer, Berlin [u.a.], pp 97–113

Schnabel F, Pastewski N, von Geibler J et al (2010) Transitions towards sustainable innovations: linking resource efficiency and new technologies. In: 15th International conference sustainable innovation 2010

Schwing D (2011) Veranstaltungsreihe e-motion. Stuttgart: INCOVIS—http://www.incovis.com/unternehmen/veranstaltungen/e-motion. Accessed 1 Aug 2011

Seiffert U, Walzer P (1989) Automobiltechnik der Zukunft. VDI-Verlag, Düsseldorf

Smith A, Stirling A, Berkhout F (2005) The governance of sustainable socio-technical transitions. Res Policy 34:1491–1510

Summerton J (1994) Changing large technical systems. Westview Press, Boulder

Van der Brugge R, van Raak R (2007) Facing the adaptive management challenge: insights from Transition Management. Ecol Soc 12(2):33–47. http://www.ecologyandsociety.org/vol12/iss2/art33/ES-2007-2227.pdf. Accessed 27 Apr 2011

Verband Region Stuttgart (2011) Sitzungsvorlage Nr. 46/2011. Regionalversammlung am 20.07.2011. Verband Region Stuttgart, Stuttgart. www.region-stuttgart.org/vrs/download.jsp?docid=10960. Accessed 1 Aug 2011

Voß J-P, Smith A, Grin J (2009) Designing long-term policy: rethinking transition management. Policy Sci 42(4):275–302

VROM (2001) Where there's a will there is a world. In: 4th National environmental policy plan - summary. Working on sustainability. Spatial Planning and the Environment Ministry of Housing, The Hague. http://www.ocs.polito.it/biblioteca/ecorete/nepp4.pdf. Accessed 21 Apr 2011

Yücel G, Chiong Meza CM (2008) Studying transition dynamics via focusing on underlying feedback interactions. Comput Math Org Theory 14(4):320–349

Zhao J (ed) (2011) Key research areas of ecological and environmental science & technology. Springer, Berlin

Zimmermann A Dr. (2006) Instrumente zur Akteurs Analyse. 10 Bausteine für die partizipative Gestaltung von Kooperationssystemen. GTZ, Eschborn. http://www.gtz.de/de/dokumente/de-SVMP-Instrumente-Akteursanalyse.pdf. Accessed 26 May 2011

Author Biography

Alanus von Radecki is a researcher at Fraunhofer Institute for Industrial Engineering (IAO) in Stuttgart—Germany. He studied Sociology and Cultural Anthropology at the University of Freiburg and received his Master's degree in 2006. After three years work as Knowledge Manager for a German IT-Company he completed the international master course: M.Sc. "Environmental Governance" at the University of Freiburg in 2011. His projects include topics of urban electric mobility, sustainability mobility for communities, governance concepts for sustainable transitions and multi-stakeholder integration for sustainable transitions of cities. At current he is project manager of a global innovation network for sustainable cities at Fraunhofer IAO.

Assessment of CO_2-Emissions from Electric Vehicles: State of the Scientific Debate

Jürgen Gabriel, Philipp Wellbrock and Marius Buchmann

Abstract The purpose of our work is to summarize and contextualize the on-going discussion about the ecological effects of electric vehicles. Key driver for the introduction of electric vehicles in Germany, besides resource scarcity, is the necessity to reduce CO_2-emissions in the transportation sector. Even though the media already proclaims the ecological supremacy of electric vehicles over the conventional combustion engine, the scientific discussion still tries to find an answer to the question: Will the use of electric vehicles as substitutes for conventional combustion engines significantly reduce the CO_2-emissions in the transportation sector? Well known experts from Germany as well as European and American scientists picked up this task and tried to find the solution. However, as at the moment less than 5,000 electric vehicles are currently driving on German roads, the database is very limited and therefore the studies need to deduce estimates for key determinates for the emission comparison, which leads to a broad variety in the results. Within our study we compare the most important and up-to-date studies with respect to the German market to derive the potential outcomes of an emission comparison. Resulting from this study we will be able to contextualize the test drive within the model region Bremen/Oldenburg. Furthermore, the data from the model regions could be used to verify the scope of the different studies and the reliability of the different estimates used. Finally, we will be able to draw a realistic picture of the potential ecological effects of electric vehicles.

List of Abbreviations

BEV   Battery-electric vehicles
CO_2 Carbondioxid
CV   Combustion vehicles
CHP   Combined Heat and Power
ETS   European Emission Trading System

J. Gabriel (✉) · P. Wellbrock · M. Buchmann
Bremer Energie Institut, College Ring 2/Research V 28759 Bremen, Germany
e-mail: gabriel@bremer-energie-institut.de

GHG   Greenhouse gas
NEDC   New European Driving Cycle
RE   Renewable energy
TtW   Tank to wheel
WtW   Wheel to wheel

1 Introduction

In the discussion about electric vehicles, a common argument is that their dissemination could substantially reduce greenhouse gas (GHG) emissions in the transport sector. However, it is common sense in the scientific debate that electric vehicles cannot be regarded as "emission-free", just because they do not emit any GHG during operation. Still, it is discussed how many CO_2 emissions are related to the operation of an electric vehicle.

In recent years, several attempts have been made to quantify the GHG effect of replacing a conventional combustion vehicle with a battery-electric one (e.g. Blesl et al. 2009; Wietschel and Bünger 2010). It becomes clear that such comparison depends on a range of parameters. As many of the studies apply different parameters it is quite difficult to compare the results and to get a clear picture of the GHG effect of electric vehicles. To allow a better comparison of the different studies we want to shed light on the dark and thereby enable the general public to discuss on a common ground about CO_2 emissions in the context of the electrification of the transportation sector. Therefore, this paper aims to give an introduction to these parameters, how several recent studies have set them and how this influences the resulting emission balance. In chapter "Mobility Scenarios for the Year 2030: Implications for Individual Electric Mobility", options are discussed how (virtually) emission-free electric mobility can be achieved through the use of additional renewable energy.

1.1 Scope and Objectives

This paper concentrates on comparing emissions of

1. battery-electric vehicles (BEV) and
2. combustion vehicles (CV).

We chose this narrow focus to directly address the differences in GHG emissions between the combustion vehicles and electric mobility. Other concepts, like hybrid or plug-in hybrid-electric cars, are always dependent on fossil based inputs like gasoline. Therefore, the GHG emissions of hybrid vehicles are only partially based on the electric drive train. However, for the comparison it is important to make a clear differentiation between the emissions from the electric drive train and the combustion engine. By comparing the two technologies independent from each other we can derive the maximum difference in GHG emissions between them. Based on these results it is possible to compare the two technologies. In addition, the results can give an insight into the potential of hybrid cars to effect GHG emissions as the emissions from hybrid concepts will always be located somewhere between the two independent technological solutions of battery electric and combustion vehicles.

Furthermore, commercial vehicles and motorcycles will not be included. Also, only emissions relevant for the anthropogenic climate change are covered; other pollutants and noise emissions are neglected.

The paper further concentrates on emissions that result from the operation of the vehicle, including the supply chains of energy sources, the so called "well-to-wheel" emissions. The well-to-wheel perspective is the standard approach in this context and therefore allows comparing the different studies. Emissions generated during construction, delivery, maintenance or disposal of the vehicle will not be considered—such a life cycle assessment requires a complex methodological framework and would exceed the scope of this paper.

The comparison focuses in the level of the individual vehicle, not on the total cumulative effect of a certain distribution of electric vehicles. Alternative ways of looking at the net emission balance of electric vehicles, which result from the legislative environment and may only become relevant given a certain distribution of BEVs, would exceed the scope of this paper and cannot be addressed here.

1.2 Vehicle Segments

Today or in the near future, the available battery-electric passenger cars mostly belong to the small or micro car segment, with some compact class vehicles.[1] This is mainly due to the limited capacity of current batteries. The resulting short range of typically 100–200 km favors the use of BEV as city or commuter vehicles. Larger and therefore heavier BEV with the same range would also be far more expensive, due to the larger storage capacity required. Of the eight examined studies, three focus on a single car segment. The remaining five studies assess both small/micro cars and compact cars.

---

[1] According to the classification of the German Federal Motor Vehicle Office (Kraftfahrt-Bundesamt), micro cars include the Smart Fortwo or the Fiat 500, examples of small cars are the VW Polo and Opel Corsa, and typical compact cars are the VW Golf and Opel Astra.

1.3 The Reviewed Studies

The following studies were selected and reviewed by the authors of this paper in order to provide a comprehensive view on the issue of BEV/CV emission comparison. The selection of the different studies was based on several criteria. First, studies had to refer explicitly to the situation in Germany and contain assumptions on the key parameters introduced in chapter "Socio-Economic Aspects of Electric Vehicles: A Literature Review". Furthermore, the studies had to apply the Well-to-Wheel approach and should have been published between January 2007 and September 2011. Eight studies were selected based on the criteria (Blesl et al. 2009; Engel 2007; Helms et al. 2010; Horst et al. 2009; Richter and Lindenberger 2010; Wietschel and Bünger 2010; Renewbility 2009a, b; Helmers 2010).

2 Parameters of Emission Comparison

The following chapter describes the differences between the Battery-electric vehicle (BEV) and the Combustion vehicle (CV). The specific content analysis focuses on the consumption values of the two vehicle types.

2.1 Overview

The comparison of combustion and battery-electric vehicle emissions depends on a set of key parameters, which are shown in Fig. 1.

The first two parameters determine the baseline scenario of the comparison, i.e. the emission balance of a conventional combustion vehicle. The total GHG balance of driving a CV consists of two factors: Firstly, direct emissions from burning the fossil fuel (so-called tank-to-wheel (TtW) emissions). Secondly, emissions caused by fuel production and provision (so-called well-to-tank (WtT) emissions). Together, these two parameters determine the overall or "well-to-wheel" emissions.

Unlike CVs, BEVs usually do not emit greenhouse gases during operation. Because of this, they are sometimes referred to as "zero-emission cars". However, the generation of electricity for charging the BEV is always associated with a certain emission of greenhouse gases. Analogous to an observation "from the oil well to wheel" (well-to-wheel) in a combustion vehicle, these supply chain emissions have to be included when investigating the actual greenhouse effect of using electric vehicles.

This means that two parameters are needed to calculate BEV emissions per km: Firstly, the gross amount of charging energy needed for driving one km, including charging losses etc. (so-called plug-to-wheel consumption). Secondly, the gross amount of GHG that was emitted producing that amount of electricity.

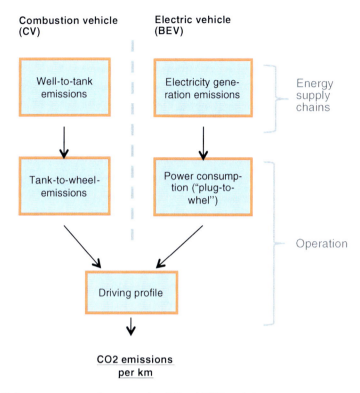

Fig. 1 Relevant parameters for comparing CV and BEV emissions

As consumption varies between different driving situations both for CV and BEV (e.g. urban or motorway driving), assumptions have to be made regarding the actual driving profile. This profile should of course be identical for both CV and BEV in order to compare the real world GHG balance. For comparing the different studies and for creating corridors, these driving profiles are of particular interest, as they contain assumptions about the actual usage of BEV and CV in a certain segment, which can significantly affect the emission balance.

2.2 Combustion Vehicle (CV) Emissions

CO_2 emissions per kilometer are calculated based on fuel consumption and fuel type. Official figures usually include only direct emissions resulting from the combustion of fuel (so-called tank-to-wheel emissions). However, the production, processing and provision of fossil fuels also cause significant greenhouse gas emissions. These well-to-tank emissions amount to approximately 19 % of tank-to-wheel emissions for diesel vehicles, and to 15 % for petrol vehicles (2007).

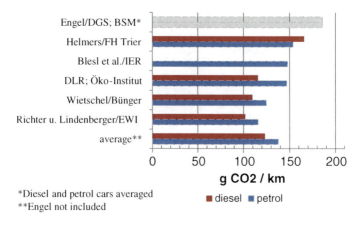

Fig. 2 Well-to-wheel CO$_2$ emissions, combined, for small/micro CV

If the current admixture of biofuels is taken into account, values are reduced to approx. 15 % for diesel and 12 % for petrol vehicles.

Figure 2 shows the combined consumption figures assumed for small/micro CV with petrol and diesel engine, and the average for both fuel types. Figure 3 shows the same figures for compact cars. As can be seen, not all studies consider both fuel options for each vehicle segment. Engel (2007) does not treat petrol and diesel cars separately. Consequently, that value is excluded from the calculation of averages.

As can be seen, assumed values vary substantially, especially for small/compact cars. Decisive factors are the choice of the reference CV and the method used to determine fuel consumption and emissions. Some of the reviewed studies use existing car models for comparison. In the small/micro segment, the usual reference is the Smart Fortwo, which is available with diesel and gasoline engines, but also with battery-electric propulsion. For the compact segment, the VW Golf is a common reference.

The remaining studies assume an average car for each segment, based on the official registration data in a certain year (e.g. Engel 2007: 61); or, like Pehnt (Pehnt 2010: 6), on other data sources.

However, even if the same car is taken as a reference, figures vary substantially: Richter and Lindenberger (2010) and Helmers (2010: 574) both take the Smart Fortwo Diesel as a reference in the small/micro segment, but assume very different emissions (102–166 g, see Fig. 2). The reason is that the former study put down the official manufacturer's figure, while the latter uses a test result by an automobile magazine.[2]

---

[2] Differences in well-to-tank emissions can be neglected: Since (Helmers 2010: 574) only observes TtW emissions, WtT emissions have been added based on (JRC 2007), taking into account the current biofuels share in German fuel. (Richter and Lindenberger:64) also consider the biofuels share in the calculation of WtT emissions.

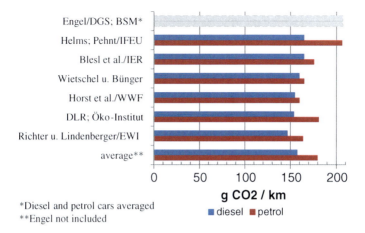

Fig. 3 Well-to-wheel CO_2 emissions, combined, for compact CV

Manufacturers' consumption figures in the EU are based on the new European Driving Cycle (NEDC). It includes urban, extra-urban and motorway driving, yet it is criticized by experts as unrealistic (e.g. Helmers 2010: 573–574)—real-world values often significantly exceed the official figures. For example, the NEDC does not include additional loads like air conditioning. Several studies therefore explicitly assume higher figures—for example (Blesl et al. 2009: 32) add 10 %. Helmers (2010: 574) adds 17 % for petrol engines, and, as mentioned above, as much as 60 % for the Smart diesel.

Naturally, CV consumption also depends strongly on the underlying driving profile—especially the assumed share of urban driving strongly influences the emission balance. Engel (2007), for example, takes the average consumption of newly registered cars in 2005 for each segment and converts the urban and extra-urban figures to a specific combined driving profile (60 % urban/40 % extra urban for small cars, 40/60 for compact cars). This method yields a significantly higher combined consumption compared to the NEDC, especially in the small car segment.

Helms (2010: 120) differentiate two driving profiles (urban and average). For this comparison, the "urban" profile (70 % urban, 20 % extra-urban, 10 % motorway driving) was chosen.

Blesl (2009) even calculate combined consumption figures for three different driving profiles. For this comparison, the "commuter" profile was chosen (1/3 urban driving, 2/3 extra-urban driving), as this profile is also used to calculate combined BEV consumption (Blesl et al. 2009: 33; 53).

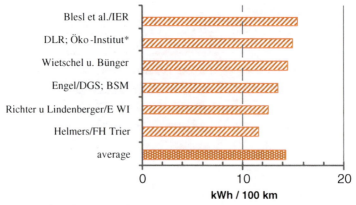

Fig. 4 Combined consumption of small/micro BEV, as given in reviewed studies

2.3 Battery-Electric Vehicle (BEV) Consumption

The consumption of an electric vehicle is measured in kWh per 100 km. Figs. 4 and 5 show the values assumed by the reviewed studies. While variations are smaller than for CV emissions, figures still differ notably.

2.3.1 Origin of Consumption Figures

As is the case for CV emissions, the examined studies use different sources for determining BEV consumption. In the majority, a "typical", i.e. hypothetical BEV is assumed by averaging consumption data of several available or announced vehicles of a certain segment. Other studies use manufacturer data or real-world test results (Helmers 2010: 574) of a specific model.

Another factor that can lead to variations is the driving profile assumed when calculating combined emissions (in analogy with CV emissions, see also section "Management Perspectives" in Assessment of CO_2-emissions from Electric Vehicles: State of the Scientific Debate). Unlike the CV, consumption of a BEV per 100 km is lowest for urban driving, somewhat higher for extra urban driving and highest at constant high speeds. This is due to low consumption in stop-and-go traffic and to best powertrain efficiency at part load, factors that negatively affect consumption in a CV.

Regarding driving profiles, four studies just state combined values without commenting on driving profiles. Those studies defining a driving profile for CV also apply this to the same-sized BEV and give detailed comsumption figures for different driving situations.

Notably, two studies include figures for motorway driving—in (Helms et al. 2010: 120), 25 kWh/100 km is assumed for compact cars, (Renewbility 2009a)

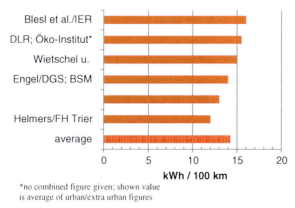

Fig. 5 Combined consumption of compact BEV, as given in reviewed studies

states 24 kWh/100 km for small and 28 kWh/100 km for compact vehicles.[3] Hence, if a hypothetical driving profile included extensive motorway driving, combined consumption would be substantially higher than the values given above.

2.3.2 Charging Losses and Additional Loads

When charging a BEV, certain charge losses occur, i.e. not all supplied energy is stored in the battery, as a portion is converted into heat and is lost. In three of the analyzed studies it is not clear whether charging losses are included in the consumption figures; the remaining five studies explicitly take them into account. In two cases a 15 % loss is assumed (Engel 2007: 32; Horst et al. 2009: 32); Helms et al. (2010: 115) estimate 10 %. Two studies give no exact value (Renewbility 2009a: 112), or a range of 5–40 % (Helmers 2010: 575). The latter study also points to the large potential impact that an inefficient charging technology may have on the emission balance of a BEV (ibid.).

Another important point in estimating the consumption of BEV is whether additional power consumers on board the vehicle are considered. Unlike in a CV, BEVs barely create excess heat that could be used for heating the vehicle. Heating in a BEV usually works electrically, resulting in significantly higher energy consumption during cold season. Also, air conditioning, ventilation or headlights can affect energy consumption proportionately more than in a CV. These effects are only quantified by Helms et al. (2010: 115–117), they expect markups of up to 20 % over the NEDC. In other studies, which set consumption values on the basis of real-world test results, the effect of other consumers might also be included to some extent.

From this point of view, the average combined consumption figures (14.3 and 19.4 kWh/100 km, respectively) derived from the examined studies may be

---

[3] This study does not state combined figures, and for the comparison in this chapter urban and extra-urban figures have been averaged.

somewhat underestimated. However, for the lack of long-term real-world figures, a combined consumption of 15 kWh/100 km for small and 20 kWh/100 km for compact cars is assumed in the development of a synthesis in section "Economic Conditions" in Assessment of CO_2-emissions from Electric Vehicles: State of the Scientific Debate.

2.4 Power Generation Emissions

Analogous to an observation from well-to-wheel for a CV, emissions that occurred during power production have to be included in the case of the BEV. To calculate emissions per km, it has to be investigated how much greenhouse gas was emitted producing the necessary amount of electricity.

In fact, the emission factor of electricity generation (measured in grams of CO_2 equivalent per kWh of electrical energy) is the determining factor for the emission balance of electric vehicles. The calculation is usually done taking into account the specific supply chains, i.e. the provision and conversion of primary energy sources and the transmission of electricity, but also including the proportionate emissions from the construction of the respective power plants. This explains why even for wind energy a certain amount of CO_2 emissions is assumed.

In most cases, BEVs are charged from the general electricity grid. To determine the related emissions, there are different methodological approaches that can lead to significantly different estimates.

2.4.1 Calculation Based on Power Generation Mix

The simple option regards the CO_2 emissions per kWh according to the average electricity mix in the relevant year. For Germany, this figure is published regularly by the Federal Environmental Agency (e.g. UBA 2009: 1); in 2009 this figure was at 565 g/kWh (ibid.). Additionally, some studies assess other hypothetical options, such as a purely fossil energy mix, a renewable mix or a mix of renewable and CHP electricity, with specific emission factors.

2.4.2 Calculation Based on the Marginal Power Plant

In the opinion of several authors, the power generation mix approach is inadequate: The use of electric vehicles instead of internal combustion vehicles results in an additional demand of electricity—every time an electric vehicle is connected to the grid for charging. Hence, one has to consider the specific emissions of the power plant that is powered up during this period to meet the additional demand for electricity.

This so-called marginal power plant typically has a much higher intensity of CO_2 than the average electricity mix: Electricity from renewable energies is fed into the grid with priority (in Germany), so usually there is no renewable stand-by reserve. Also the low-emission nuclear plants tend to be operated at full capacity, whenever possible. Typical marginal power stations would be medium load power plants such as black coal-fired plants, as well as peak load plants, such as gas turbine plants.

Marginal power plant emissions are also influenced by other factors: Different loading scenarios (timing, duration, etc.), market distribution of BEV, a changing power plant structure and the implications of an European energy market make it difficult to determine the marginal plant and its emissions, especially for the future. The only study calculating the effects of different load scenarios on CO_2 emissions is Pehnt et al (2011: 5–9). Given the unregulated charge of 10–12 million BEVs, it expects an increase in peak load by 12 %. An intelligent load management with nightly charging could, however, contribute to the integration of renewable energy sources—as renewable energy plants may remain connected at night which otherwise would have to be switched off. At present, though, black coal-fired power plants usually represent the marginal power plant, with distinctly higher specific emissions than the marginal power plants at daytime (mostly gas-fired power plants).

Notably, only two of the reviewed studies explicitly address the perspective of marginal power plants in this context (Horst et al. 2009: 33; Pehnt et al. 2011: 5–9).

2.4.3 Alternative Perspectives on Emission Effects

Alternative ways of looking at the net impact of power generation emissions take into account the conditions of the European Emission Trading System (ETS) (Horst et al. 2009: 32–37; Richter and Lindenberger 2010: 70–71) and the EU targets for renewable energy usage (Pehnt 2010: 8–9). Discussing these approaches in detail would exceed the scope of this paper, further information can be found in the referenced documents. Options for and implications of charging BEVs with 100 % renewable electricity are discussed in chapter "Mobility Scenarios for the Year 2030: Implications for Individual Electric Mobility".

2.5 Synthesis and Outlook

With typical BEV consumption and CV emissions derived from the reviewed studies (see section "Usability" in Assessment of CO_2-emissions from Electric Vehicles: State of the Scientific Debate), it is now possible to compare the resulting CO_2 emissions per km for different emission factors of power generation. As the year of reference for the German power mix varies between the reviewed studies, it is not considered reasonable to average the given figures. Instead, the value predicted for 2010 by UBA (UBA 2009) is assumed.

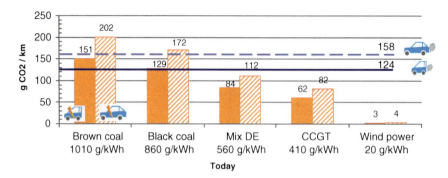

Fig. 6 BEV emissions for various power generation options, compared to CV WtW emissions (today)

Figure 6[4] shows these values compared to typical CV emissions, both for small/micro and compact cars. It can be seen that BEV emissions actually exceed those of a same-sized combustion vehicle when charged with electricity from a coal-fired power plant (which represents the marginal power plant at night, see section "Standardization" in Assessment of CO_2-emissions from Electric Vehicles: State of the Scientific Debate). Assuming the average generation mix, BEVs produce 25–30 % less emissions than CV. For gas-fired power plants, emissions are only 50 % of the CV figures.

It should be kept in mind that these parameters may vary significantly depending on compared vehicles, actual driving profiles et cetera.

Table 1 summarizes the findings of the individual studies regarding a charge based on the power generation mix. This illustrates how the setting of parameters influences overall results, with predicted emission reductions ranging from 3 % to over 50 %. However, it can also be noted that five of the eight reviewed studies assume similar savings between 20 and 30 %.

2.6 Outlook

In total, four studies attempt a prognosis of how the emission balance will develop in the future. Pehnt et al (2011: 7) only model power generation emissions, without assuming changes in CV emissions or BEV consumption.

---

[4] Assumptions and data sources:

- BEV consumption: small/micro car 15 kWh/100 km; compact car 20 kWh/100 km (derived from reviewed studies)
- CV WtW emissions: small/micro diesel car 124 g/km; compact diesel car 158 g/km (derived from reviewed studies)
- DE electricity mix 2010 according to (UBA 2009)
- Power plant emission factors according to (Öko-Institut 2010)

Table 1 Results of reviewed studies for mix-based charging

Study	Power generation emissions (g/kWh)	Vehicle segment	CV emissions, WtW (g CO_2/km)	BEV emissions (g CO_2/km)	BEV balance
DGS/bsm	650	Small/micro	186 (di./pet.)	91	−51 %
		Compact	207 (di./pet.)	111	−46 %
EWI[a]	596	Micro	102 (diesel)	79	−22 %
		Compact	147 (diesel]	122	−17 %
Fraunh. ISI/RWE	527	Micro	110 (diesel)	79	−28 %
Helmers	600	Micro	148 (petrol)	71	−52 %
IER	624	Micro	133 (petrol)	100	−25 %
		Compact	165 (diesel)	127	−23 %
IFEU[b]	630	Compact	165 (diesel)	131	−21 %
Öko-Institut/DLR	596	Small/micro	116 (diesel)	86	−26 %
		Compact	154 (diesel)	110	−29 %
WWF	625	Compact	155 (diesel)	150	−3 %

[a] Driving profile: commuter (1/3 urban, 2/3 extra urban)
[b] Driving profile: 70 % urban, 20 % extra-urban, 10 % motorway (Helms et al. 2010: 120)

Of the examined studies, three take assumptions about the future development of CV consumption. Herein, two studies refer to the EU regulation 443/2009 (EU 2009), which for 2020 establishes a mandatory fleet limit of 95 g/km (tank-to-wheel), and adopt this figure as a guideline (e.g. Renewbility 2009b: 29).

Regarding changing BEV consumption, only Blesl et al. (2009) and Richter and Lindenberger (2010) take assumptions. While the former expects a decline in small BEV consumption by 20 % until 2030, the latter only assumes 8 % less.

Predictions about the future development of the power generation mix are included in all four studies that model BEV emissions in the medium-term. Herein, some studies, like Wietschel and Bünger (2010: 19) and Renewbility (2009a: 189) rely on prognoses issued by the German Environment Ministry (e.g. BMU 2009). Blesl et al. (2009: 44) refer to the scenarios developed by IER/RWI/ZEW (IER/RWI/ZEW 2009), while Richter and Lindenberger (2010: 57–58) take own assumptions.

Figure 7[5] summarizes the prognoses made in the reviewed studies concerning CV emissions and compares them to BEV emissions. These are based on a projection of power generation emissions by IER/RWI/ZEW (IER/RWI/ZEW 2009: 104),[6] which lies well within the range of values assumed by the other sources mentioned above.

---

[5] Assumptions and data sources:

- BEV consumption: 15/20 kWh/100 km (small/compact car, derived from reviewed studies)
- Future CV WtW emissions (small petrol car/ compact diesel car): average derived from reviewed studies
- DE electricity mix: for 2010 based on (UBA 2010); for 2020 and 2030 based on the reference scenario of (IER/RWI/ZEW 2009: 104)

[6] Figures are based on the so-called reference scenario, which assumes a fade-out of nuclear energy according to the 2002 revision of the Nuclear Energy Law (IER/RWI/ZEW 2009: 4).

Fig. 7 BEV emissions for projected power generation mix, compared to projected CV WtW emissions (2010–2030)

BEV consumption is assumed to remain constant. The result shows a growing advantage for BEVs because of substantially lower power generation emissions—CV emissions decline at a lower rate and stagnate towards 2030 (note that BEV emission projections are based on the average generation mix, not the marginal power plant perspective; compare section "Standardization" in Assessment of CO_2-emissions from Electric Vehicles: State of the Scientific Debate).

3 Charging with Additional Renewable Energy

In the face of the significant emissions related to fossil-generated electricity and hence the current power generation mix, using renewable energy for charging is the only way to achieve virtually emission-free electromobility. In fact, five of the reviewed studies assess one or more options for power generation from renewable energy sources.

On closer inspection of the power sector's regulatory framework, some issues complicate the use of green electricity. The key challenge is to guarantee an actual, *additional* production of renewable electricity, instead of just requiring a portion of the existing renewable production.

3.1 Construction of New RE Plants to Compensate BEV Energy Demand

One possibility, according to Pehnt (2010: 12), would be to tie the sale of a BEV to the construction of new RE capacity, either directly or through a special fund.

Owners or manufactures could either be obliged to contribute, or be motivated through certain incentives.

However, if a renewable energy power plant is built "for compensation", it may not simply sell its production under the German RE feed-in tariff—in that case, power generation could not be regarded as additional: If the plant can operate profitably under the feed-in tariff regime, most likely it would have been built anyway, also without the BEV background (Pehnt 2010: 9).

An additional production would occur if the newly built plant directly supplied charging electricity to BEVs and sold only the remaining production under the feed-in tariff or on the free market for green power. However, this would not be profitable in the medium term, manufacturers or owners would therefore face additional costs compared to charging with the regular grid mix.

3.2 Purchase of Green Power Products

If the BEV owner uses a green power product to supply the necessary electricity, the additionality depends on the conditions of the offer: Certain products are only "green washed" through the acquisition of cheap Renewable Energy Certificate System (RECS) certificates, e.g. from foreign hydro-electric plants. Here, no new RE capacity is created, and the actual power demand is met by the next best power plant according to the merit order. Certified green power products, however, must include a certain percentage of electricity from new, non-depreciated power plants; some providers also commit themselves to investing in new facilities. In such a case, it can be assumed that at least some renewable energy is produced additionally (compare Richter/Lindenberger 2010: 71). Compared to the use of conventional grid power, BEV owners would have to bear higher costs.

3.3 Adaptation of Renewable Energy Targets

A third alternative aims at the mandatory targets for expanding renewable power generation, set by the German federal government: The amount of electricity used for electric vehicles could be deducted from the total renewable energy production—regardless of its actual origin. In order to still achieve the renewable energy targets, incentives would have to become effective that lead to additional RE production. In this case, the general public would bear the cost of a CO_2-neutral integration of electric vehicles. However, as Pehnt (2010: 14) points out, this option may not have the desired effect, as the renewable energy targets may already be exceeded given the current development (according to BMU 2009).

4 Conclusion

The main result of this review, which it shares with all the analyzed studies, is that battery-electric vehicles have somewhat lower well-to-wheel emissions than CV, when charged based on the average German power generation mix.

However, it should be kept in mind that the comparisons in the reviewed studies rely on a set of assumptions, namely that the reduced usability of a BEV (regarding range, top speed) is not relevant, driving profiles do not include larger shares of high speed driving, and that extra consumers like heating or air conditioning do not raise consumption figures significantly.

Also, the marginal power plant perspective implies that, at least for overnight charging, a much higher emission factor may apply, leading to emissions that may exceed those of a CV.

These limitations regarding the emission balance based on fossil power generation underlines the importance of charging BEVs with renewable electricity, which is the only way to guarantee virtually emission-free electric mobility.

Therefore, along with the technical development and market penetration of electric vehicles, it is necessary to create an adequate regulatory framework. Its goal is to ensure and facilitate the creation of sufficient renewable generation capacity to cover the additional energy demand of BEVs. Also, it has to be discussed whether manufactures, buyers or the general public should bear the extra costs of a climate-friendly integration of electric mobility.

References

Blesl et al (2009) Entwicklungsstand und Perspektiven der Elektromobilität. http://elib.uni-stuttgart.de/opus/volltexte/2010/ 5218/pdf/Elektromobilitaet_Endbericht_20100322.pdf. Accessed 21 June 2011

BMU (2009) Bundesministerium für Umwelt, Naturschutz und Reaktorsicherheit (2009) Langfristszenarien und Strategien für den Ausbau erneuerbarer Energien in Deutschland unter Berücksichtigung der europäischen und globalen Entwicklung—Leitszenario 2009. http://www.bmu.de/files/pdfs/allgemein/application/pdf/leitszenario2009_bf.pdf. Accessed 21 June 2011

Engel (2007) Plug-in Hybrids. Studie zur Abschätzung des Potentials zur Reduktion der CO_2-Emissionen im PKW-Verkehr bei verstärkter Nutzung von elektrischen Antrieben im Zusammenhang mit Plug-in-Hybrid Fahrzeugen. München

EU (2009) REGULATION (EC) No 443/2009 setting emission performance standards for new passenger cars as part of the Community's integrated approach to reduce CO2 emissions from light-duty vehicles 23.04.2009

Helmers (2010) Bewertung der Umwelteffizienz moderner Autoantriebe—auf dem Weg vom Diesel-Pkw-Boom zu Elektroautos, in: Umweltwissenschaften und Schadstoffforschung 22/5, 2010, p. 564-578. http://www.springerlink.com/content/c3hkxq73812x4t26/fulltext.pdf. Accessed 28 June 2011

Helms et al (2010) Electric vehicle and plug-in hybrid energy efficiency and life cycle emissions. In: Empa (ed) Tagungsband des 18. Transport and Air Pollution Symposiums, Dübendorf, 18.-19.5 2010. http://www.ifeu.de/verkehrundumwelt/pdf/Helms%20et%20al.%20(2010)%20 Electric%20vehicles%20(TAP%20conference%20paper)%20final.pdf. Accessed 21 June 2011

Horst et al (2009) Auswirkungen von Elektroautos auf den Kraftwerkspark und die CO_2-Emissionen in Deutschland, WWF Deutschland (ed), 2009. http://www.wwf.de/downloads/publikationsdatenbank/ddd/30496. Accessed 21 June 2011

IER/RWI/ZEW (2009) Die Entwicklung der Energiemärkte bis 2030—Energieprognose 2009. http://www.rwi-essen.de/media/content/pages/publikationen/rwi-projektberichte/PB_Energieprognose-2009.pdf. Accessed 28 June 2011

JRC (2007) Joint Research Centre/EUCAR/CONCAWE, Well-to-Wheels analysis of future automotive fuels and powertrains in the European context, Version 2c. Ispra

Öko-Institut (2010) Globales Emissions-Modell Integrierter Systeme (GEMIS), Version 4.6. http://www.oeko.de/service/gemis/de/index.htm. Accessed 21 June 2011

Pehnt (2010) Elektromobilität und erneuerbare Energien, in: Müller (ed.): 20 Jahre Recht der Erneuerbaren Energien. Tagungsband der 7. Würzburger Gespräche zum Umweltenergierecht, Würzburg, 13.-14.10.2010. http://www.ifeu.de/verkehrundumwelt/pdf/Pehnt%20(2010)_%20Erneuerbare%20Energien%20und%20Elektromobilitaet%20final.pdf. Accessed 21 June 2011

Pehnt et al (2011) Elektroautos in einer von erneuerbaren Energien geprägten Energiewirtschaft, in: Zeitschrift für Energiewirtschaft. http://www.springerlink.com/content/a156165p51827682/fulltext.pdf. Accessed 21 June 2011

Renewbility (2009a) Stoffstromanalyse—Nachhaltige Mobilität im Kontext erneuerbarer Energien bis 2030. Endbericht Teil 1 - Methodik und Datenbasis. http://www.renewbility.de/fileadmin/download/endbericht_renewbility_teil1V1.pdf. Accessed 21 June 2011

Renewbility (2009b). Stoffstromanalyse—Nachhaltige Mobilität im Kontext erneuerbarer Energien bis 2030. Endbericht Teil 2 —Szenarioprozess und Szenarioergebnisse. http://www.renewbility.de/fileadmin/download/endbericht_renewbility_teil2V1.pdf. Accessed 21 June 2011

Richter and Lindenberger (2010) Potenziale der Elektromobilität bis 2050—Eine szenarienbasierte Analyse der Wirtschaftlichkeit, Umweltauswirkungen und Systemintegration. Köln

UBA (2009) Umweltbundesamt Entwicklung der spezifischen Kohlendioxid-Emissionen des deutschen Strommix 1990–2007, April 2009, updated March 2011. www.umweltbundesamt.de/energie/archiv/co2-strommix.pdf. Accessed 21 June 2011

Wietschel and Bünger (2010) Vergleich von Strom und Wasserstoff als CO_2-freie Endenergieträger—Studie im Auftrag der RWE AG. http://isi.fraunhofer.de/isi-de/e/download/publikationen/Endbericht_H2_vs_Strom-final.pdf. Accessed 02 July 2011

Author Biographies

Jürgen Gabriel is researcher at the Bremen Energy Institute since 2002. After completing his economics degree, he worked at the University of Dortmund and graduated with the topic on "forecast regional demand for housing." Jürgen Gabriel has also practical experience in urban development and planning and at a large municipal energy company. His research interests are liberalized energy markets with focus on conditions of network operations, the integration of renewable energy into the power system as well as the macro-economic effects of different energy economics or energy policy developments.

Philipp Wellbrock studied geography and is researcher at the Bremen Energy Institute. His research focuses are in the area of the future of electricity generation in the European comparison, the impact of European investment conditions of employment in the German electricity sector or the evaluation of the impact of electric vehicles.

Marius Buchmann gained a BA in Economics and Law at the University of Erfurt and studied the Master program "Sustainability Economics and Management" at the University of Oldenburg. Marius Buchmann has also experience in the management consulting. His research focus is on user integration and energy efficiency in buildings.

How to Integrate Electric Vehicles in the Future Energy System?

Patrick Jochem, Thomas Kaschub and Wolf Fichtner

Abstract Main challenges within the energy system of tomorrow are more volatile, less controllable and at the same time more decentralized electricity generation. Furthermore, the increasing research and development activities on electric vehicles (EV) make a significant share of electric vehicles within the passenger car fleet in 2030 more and more likely. This will lead to a further increase of power demand during peak hours. Answers to these challenges are seen, besides measures on the electricity supply side (e.g. investing in more flexible power plants or storage plants), in (1) grid extensions, which are expensive and time consuming due to local acceptance, and in (2) influencing electricity demand by different demand side management (DSM) approaches. Automatic delayed charging of electric vehicles as one demand side management approach can help to avoid peaks in household load curves and, even more, increase the low electricity demand during the night. This facilitates integrating more volatile regenerative power sources, too. Bidirectional charging (V2G) and storing of electricity extends the possibilities to integrate electric vehicles into the grid. But, comparing electricity storage costs and availability of electric vehicles with costs and technical conditions of other technologies leads to the conclusion, that vehicle to grid (V2G) is currently not competitive—but might be competitive in the future, e.g. within the electricity reserve market. In summary, the chapter gives an overview of the future electricity market with the focus on electric vehicles and argues for automatic delayed charging of electric vehicles due to economic and technical reasons.

P. Jochem (✉) · T. Kaschub · W. Fichtner
Karlsruhe Institute for Technology (KIT), Institute for Industrial Production (IIP), Hertzstr. 16, Building 06.33 76187 Karlsruhe, Germany
e-mail: jochem@kit.edu

T. Kaschub
e-mail: kaschub@kit.edu

W. Fichtner
e-mail: wolf.fichtner@kit.edu

1 Introduction

The energy system of today is faced with advancing challenges: more volatile, less controllable and at the same time more decentralized electricity generation. Simultaneously, the political objective is to increase energy efficiency [see e.g. proposal for the directive on energy efficiency (EC 2011)] and to reduce greenhouse gas emissions (e.g. Erdmenger et al. 2009). From the German perspective these challenges might be tremendous due to the ambitious aim of CO_2 emission reduction of up to 80 % in 2050 compared to 1990 levels (e.g. Nagl et al. 2011; BMU 2010) and the nuclear phase-out until 2022 (German Atomic Energy Law—AtG).

The development of the electricity generation and its composition in Germany until 2050 illustrates these challenges (Fig. 1). While the share on conventional power generation (coal, nuclear, natural gas, and oil) was above 80 % in 2008, it is supposed to be below 50 % in 2050 (Nagl et al. 2011; EWI et al. 2010).

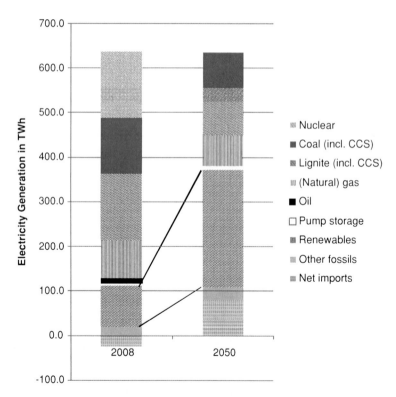

Fig. 1 Electricity generation in Germany in 2008 and 2050 by fuels in TWh. *Source*: EWI et al. (2010)

This increase in power generation from renewable sources, together with the trend to heat driven[1] domestic combined heat power plants (CHP), leads to a higher feed-in of "uncontrolled" electricity generation in the lower voltage grid levels (distribution grid). Beside these challenges a significant share of electric vehicles (EV) on the passenger car fleet in 2030 is highly probable (Pehnt et al. 2011). This leads to a further increase in electricity demand (energy and power)—especially during peak hours and, again, on the low voltage grid (Pollok et al. 2009). Hence, on the one hand a more complex energy supply can be observed, and on the other hand an increased electricity demand with higher peak loads in the evening hours is expected.

However, a modified charging process of EV might help to tackle these challenges: (1) by controlled unidirectional charging and (2) by controlled bidirectional charging (Kempton and Tomić 2005b). This allows reversing the previously unchanged principle of energy economics, which states that electricity supply usually adjusts to the more or less price-inelastic electricity demand. With the price-elastic charging of EV, however, the electricity demand becomes more variable and helps to adjust the electricity demand to the inflexible but volatile future electricity supply with a high share of renewable energy.

The structure of this chapter is as follows: Section 2 introduces the issue. Section 3 illustrates an answer to these challenges in terms of the demand side management (DSM) approach. This gives an incentive to charge EV whenever the energy system contains "abundant" electricity. In order to assess the corresponding profitability, the related costs are estimated, which contains battery depletion costs and other related costs of EV to offer auxiliary services to the grid (including vehicle to grid—V2G). These costs are estimated in Sect. 4, which ends with a competitiveness assessment of vehicle to grid to other storage technologies. Section 5 concludes.

2 Challenges

When considering the challenges in the energy system of tomorrow, the question remains whether electric mobility is boon and bane for these developments. Obviously, EV generate an additional energy demand within the low voltage grid, which is at the same time less time critical than the usual electricity demand of households (see below). The additional load might lead to shortages within the grid. Currently, two solutions are seen to prevent the lower voltage grid levels from this hazard:

- Grid extensions, which are however expensive and time consuming due to local acceptability, and
- Smart grids, in terms of demand side management (DSM) with local storage technologies (e.g. batteries).

---

[1] Currently, most domestic combined heat power plants are heat driven—their operation time depends on the heat demand of the household and not on the electricity demand.

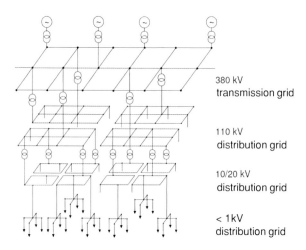

Fig. 2 Different grid levels in Germany (*Source*: Oswald 2009)

2.1 Impact on the Grid

In Germany four different grid levels can be distinguished (Mez 1997): The 380 kV grid level, the transmission grid, is responsible for the national balancing of electricity and serves as feeding point for most conventional power plants (see Fig. 2). The distribution grid on 110 kV and 10 or 20 kV level distributes electricity within a given region and delivers electricity to industrial consumers. The low voltage grid (in Germany usually 0.4 kV) gives access to private households. This transformation results due to the fact that the Ohmic resistance decreases quadratically with higher voltage by constant transport capacity.[2] Hence, for longer distances a transformation to high voltage is reasonable.

The charging of a single EV is generally located in the low voltage distribution network. The charging process of an EV at home with 11 kW is displayed in the Fig. 3. The exemplary, but empirical, load curve of this single household over the period of one day shows the load of an EV between 9 p.m. and 10:15 p.m. The charging of the EV nearly doubles the household electricity consumption of this day and triples the peak power. Considering, that there could be several EV charging uncontrolled in the neighborhood on that evening, there will be an overload at the local transformer due to the simultaneity of this high load peaks. A local power outage might be the result. Unfortunately, the prediction of this hazard is complex as the spatial probability for EV penetration rates differ strongly and German low voltage grids are rather heterogeneous (Stöckl et al. 2011). In some

---

[2] The corresponding formula is $P_v = 3 \cdot R \cdot \frac{P^2}{U^2}$.

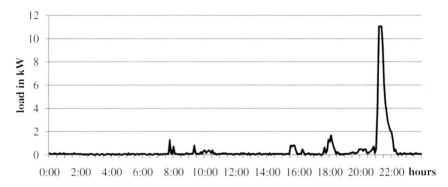

Fig. 3 Household load curve with an EV charging at 9 p.m.

grid areas, especially in cities, the average load per household is very low—sometimes even below 1.2 kW (Pollok et al. 2010). Nevertheless, most current low voltage grids seem to cope with the additional load of normal distributed charging in the evening hours and the forecasted penetration rates of EV. From the current perspective the high penetration rate of domestic photovoltaic systems (with per se high simultaneity) in Germany seems to be more challenging.

Looking at higher distribution network levels, the situation is similar. Kaschub et al. (2010) analysed the impact within a representative urban grid in Germany supplying 25,000 households on the 10 kV level. A penetration rate of 100 % EV without special charging infrastructure and instant charging at 3.5 kW would lead to a significant load increase, particularly in the evening hours (cf. Fig. 4).[3] With regard to the infrastructure, transformers seem to be a bottleneck for the German low voltage grid, especially for high charging rates (Pehnt et al. 2011). These effects do highly depend on the network design and its number of households (Stöckl et al. 2011).

In the transmission network, the impact of EV is rather low due to the relatively little additional electricity demand compared to overall electricity consumption—this is also true for the power plant portfolio (Heinrichs et al. 2011 and Pehnt et al. 2011). However, other political framework conditions, e.g. the increasing volatile electricity supply by renewable energies, in particular wind, have a much stronger impact on the transmission grid (e.g. DENA 2005, 2010).

The previously offered solution for bottlenecks in the electricity grid was to expand the current grid, which is however laborious and costly (DENA 2010; BDEW 2011). This results not only from the technical costs, but equally from the local resistance of the population. In the following, we therefore concentrate on the new demand side management approach—even though both approaches do not exclude each other.

---

[3] This is confirmed for many other countries by other studies e.g. Göransson et al. (2009), Davies and Kurani (2010), Waraich et al. (2009), Weiller (2011), Hartmann and Özdemir (2010), van Vliet et al. (2011), or Leitinger and Litzlbauer (2011).

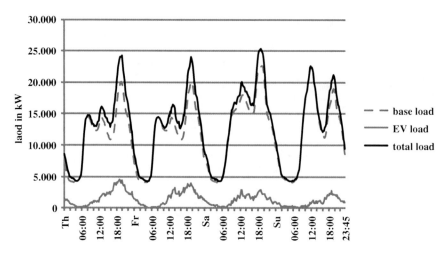

Fig. 4 Load curve of an urban district with uncontrolled EV charging (*Source*: Kaschub et al. 2010)

2.2 Smart Grid and Demand Side Management (DSM)

Smart grids, with direct communication between energy supplier and consumer, allow optimizing the coordination of electricity demand and supply in an efficient manner. This includes demand side management approaches influencing the electricity user to adapt its electricity demand over time: e.g. through dynamic tariffs (see Hillemacher et al. 2011). The potentials in average households are usually small due to the users' low price elasticity. However, EV increase these potentials substantially: the energy consumption in an average household will almost double and the parking time is by far long enough to allow a load shift without constraining the users. Hence, if users allow automatic delayed charging, the increase of peaks can be avoided ('peak shaving') and, additionally, the low electricity demand during the night can be increased, which leads to an increased efficiency of baseload power plants ('valley filling') (cp. Fig. 5). Even more, in stormy nights a significant share of superfluous electricity in the grid could be absorbed by EV.

The degree of freedom and the effectiveness of demand side management can be enhanced by implementing stationary storage systems in the distribution grid. This is an alternative to vehicle to grid, but could be more expensive, since the battery is only used for storing electricity.

Hence, with demand side management a reduction of the load is feasible, but does not fully replace the grid extension approach—especially in regions where load shift potentials and its acceptability are small.

Demand side management can be implemented by highly complex and intelligent domestic control boxes (smart grid technologies), or alternatively initially by the already implemented basic technology ripple control (a signal from the

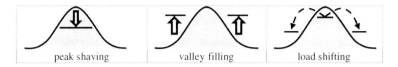

Fig. 5 Selected demand side management strategies

electricity provider through the low voltage grid) as e.g. used for night storage heating. The significant load shift potential by EV might help to overcome the challenges in the future energy system. However, from a socio-demographic view, this potential is highly uncertain. Users might not allow any external control or only use instant-charging for maximum flexibility. Therefore, the technical potential is significantly influenced by personal constraints, which are hardly predictable. Even though some first questionnaires about the participation of demand side management with EV already exist (e.g. Bunzeck et al. 2011; Paetz et al. 2011), experience in the real environment is still missing.

2.2.1 Load Shift Potentials

Defining the load shift potential (LSP) of EV is not as trivial as it seems. For one EV it is rather straight forward: The car arrives with a certain state of charge (SoC) of its battery and a certain future point in time, where it will depart again with a certain desired state of charge. In between the charging process can be allocated arbitrarily—respecting the user requirements. For several EV with different arriving and departing times as well as charging rates and state of charge, the issue is getting complex.

To allow a simple estimation, we assumed two extreme behaviors that are idealized in the following:

- recharging the battery as early and as full as possible, and
- recharging the battery as late and as little as possible.

The first charging strategy leads to an always fully charged battery and therefore highest flexibility, but at the same time no load shifting potential. The second charging strategy allows the maximum load shifting potential but no flexibility in using the car unexpectedly earlier or for a longer distance (cf. Fig. 6). Both strategies might be seldom and real-life behavior will be somewhere between these extremes.[4]

With these two extreme charging strategies it is possible to generate fictive charging curves during one week, based on the representative trip data of current passenger cars from the study Mobility in Germany 2008 (Infas and DLR 2008).

---

[4] Especially for the second strategy an instant charging to a certain state of charge (SoC) might be meaningful (e.g. to guarantee a trip to the next hospital).

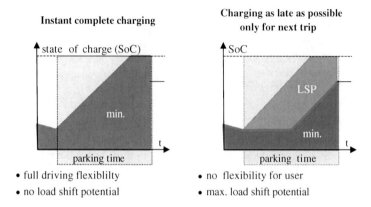

Fig. 6 Extreme charging strategies to estimate the load shifting potentials (LSP) of EV

From the study database we selected all single car trips in German city districts together with the arrival and the following departing time, as well as the required energy resulting from the last and following trip.[5] Furthermore, we assumed an average battery capacity depending on the car segment, a charging rate of 3.5 kW and that vehicles are plugged-in whenever they are parked at home. The resulting sum of all EV load shift potentials shows the maximum technical load shift potential based on our assumptions and is indicated with the blue area in Fig. 7. It shows a theoretical load shift potential of 40–90 % of the available battery capacity depending on the time and day of week. During the night, the potential is greatest and during midday lowest. The green line shows the theoretical minimum state of charge required to accomplish all trips. In reality this minimum line would be higher, and, hence, the real load shift potential smaller.

Other studies on load shifting potentials confirm our results (Göransson et al. 2009; Waraich et al. 2009; Davies and Kurani 2010). However, the real load shifting potential might strongly differ from these calculations and depends on the acceptance of the customer and the underlying smart charging principle, which is explained in the following.

3 Smart Charging

The technical possibilities for smart charging depend on the one hand on the charging infrastructure at the parking places, and on the other hand on the information and communication technologies (ICT) implemented. Without special charging infrastructure the communication and billing requirements have to be

---

[5] We assumed average electricity consumption depending on the car segment (small, medium, large).

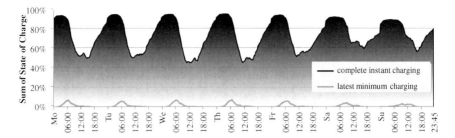

Fig. 7 Technical load shift potential of the EV fleet

inside the car. In case of charging stations with implemented smart charging capabilities, this equipment has to be located there. Furthermore, a communication between the charging station and the car is mandatory. These communication capabilities are still in standardization process (DIN IEC 62196).

Depending on the existing infrastructure three main charging scenarios can be defined:

- Charging only at the household plug at home, i.e. in Germany at 3.5 kW
- Charging only at home and at the workplace at e.g. 3.5, 11 or 22 kW
- Charging everywhere with up to 43 kW.

Moreover, the operator of smart charging is not defined yet. This may be the power transmission system operator (TSO), the power distribution system operator (DSO), or another stake holder, e.g. a third entity representing the objectives of the utilities. Depending on the objective for smart charging, each operator might have advantages and disadvantages.

Additionally, there are several possibilities to implement the smart charging process. There could be a binding control signal like the already mentioned ripple control for night storage heating, or a voluntary incentive scheme. Dynamic tariffs, which are currently of high interest in research, mainly focused on time variable tariffs, are promising representatives. An already existing dynamic tariff is the Time of Day (TOD) price (e.g. HT/NT-tariff in Germany) with two fixed levels, one for the daytime and one for the nighttime.

Simulations show that smart charging principles help to avoid increasing load peaks of uncontrolled charging and to raise the low electricity demand during night time (valley filling). Comparing Fig. 8 with Fig. 4 shows how the smart charging scenario in an urban district with low electricity prices during the night leads to significant load shifts into cheaper night time, and hence fulfills the principles of peak shaving and valley filling (cf. Kaschub et al. 2010). In order to prevent a charging peak at the price leap at 9 p.m., the cars are divided into early, later and latest charger.

Hence, through smart charging of EV the electricity grid is released and the supply side of the market gains stability. But is this profitable for the vehicle user?

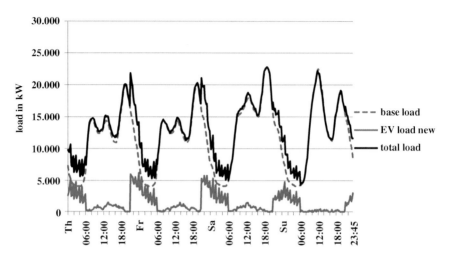

Fig. 8 Load curve of an urban district with controlled EV charging (*Source*: Kaschub et al. 2010)

For this the user might ask whether the spread of the dynamic tariff is sufficient to cover the following costs:

- the implementation costs of the information and communication technologies infrastructure (i.e. smart meter), and
- the additional degradation of the battery (which can be neglected as far as the maximum charging rate and numbers of charging cycles remains equal).

The following example should illustrate these considerations: In average about 3,000 kWh/a charging energy per car seems realistic.[6] Assuming that charging at 15 ct/kWh (night tariff) instead of 25 ct/kWh (day tariff) is always possible, which leads to savings of 300 €/a. The costs for charging infrastructure at home are between 150 and 350 € and communication hardware at max. 100 € (Kley 2011). Hence, while investing[7] less than 100 €/a, the car user can benefit from a considerable revenues of about 300 €/a. This confirms the results from Szczechowicz et al. (2011). Therefore, the economic question alone does not seem to be the obstacle in order to cut the high technical potential for load shifting through smart charging in praxis—its acceptance, however, is still vague (Paetz et al. 2011).

In using the battery of the car not only as energy storage for mobility, but also for the grid, is referred as "Vehicle to Grid" (V2G). It includes on the one hand the controlling of the charging time and charging rate as well as the battery discharging (and hence feeding back into the grid) to allow optimal and cost-efficient

---

[6] In Germany the average mileage of a car is 12,000 km and the average consumption of battery EV is 0.25 kWh/km, which results in a required energy of 3 MWh/a.

[7] In general the depreciation time for vehicle investment in Germany is 6 years (BMF 2010).

grid stability. On the other hand the mobility autonomy of the vehicle owner might be further reduced. Moreover, the additional battery use accelerates its degradation, which might lead to extra costs if an additional battery replacement during the vehicle's life time is required.

4 Storage Technologies in Comparison to Vehicle to Grid

In the past, energy storage in the power grid was mainly accomplished by pump storage power plants. Only few compressed air energy storage power plants (CAES) or stationary battery storage systems (BESS) are in operation (e.g. EAC 2011). These alternatives operate either as demonstration projects or in special system setups. The economic competitiveness of these storage systems is dependent on several parameters. Use cases reach from long term storage for up to several weeks, down to short term storage for only several minutes or seconds, which has a significant impact on the number of cycles and hence on the costs and potential benefits of these technologies. When calculating business cases for vehicle to grid it is inevitable to know the alternatives available: Where other technologies are more economically efficient, vehicle to grid will not be applied. Therefore, in the following a short introduction in alternative technologies is given.

4.1 Overview of Storage Technologies

The storage technologies can be classified in energy and power storage technologies or based on its physical storage technique (mechanical, electrical, electricchemical, etc.). This classification is done based on parameters that are listed in Table 1. The specifications often include a broad range and indicate that the listed storage type is representing a couple of subgroups.

Table 1 shows, that the Li-ion high energy (Li-ion HE) technology has no outstanding parameter compared to other technologies. This indicates that the combination of the parameters is decisive. For Li-ion batteries especially the combination of energy density, price projection and usability in large battery packs make them interesting for mobile devices and vehicles.

Pump storage power plants remain the most efficient large storage system due to their high cycle capability and therefore unrivalled low specific energy costs (3.5 ct/kWh, VDE 2008), combined with very low self-discharge for stationary applications. But the geographical requirements limit capacity extension. Nevertheless, several projects are in planning for the coming years (e.g. Deane et al. 2010). Only for flat landscapes other technologies could be advantageous in the near future when transmission grid capabilities are insufficient. In the following, we outline the vehicle to grid approach to allow a more sophisticated comparison between vehicle to grid and pump storage thereafter.

Table 1 Characteristics of different power and energy storage technologies

Type	Cycles	E-density in Wh/kg	P-density in W/kg	Invest in EUR/kWh	self-discharge
Pump storage	Very high	~1	–	600–3 k	Low
CAES	Very high	2 kWh/m^3	–	400–800	Low
Lead-Acid	<1 k	20–30	80–300	100–250	Low
NiMH	eq. Li-ion	50–80	200–1.5 k	750–1.5 k	High
Li-ion (HE)	2.5–7.5 k	120–180	80–300	450–1.2 k	Low
Li-ion (HP)	2–5 M	60–140	200–2 k	1–2 k	Low
Ultracapacitors	500 k–2 M	2.2	1.4 k	1–2 k	High
Flywheels	>5 M	3.7–11.1	180–1.8 k	4.5–5 k	High
NaS	1.5–4.5 k	90–110	100–120	250–3 k	Low
Redox-flow	5–15 k	20–65	–	300–1 k	Low

CAES Compressed Air Energy Storage; *H* High Power; *HE* High Energy
Source: IIP (2011)

4.2 Relevant Markets for Vehicle to Grid (V2G)

If the number of registered EV does further increase, the vehicles, while plugged-in, can be interpreted as a large virtual electricity storage system. This requires a bidirectional electricity link and an intelligent management system for all plugged-in EV. If one million EV are assumed, an average available battery capacity of about 10 kWh per EV leads to a storage capacity of 10 GWh—which is about the capacity of two large pump storage power plants. The corresponding capacity amounts to 3.5 GW (or 22 GW for 22 kW charging) respectively—more than three large power plants at nominal load. But, would this storage for electricity with EV batteries be competitive to other technologies? Is the acceptance of vehicle to grid assured? In order to give an answer to the techno-economic question of profitability, two issues are relevant:

1. the potential gain through vehicle to grid, which is highly dependent on the time-specific (without usage) and the usage-specific depletion cost of the battery, and
2. the electricity prices and storage requirements from the grid perspective.

Furthermore, the additional costs for the infrastructure (components for the bidirectional link) are to be defined.

1. The corresponding costs of the battery highly influence the profitability within these markets. Due to little experience with this fast developing technology a well-founded cost estimate is highly unlikely—especially for the given context (battery type, usage etc.). Nevertheless, first estimates are used in the chapter below to allow a preliminary forecast of the relevance of vehicle to grid in the future electricity grid.

2. Electricity storage technologies can participate mainly in two different markets[8]:

   a. Short-term electricity trading market (in particular day-ahead and intraday market).
   b. Control reserve markets for stabilizing the grid if electricity supply or demand deviates from its forecast at short notice.[9]

The corresponding requirements for the markets and therefore the resulting chances for each technology, differ strongly. The participation within the short-term electricity trading market (a) is rather a future opportunity for vehicle to grid: Only a huge share of volatile and non-controllable wind energy makes an integration of this technology into the day-ahead market reasonable. The great amounts of energy and long-time horizons require a huge number of highly reliable participating car owners, as well as a good knowledge of the trip scheduling: If by chance one day all participants make a long trip, the storage capacity would be zero—even though this is not a very likely scenario (see Kaschub et al. 2011). Nevertheless, the control reserve market (b) seems much more convenient, as its advantage is twofold: Less energy and power is required and even the provision of capacity is funded ("procurement fee"). For example, the current average retrieved capacity in the German secondary control reserve (SCR) market is about 450 MW (Regelleistung 2011)—less than half of a large power plant at nominal load.[10] Hence, due to the procurement fee and energy price, a certain amount of parked and plugged-in vehicles lead to a decrease in the average electricity costs for each participant.

Besides the SRM two other reserve markets are implemented: The UCTE-wide primary and the tertiary control reserve market (see Fig. 9) (UCTE 2009). In case of instable frequency within the grid, first (within seconds) the primary control reserve (provided directly within the power plant) is automatically activated to provide positive or negative power reserve. After about half a minute, other storage technologies with longer ramp-up times can undertake this operation within the secondary control reserve (e.g. pump storage power plants). Thereafter the tertiary control reserve has to take over for up to several hours, which is in general achieved by additional power plants. Due to the smaller electricity demand in the tertiary control reserve (BNetzA 2011: 108ff), this market is not considered at this stage, but could also be relevant for vehicle to grid in the future. In the following we concentrate on the secondary control reserve.

---

[8] Other relevant use cases could be focused on the own household or a micro-grid, where no market is influenced and the benefit is localized in the own household or community.

[9] In Germany, however, it is currently not possible to integrate the vehicles due to the organizational and technical requirements. e.g. the smallest bid is 5 MW (Regelleistung 2011).

[10] The average weekly energy provided is about 80 GWh (Regelleistung 2011) with a potential capacity offer of 3.2 GW (Regelleistung 2011).

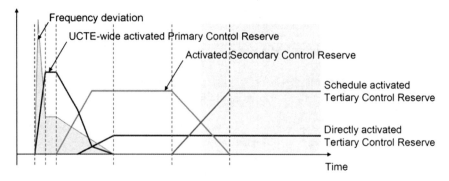

Fig. 9 Principle frequency deviation and subsequent activation of control reserves (*Source*: UCTE 2009)

4.3 A Techno-Economic Analysis of Vehicle to Grid Competitiveness

Today the usage of existing Li-ion batteries seems far from being competitive in the energy storage market. But the increasing need for electricity from the control reserve market might lead to rising electricity prices and together with the declining costs of Li-ion batteries (RB 2010) the situation might change significantly in the future.

As depicted above, the costs for electricity storage in automotive Li-ion batteries consist of time-specific (without usage) and the usage-specific depletion costs of the battery. The time-specific costs depend on the depletion due to aging of the battery and the self-discharge rate. The usage-specific costs depend on the depletion due to its usage and the corresponding conditions (temperature, depth of discharge (DoD), number of cycles, charging rate (C-rate), etc.). Figure 10 illustrates the bandwidth of specific prices of some current battery technologies depending on the applied depth of discharge. The Li-ion batteries seem unrivalled at these low DoD, but have a broad price-range from 0.01 to 0.14 €/kWh.

To allow a first estimation of the competitiveness of vehicle to grid, we illustrate a simple and optimistic calculation in the following, assuming an average vehicle battery of 20 kWh and investments of 8,000 € (Kalhammer et al. 2007). Time-specific costs depend mainly on the DoD. For the use-specific cost of the battery with 400 €/kWh specific investment costs[11] and 20 % DoD a price of 0.05 €/kWh can be derived from Fig. 10.[12] Other (ICT) hardware system costs amount to about 500 € (see Kley 2011), which equals about 85 €/a.

---

[11] Today prices for Li-ion batteries are more likely at 600 €/kWh (Jochem et al. 2011), although severe data is not given.

[12] The C-rate is neglected here, as charging rates at 3.5 kW are much lower compared to usual driving cycles and thus causes hardly any additional costs. Other additional costs are not measurable or not known yet.

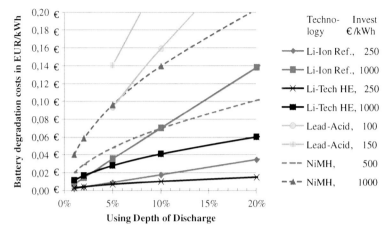

Fig. 10 Minimum battery degradation costs depending on the depth of discharge

After having identified the underlying costs, we now focus on the identification of possible revenues in the secondary control reserve market. The secondary control reserve market is characterized by an erratic demand. In Fig. 11 an example is given for an arbitrary week in the German secondary control reserve market. The demand of positive reserve capacity is changing from zero up to 3 GW (Regelleistung 2011). Most of the time the demand is at low capacity: 70 % below 100 MW, 90 % below 500 MW, and less than 5 % of the time above 1,000 MW. The amount of energy within the positive reserve amounts to 30 GWh in this week. Within regional markets the quotation is even more erratic. An example is the region within the EnBW grid, where almost no positive secondary reserve was requested in the week between July 18th and 24th, 2011.

Because negative secondary reserve energy can also be provided through controlled unidirectional charging, the following calculation focuses on the revenues with positive reserve energy provided through vehicle to grid. At first glance, providing 2 GW of peak power, about 600,000 EV with a 3.5 kW charging rate seem to be sufficient to supply the required power. On second glance however, the amount of vehicles is far too small with respect to the requested energy. If we still assume a provision of 20 % DoD for average 20 kWh traction batteries for vehicle to grid services in the considered time period, we would need about 5 million plugged-in vehicles (12 % of German passenger car fleet). In reality these numbers should be much higher in order to guarantee the grid services and allowing unforeseen trips at the same time. Assuming a supply of 2 GW with an

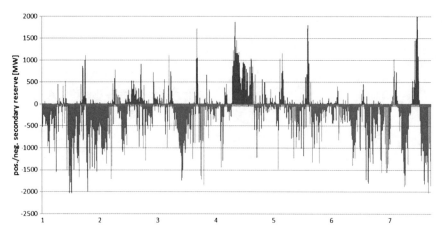

Fig. 11 Released secondary control reserve energy in Germany from October 24th to 30th 2011 (*Source*: Regelleistung 2011)

average price of 1,000 €/MW (per week) for procurement,[13] plus an energy price of about 130 €/MWh (Regelleistung 2011), the resulting revenue with a demand of 1.5 TWh reserve energy per year would amount to 5.75 million € per week, which equals (in an optimistic calculation) to 1.20 € per vehicle and week, or about 60 € per vehicle and year.

The corresponding costs can be calculated in assuming the 85 € p.a. depicted above for the required hardware and the battery depreciation costs of 50 €/MWh. Together with the required fleet of 5 million vehicles the costs per vehicle and year are:

$$85\,\text{\texteuro}+\frac{\left(1.5\,\text{TWh}\times 50\,\frac{\text{\texteuro}}{\text{MWh}}\right)}{5,000,000\,\text{EV}}=100\,\frac{\text{\texteuro}}{\text{a}}. \quad (1)$$

The corresponding revenues per vehicle and year amount to

$$\frac{1.5\,\text{TWh}\times 130\,\frac{\text{\texteuro}}{\text{MWh}}+1,000\,\frac{\text{\texteuro}}{\text{MW}}\times\frac{365}{7}\times 2\,\text{GW}}{5,000,000\,\text{EV}}=59.86\,\frac{\text{\texteuro}}{\text{a}} \quad (2)$$

Consequently, the resulting profit of this estimation is negative (40 € per vehicle and year).

Hence, vehicle to grid for positive reserve energy in the German secondary control reserve market seems not to be profitable today. We should, however, keep in mind that the underlying prices are highly volatile and might develop in the

---

[13] The procurement period is reduced from one month to one week per bid, since June 2011. Price fluctuation is still very high and prices between 800 and 1,500 are common; but deviations thereof in both directions numerous. This volatility holds equally for the energy price.

coming years in favor of vehicle to grid (e.g. decreasing battery costs and increasing market prices in the secondary control reserve market). Furthermore, if we assume a longer lifetime of the battery than the vehicle, the depreciation price of the battery is zero. This would lead to a smaller loss of about 25 € per vehicle and year. In conclusion we might say that vehicles with low mileage but "overdesigned" batteries may be dedicated to provide ancillary services in the future, when battery prices are lower and market prices higher.

Other studies confirm our results. e.g. Dallinger et al. (2011) concludes that under today's condition in Germany vehicle to grid is only profitable for primary control and negative control reserve in the secondary control reserve and tertiary control reserve markets, whereas it is not profitable for positive control. However, according to Kempton and Tomić (2005a), vehicle to grid with ancillary services has been already profitable in the US in 2005. This was confirmed by Tomić and Kempton (2007) for the EV "Think City", using different charging rates.

Furthermore, Hill et al. (2012) showed for an extended-range EV (EREV) fleet with a sensitivity analysis of the influencing factors on profitability of vehicle to grid in the US reserve market, that the significance of each variable differs and is strongly case specific. In their analysis they show—despite the uncertainties in technical and market development—two profitable market conditions.

Another possibility to evaluate the economy of market participation is the comparison to other storage technologies. For short term storage applications in the transmission grid (about one cycle per day) the competitor technology is pump storage. The specific costs for pump storage are between 0.03 and 0.04 €/kWh (VDE 2008). These costs can be compared to the battery usage costs through additional degradation (Fig. 10). When only 20 % DoD are assumed with investment of 600 €/kWh, the costs are between 0.05 and 0.08 €/kWh. The battery technology is therefore obviously not economic nowadays—but at the same time in a similar league with other storage technologies. Through intensive research and innovation in Lithium-ion technology (RB 2010), a cost development down to 250 €/kWh$_{invest}$ is realistic before 2030 (which is a target for the *Competence E* research project at KIT). This leads, together with the assumption of 20 % DoD, to costs of 0.02–0.04 €/kWh, which enables a profitable application. Increased cycle stability can reduce these costs further below 0.02 €/kWh. In this case the Li-ion technology is competitive to pump storage.

5 Conclusions

This chapter outlines the impact of EV on the power grid—especially the impacts on different grid levels are outlined, with lower grid levels seeming to be the most challenging. Especially the contribution to an increased load peak in the evening hours is seen as a drawback in the energy system. Two possible solutions are: (1) grid extensions or (2) demand side management approaches. Within the latter we indicated, that the load shift potential of private households multiplies, if the household owns an EV. The average German household would roughly double its energy demand and even more than triple its power demand (when using faster charging rates). The high load shift potential derives from the high power flexibility, comprehensive energy storage possibilities (batteries) and long parking times of more than 12 h at home or even 23 h per day on all parking sites. This, together with the fact that most charging processes will occur in the evening, where already today a peak in the lower grid levels is observed, turns a delayed charging of EV to an essential milestone for a successful EV market penetration.

In order to give a short contribution to the ongoing discussion on vehicle to grid, we estimated potential profits of vehicle to grid in the current German secondary control reserve market for positive control reserve. It has been shown that in the current market situation vehicle to grid is not competitive. But several assumptions affect the results of the calculation considerably such as battery costs, battery lifetime (which particularly depends on the number of cycles and depth of discharge), costs for vehicle to grid infrastructure, market prices, as well as market and regulatory conditions.

In the future market, the increasing non-controllable volatility in German electricity generation might (even with simultaneously improved weather forecasts) raise the demand in the reserve market, and hence lead to an increasing market price. Furthermore, from the current perspective it seems that the number of EV will increase and the battery costs will decline. This might lead to increasing vehicle to grid profits in the future.

Besides these economic constraints however, legal aspects, user's acceptance, the reliability (of thousands of EV) as well as the future technical development, make profits of vehicle to grid still uncertain in the future electricity system. Yet, a controlled charging is necessary to allow high penetration rates of EV without jeopardizing the low voltage grid and to foster higher shares of renewable electricity generation. Hence, controlled charging converts electric mobility from a gatecrasher to a welcome guest.

References

Barenschee ER (2010) Energiespeicherung und Lithium-Ionentechnologie. In: Proceedings of senior expert chemists annual meeting, Bitterfeld
BDEW (German Association of Energy and Water Industries) (2011) Abschätzung des Ausbaubedarfs in deutschen Verteilungsnetzen aufgrund von Photovoltaik- und Windeinspeisungen bis 2020. Berlin
BMF (German Federal Ministry of Finance) (2010) AfA-Tabelle für die allgemein verwendbaren Anlagegüter. Bonn
BMU (Federal Ministry for the Environment, Nature Conservation and Nuclear Safety) (2010) Langfristszenarien und Strategien für den Ausbau der erneuerbaren Energien in Deutschland bei Berücksichtigung der Entwicklung in Europa und global—"Leitstudie 2010". Berlin
BNetzA (German Federal Network Agency) (2011) Monitoringbericht 2011. Bonn

Bunzeck I, Feenstra CEJ, Paukovic M (2011) Preferences of potential users of electric cars related to charging—a survey in eight EU countries. Deliverable D3.2 of the Grid for Vehicles Project within the FP7 program of the European Commission, Brussels

Dallinger D, Krampe D, Wietschel M (2011) Vehicle-to-grid regulation reserves based on a dynamic simulation of mobility behavior. IEEE Trans Smart Grid 2(2):302–313

Davies J, Kurani KS (2010) Households' plug-in hybrid electric vehicle recharging behavior: observed variation in households' use of a 5 kWh blended PHEV-conversion. In: Working paper UCD-ITS-WP-10-04, Institute of Transportation Studies, University of California, Davis

Deane JP, Gallachóir BPÓ, McKeogh EJ (2010) Techno-economic review of existing and new pumped hydro energy storage plant. Renew Sustain Energy Rev 14:1293–1302

DENA (German Energy Agency) (2005) Energiewirtschaftliche Planung für die Netzintegration von Windenergie in Deutschland an Land und Offshore bis zum Jahr 2020 (DENA Netzstudie 1). Berlin

DENA (German Energy Agency) (2010) Integration of renewable energy sources in the German power supply system from 2015–2020 with an Outlook to 2025—dena Grid Study II. Berlin

EAC (Electricity Advisory Committee) (2011) Energy storage activities in the United States electricity grid. Washington, DC

EC (European Comission) (2011) Proposal for the directive of the European Parliament and of the council on energy efficiency and repealing directives 2004/8/EC and 2006/32/EC, SEC(2011) 779 final, 22/06/2011. Brussels

Erdmenger Ch, Lehmann H, Müschen K, Tambke J, Mayr S, Kuhnhenn K (2009) A climate protection strategy for Germany—40 % reduction of CO_2 emissions by 2020 compared to 1990. Energy Policy 37:158–165

EWI (Institute of Energy Economics at the University of Cologne), GWS (Institute of Economic Structures Research) & prognos (2010) Energieszenarien für ein Energiekonzept der Bundesregierung. Basel, Cologne, Osnabrück

Göransson L, Karlsson S, Johnsson F (2009) Plug-in hybrid electric vehicles as a mean to reduce CO_2 emissions from electricity production. In: Proceedings of EVS24, Stavanger

Hartmann N, Özdemir ED (2010) Impact of different utilization scenarios of electric vehicles on the German grid in 2030. J Power Sources 196(4):2311–2318

Heinrichs H, Eßer-Frey A, Jochem P, Fichtner W (2011) Zur Analyse der langfristigen Entwicklung des deutschen Kraftwerkparks—Zwischen europäischem Energieverbund und dezentraler Erzeugung. In: VDI-GET (ed) Optimierung in der Energiewirtschaft, VDI Publisher, Düsseldorf

Hill DM, Agarwal AS, Ayello F (2012) Fleet operator risks for using fleets for V2G regulation. Energy Policy 41:221–231

Hillemacher L, Eßer-Frey A, Fichtner W (2011) Preis- und Effizienzsignale im MeRegio Smart Grid Feldtest—Simulationen und erste Ergebnisse. In: Proceedings of IEWT (Internationale Energiewirtschaftstagung), Wien

IIP (Institute for Industrial Production) (2011) Internal database of electricity storage technologies

Infas (Institute for Applied Social Sciences), & DLR (German Aerospace Centre) (2008) MID—Mobilität in Deutschland 2008. Berlin

Jochem P, Feige J, Kaschub T, Fichtner W (2011) Increasing demand for battery applications. In: Proceedings of 6th international renewables energy storage conference and exhibition, Berlin

Kalhammer FR, Kopf BM, Swan DH, Roan VP, Walsh MP (2007) Status and prospects for zero emissions vehicle technology—report of the ARB independent expert panel 2007. California Environmental Protection Agency—Air Resources Board, California

Kaschub T, Mültin M, Fichtner W, Schmeck H, Kessler A (2010) Smart charging of electric vehicles in the context of an urban district. In: VDE congress E-mobility, Leipzig

Kaschub T, Jochem P, Fichtner W (2011) Integration von Elektrofahrzeugen und Erneuerbaren Energien ins Elektrizitätsnetz - eine modellbasierte regionale Systemanalyse. In: 7. Internationale Energiewirtschaftstagung, TU Wien

Kempton W, Tomić J (2005a) Vehicle-to-grid power fundamentals: calculating capacity and net revenue. J Power Sources 144(1):268–279

Kempton W, Tomić J (2005b) Vehicle-to-grid power implementation: from stabilizing the grid to supporting large-scale renewable energy. J Power Sources 144(1):280–294

Kley F (2011) Ladeinfrastrukturen für Elektrofahrzeuge. Frauenhofer Publisher, Karlsruher Institut für Technologie, Karlsruhe

Leitinger C, Litzlbauer M (2011) Netzintegration und Ladestrategien der Elektromobilität. Elektrotechnik und Informationstechnik 128(1–2):10–15

Mez L (1997) The German electricity reform attempts: reforming co-optive networks. In: Midttun A (ed) European electricity systems in transition. Elsevier, Amsterdam, pp 231–252

Nagl S, Fürsch M, Paulus M, Richter J, Trüby J, Lindenberger D (2011) Energy policy scenarios to reach challenging climate protection targets in the German electricity sector until 2050. Utilities Policy 19:185–192

Oswald BR (2009) Optionen im Stromnetz für Hoch- und Höchstspannung: Freileitung/Erdkabel, Drehstrom/Gleichstrom. In: Proceedings of Netz-Event 14. Mai 2009: Freileitung/Erdkabel

Paetz A-G, Dütschke E, Schäfer A (2011) Die Last mit der Lastkontrolle. Energie & Management, 12/2011, 19

Pehnt M, Helms H, Lambrecht U, Dallinger D, Wietschel M, Heinrichs H, Kohrs R, Link J, Trommer S, Pollok T, Behrens P (2011) Elektroautos in einer von erneuerbaren Energien geprägten Energiewirtschaft. Zeitschrift für Energiewirtschaft 35(3):221–234

Pesaran A (2007) Battery choices and potential requirements for plug-in hybrids. In: National renewable energy laboratory (NREL), plug-in hybrid electric truck workshop, hybrid truck users forum, Los Angeles

Pollok T, Szszechowicz E, Matrose C, Schnettler A, Stöckl G, Kerber G, Lödl M, Witzmann R, Behrens P (2010) Electric mobility fleet test—grid management strategies with electric vehicle fleets. In: VDE-Kongress 2010, Leipzig

Pollok T, Dederichs T, Smolka T, Theisen T, Schowe von der Brelie B, Schnettler A (2009) Technical assessment of dispersed electric vehicles in medium voltage distribution networks. CIRED, 0887. Prague

RB (Roland Berger) (2010) Electro-mobility—challenges and opportunities for Europe. European Economic and Social Committee, Brussels

Regelleistung (2011) Own evaluation of data from www.regelleistung.net in the time period 12.07.11 until 11.12.11

Schäfer T (2009) Batterietechnologie: Trends, Entwicklungen, Anwendungen. In: Proceedings of 3rd Expert Forum Leipzig. Leipzig

Szczechowicz E, Pollok T, Schnettler A (2011) Economic assessment of electric vehicle fleets providing ancillary services. CIRED, 0967. Frankfurt

Stöckl G, Witzmann R, Eckstein J (2011) Analyzing the capacity of low voltage grids for electric vehicles. In: IEEE international electrical power and energy conference, Winnipeg

Tomić J, Kempton W (2007) Using fleets of electric-drive vehicles for grid support. J Power Sources 168:459–468

UCTE (Union for the Co-ordination of Transmission of Electricity) (2009) Operation handbook: policy 1: load-frequency control. V3.0 rev15 01.04.2009, Brussels

VDE (German Association for Electrical, Electronic and Information Technologies) (2008) Energiespeicher in Stromversorgungssystemen mit hohem Anteil erneuerbarer Energieträger—Bedeutung, Stand der Technik, Handlungsbedarf. Power Engineering Society (ETG). Berlin

van Vliet O, Brouwer AS, Kuramochi T, van den Broek M, Faaij A (2011) Energy use, cost and CO_2 emissions of electric cars. J Power Sources 196(4):2298–2310

Waraich RA, Galus MD, Dobler C, Balmer M, Andersson G, Axhausen KW (2009) Plug-in hybrid electric vehicles and smart grid: investigations based on a micro-simulation. In: Proceedings of 12th international conference on travel behaviour research, Jaipur

Weiller C (2011) Plug-in hybrid electric vehicle impacts on hourly electricity demand in the United States. Energy Policy 39:3766–3778

Author Biographies

Patrick Jochem is research group leader at the KIT-IIP, chair of energy economics. In 2009 he received his PhD in transport economics from Karlsruhe University. He studied economics at the universities in Bayreuth, Mannheim and Heidelberg, Germany. His research interests are in the fields of electric mobility, and ecological economics.

Thomas Kaschub is Research Associate at the KIT-IIP in the working group transport and energy since 2009. He focuses the topics electric mobility, BESS and energy system modelling. Before he studied Mechanical Engineering (Dipl.-Ing) at the Universität Karlsruhe (TH).

Wolf Fichtner is Director of the Institute for Industrial Production and the French-German Institute for Environmental Research. He is full professor and holder of the Chair of Energy Economics at KIT. His main areas of research are Energy System Modelling and the Techno-economic Analysis of Energy Technologies.

An Approach Towards Service Infrastructure Optimization for Electromobility

Tim Hoerstebrock and Axel Hahn

Abstract Electric mobiles are one of many possible answers regarding the increasing costs of fossil fuels and carbon emission reduction targets. The main contraint is that the new technology must cover the users' mobility needs which are still rising. Due to the technical restrictions of electric vehicles, a switch to electromobility is bound to the need for a sufficient charging infrastructure taking into account current and future mobility behavior as well as the characteristics of electric vehicles. This paper explains a methodology to design and assess the charging infrastructure layout without having a significant amount of electric vehicles available. This approach includes the use of the simulation tool TrIAS (Transportation infrastructure assessment by simulation). The basis is a planning model that is able to represent concepts and parameters like users' mobility patterns, regional street layout, available parking infrastructure as well as the e-car and its technical characteristics like range or battery capacity itself. This model is the input for a simulation-based analysis. The approach is applied in the region Bremen/Oldenburg to analyze the requirements towards a sufficient charging infrastructure for electric vehicles.

1 Introduction

Electromobility holds high potentials when it comes to sustainable transportation. But it has not been fully developed yet, as one can see in the short range of the vehicles and the long charging times of the batteries. Due to the mobility pattern of

T. Hoerstebrock (✉) · A. Hahn
R&D Division Transportation, OFFIS - Institute for Information Technology,
Escherweg 2, 26121 Oldenburg, Germany
e-mail: tim.hoerstebrock@offis.de

A. Hahn
e-mail: hahn@wi-ol.de

urban regions, which are characterized by short trips and long standing times (Infas and DLR 2010), these regions are suitable to successfully deploy electric vehicles in the first run. Current research deals mostly with the individual transport, investigating new ways to help the private user to overcome the current disadvantages of the electric vehicle (Arnold et al. 2010; Hoberg et al. 2010; Spath et al. 2010). One solution is an introduction of a wide-spread charging infrastructure. The scientific challenges in this context are manifold. First, the infrastructure layout has to fulfill the (often opposing) requirements of many stakeholders (e.g. electric vehicle users, energy provider, authorities, and car producers) who are involved in its installation and usage. Second, many combinations of possible positions and technical equipment of the individual charging stations must be handled and assessed including the analysis of time-sensitive bottlenecks over a given period. Therefore, the multi-agent simulation tool TrIAS was developed to cope with the complexity of charging infrastructure placement and to address the mentioned challenges. This chapter explains the methodology and the design of the integrated simulation tool TrIAS.

The remainder of this paper is structured as follows: Sect. 2 summarizes the related work on current evaluation and planning methods in logistic location planning and traffic simulation. Section 3 describes the utilized methodology and explains the input of the simulation tool (so called planning model). At the end of the chapter, an evaluation example from the model region Bremen/Oldenburg is given to show two possible assessment dimensions to evaluate a given infrastructure layout. The last section concludes the paper and gives an outlook on future work.

2 Related Work

Planning and designing layouts of points of service (POS) to satisfy customers'demands is a common logistic task. The problem of resource positioning has been widely studied and is tackled by many scientists. The Facility Location Problem (FLP) has many classifications concerning analytic, continuous, network and discrete models. An overview and classification is provided by (Zarinbal 2009). Solving the facility location problem is usually NP-hard due to their high amount of valid combinations and the high amount of constraints that must be taken into account. To be able to handle these problems, many assumptions are created to reduce the complexity. For example the Single Facility Location Problem deals with only one incapacitated facility to be installed in a given region (Moradi 2009). Azarmand and Daneshzand extend this problem by a multi-facility view (Azarmand 2009; Daneshzand 2009). What all models and approaches have in common are the sets of target functions. Many models aim to minimize the distances and the allocated transportation costs, resulting from satisfying the customers' needs by the installed facilities. Common constraint when solving these problems is to fulfill the needs of all customers.

Trying to transfer the problem described in Sect. 1 on the FLP, one can interpret the charging infrastructure as a facility and the electric vehicles as customers with a certain demand of electricity. The target function would include the minimum distance between supply and demand. This approach has different issues however. First, the points of demand and the points of service have to be modeled on a high detail level considering every possible parking and charging spot within the region. Besides the missing empirical data, the complexity would rise immensely. Second, the models consider only static (and sometimes stochastic) demand ignoring the distribution over time (e.g. demand per hour).

Next to the logistic approaches, several simulation techniques and frameworks in transport planning and management exist. These mainly address problems in traffic management (Choy et al. 2003; Hernándes et al. 2002; Ossowski and García-Serrano 1999; Vasirani 2009), signal control and optimization (Junges and Bazzan 2008; Kosonen 2003; Rochner et al. 2006; Wiering 2000), intersection management (France and Ghorbani 2003; Klügl 2001; Nagel and Schreckenberg 1992; Nunes and Oliveira 2004; de Oliveira et al. 2004), or mobility decision and distribution (Balmer et al. 2006; Gringmuth et al. 2000; Hunecke et al. 2004). In these works infrastructural changes in the network are mainly considered to be either impossible or bound to severe economic, ecologic or social constraints (France and Ghorbani 2003). Thus, classic approaches to traffic system planning (Steierwald 2005) cannot be applied and researches mainly maximize the throughput of the traffic system by intelligent systems (e.g. cooperative and adaptive signal management systems) either on a macro (society) or micro (individual) level. In this context multi-agent techniques play a major role to assess the whole throughput of the focused system and to maximize the utilization of the transportation infrastructure's capacity.

As one can see, the agent-based simulation approaches in traffic system assessment are highly available but mainly focus on the utilization optimization of the existing transportation network and its subsystems. With the penetration of new technologies (electromobility) into the established traffic systems, the question of designing and optimizing the fuel refilling infrastructure comes into the focus. Until now, an optimization was not needed (at least in Germany) since fuel shortages were neglected and the range of vehicles assumed to be unlimited. Thus, a system or methodology to assess the service infrastructure of transportation systems in general (e.g. gas stations) and the service infrastructure of electromobility in particular (e.g. charging stations) has not been designed yet. In summary, an analysis of the performance of service infrastructure over a certain period is needed (e.g. the identifying bottlenecks during the daily usage of the infrastructure). FLP can help us optimizing the positions and the equipment of the charging stations by applying appropriate (heuristic) optimization algorithms (Salhi and Gamal 2003).

But the described approaches are not flexible enough to easily integrate new aspects like future mobility patterns or technical change of the vehicles. Also, the time-sensitive analysis cannot be performed (bottleneck analysis). Therefore, a simulation based approach was chosen, taking the above methods and tools into account.

3 Methodology

In the following, the simulation based methodology to virtually assess infrastructures for electromobility will be explained. The goal of the examination is to find an optimal infrastructure layout so that the electromobility users have the least restrictions in comparison of using a conventional vehicle with a combustion engine. The additional waiting time—while charging the vehicle—and the risk of breakdowns are key measures to quantify the quality of the infrastructure layout.

Preliminary, an overview of the overall procedure to assess a scenario is given. The above mentioned key figures are dependent on many parameters regarding regional aspects, mobility movement, vehicle specifications and considered infrastructure layout. Therefore, a planning model is designed which contains these elements and their dependencies. This model is described in Sect. 3.1. To evaluate a given scenario the planning model is instantiated with a concrete set of data. The scenario created serves as input for TrIAS described in Sect. 3.2. The user extends the scenario by certain restrictions and constraints that are relevant. This can include for example assumptions about the possibility of charging the vehicle at home. TrIAS transfers the planning model into internal simulation objects and executes the simulation run(s). While executing, the simulation tool collects, aggregates, and exports relevant key figures. The user then processes this data to address his questions and problems. The outputs from the simulation include several key figures that can be evaluated by the user. Section 3.3 gives an overview of the simulation output including the calculated key figures and gives an example about the evaluation of a modeled scenario. Figure 1 shows the chapter overview.

3.1 Planning Model

The model is divided into four sections: region model, mobility pattern model, vehicle model, infrastructure model. Each section covers an area which has a significant influence on the performance of the whole infrastructure system.

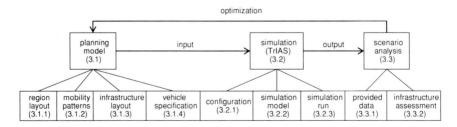

Fig. 1 Chapter overview

Fig. 2 Region model

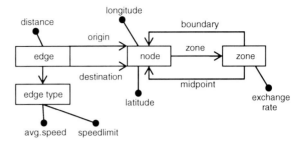

3.1.1 Region Modeling

First of all, the designated region must be defined. This is done by a graph based approach. Since a microscopic study including the detailed modeling of junctions, lanes and traffic lights is out of scope, a network as a representation of the street layout is sufficient for handling the vehicles' movement and fuel consumption. Thus, the region is designed by nodes (positions), edges (e.g. streets) and zones (e.g. districts), whereas nodes are connected by edges and certain nodes define the midpoint of a zone. The network serves as a base for the deployment of infrastructure and for the navigation and movement of the vehicles.

The procedure of separating the region into several districts with a representative node was chosen because of the available data on people's mobility habits. This data only considers the movement between districts of the region.[1] In order to represent the current market penetration of electric vehicles, an exchange rate is allocated to each zone. The edge connects two nodes and has a distance and an edge type. Edge types (e.g. living streets or highways) classify edges in terms of average speed and speed limits. Figure 2 shows the basic model which characterizes the internal representation of the network and its elements.

When it comes to instantiate the model the user has to define which zones he wants to investigate and—concerning the modeling of the street network—which level of detail he needs. For example, the consideration of only average street representations may be more efficient and target orientated than modeling each individual street.

3.1.2 Mobility Pattern Modeling

After modeling the region the mobility patterns within the region need to be addressed. As mentioned above the deployment and movement of vehicles are covered by this model and serves as a key input for the simulation. The data structure is displayed in Fig. 3.

---

[1] The data was provided by the project partners Centre for Regional and Innovation Economics (CRIE) and Senator of Environment, Building and Transportation (SUBVE).

Fig. 3 Mobility patterns

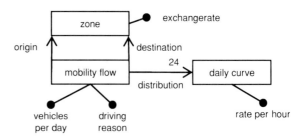

As described in 3.1.1, zones include the rate of market penetration and serve as origins and destination for vehicle traffic. Therefore, the quantitative amount of vehicles driving from one zone to another is modeled. The object mobility-flow holds those parameters. To be able to decide where to position infrastructure within a zone, each mobility-flow is also assigned a driving reason specifying the user's activity, e.g. work, shopping, or other spare time activities. This allows the user to distinguish between the availability of charging infrastructure for each driving reason and determine the zone positions of each individual charging station (private, semi-public, and public grounds). To give a realistic picture of the mobility patterns, the total amount of vehicles traveling through the network is distributed over the day. The element daily-curve represents this distribution on an hour basis.

3.1.3 Vehicle Modeling

In order to calculate consumption and emission behavior of the vehicle, its attributes including the propulsion and fuel system are modeled. This allows the user to assess the infrastructure requirements of electromobility and to determine the influence of infrastructural changes.

In the focus of Fig. 4 lies the vehicle type. This object and its concretizations (Car—FEV/Combustion) represent the several car models on the market. It has a reference to one or more propulsion types which allows the consideration of hybrid vehicles as well. The combination of the consumption behavior of the propulsion type and the used fuel determine the exhaust emissions caused by the chosen propulsion system. Each instance of the class exhaust emissions can define a unique exhaust pattern with regard the concretizations of the object fuel/tank (e.g. electricity, diesel).

3.1.4 Infrastructure Modeling

Finally, the planning model contains a description of the infrastructure located in the investigated region (see Fig. 5). At the moment the infrastructure consists of types: parking space and charging infrastructure. This differentiation is done in

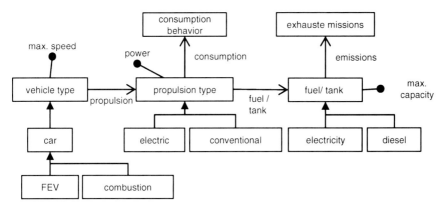

Fig. 4 Vehicle model

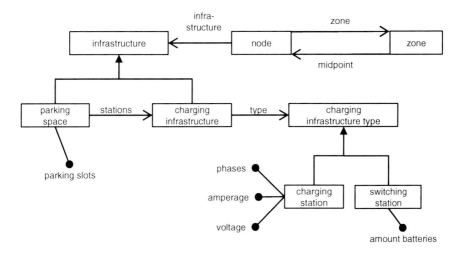

Fig. 5 Infrastructure model

order to display the different possibilities to place charging infrastructure. Since the placement of charging spots will be primarily realized on private and semi-public grounds the modeling of parking space is required (VDE 2010). Infrastructure is associated to a node and therefore indirectly part of a zone. Charging infrastructure is installed at parking spaces to satisfy the needs of electromobility users. Parking space includes parking lots, park&ride and parking garages.

The performance and behavior of the single charging infrastructure is determined by the charging infrastructure type. The types define how vehicles will be charged (e.g. by charging the battery or by switching it). The concretizations of charging infrastructure types hold several parameters according to its way of charging (e.g. the power (AC or DC) or the amount of batteries provided).

3.2 Simulation (TrIAS)

The assessment and optimization of a concrete region is a two-step process with several tasks. The first step implies the configuration of the planning model described in Sect. 3.1 and the simulation parameters. This configuration is explained in Sect. 3.2.1. The second step involves the execution of simulation runs and the analysis of their results. In order to do so, dynamic objects are required that obtain the static parameters of the planning model. An internal simulation model is the base for this transformation (see Sect. 3.2.2). The actual process of the simulation run is explained in Sect. 3.2.3.

3.2.1 Configuration

The configuration considers two parts: the configuration of the planning model (scenario) and the configuration of the simulation parameters:

- Scenario configuration: This task includes the collection of appropriate data and the instantiation of the planning model. The modeled region is mainly taken from the project OpenStreetMap which has the goal to provide a free world map (OpenStreetMap 2011). The street layout can be extracted from the XML-based export file (OSM). Since the exported street layout does not contain information about the zone definitions, this data needs to be added manually. Additionally, the file is extended by the definition of the zones. Afterwards, the planning model can convert this data into its internal objects. The other objects like the mobility patterns, the charging infrastructure types, and the placement of the individual charging stations is provided by easily editable CSV files. Each set of files defines one scenario.
- Simulation configuration: Next to its definition, the scenario is selected by the user within the simulation tool. Before loading the scenario the user selects certain options for the simulation execution. This includes the amount of simulation runs and the restrictions and constraints that are relevant to the defined scenario like assumptions about the possibility of charging the car at home.

3.2.2 Simulation Model

After the configuration of the simulation, the simulation tool instantiate an internal simulation model which holds the dynamic objects. This adoption is required to internally handle the behavior of the different elements and the interactions between them. In the following, the internal simulation model will be shown, whereas the behavior and the interactions will be discussed in 3.2.3.

Figure 6 shows the whole simulation model. As one can see, several static types from the planning model are matched with one dynamic object in the simulation

An Approach Towards Service Infrastructure Optimization for Electromobility

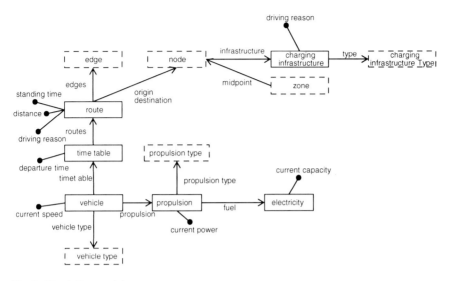

Fig. 6 Simulation model

model which holds the current values during the simulation run (see vehicle/ vehicle type, propulsion/propulsion type, charging infrastructure/charging infrastructure type). The actual values depend on the definition of the types and their concretizations, if any. The nodes and edges are directly adopted from the planning model.

Besides those direct adoptions, there is one conversion from the planning model into the simulation model concerning the mobility patterns. At the start of the simulation each mobility flow instantiate an amount of vehicles equal to the attribute 'vehicle per day' multiplied with the 'exchange rate' of the origin zone.

Moreover, each mobility flow instantiates timetables which are allocated to the vehicles. These timetables include round trip routes from the origin zone to the destination zone and a driving reason. At last each time table gets a departure time depending on the daily curve defined in the mobility patterns.

3.2.3 Simulation Run

This subsection describes how the elements behave during the simulation and which elements interact with each other. Especially the vehicle is in the center of this description, because of its high amount of behavioral control.

A scheduler within the simulation tool calculates the time of the vehicles' actions and steps them when that time has come. The individual behavior of each vehicle is determined by an internal state. Depending on this state the vehicle calls one or more actions and changes its state depending on the outcome of the fulfilled actions. The different states and their possible successors are explained in following list and displayed in Fig. 7.

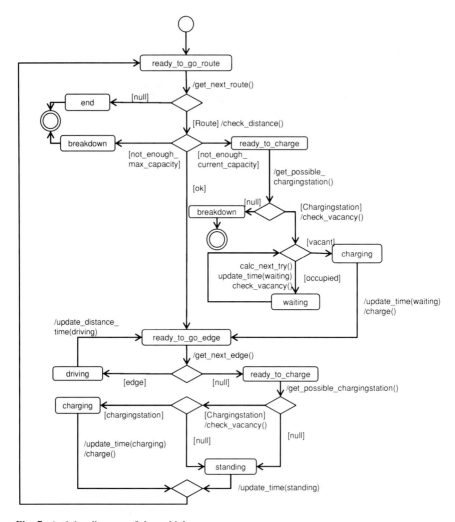

Fig. 7 Activity diagram of the vehicle

As one can see, the vehicle interacts with all different objects within the simulation model. First it interacts with its components by choosing the power of the propulsion. It interacts with the edges several times to either calculate the route distance or by driving on it. Furthermore the vehicle communicates with the destination node to find vacant charging stations and to connect to those.

An Approach Towards Service Infrastructure Optimization for Electromobility 275

3.3 Scenario Analysis

This section describes the analysis of a scenario evaluated within the model region of Bremen/Oldenburg. First, an overview of the data calculated by TrIAS is given which is followed by an analysis of the concrete scenario.

3.3.1 Provided Data

The simulation tool currently provides two sets of data. First, it generates an overview of general statistics where key figure are mapped and aggregated during the simulation run. This data is separated by number of charging places at the destination zones and the replacement rate of the conventional vehicles by electric vehicles. Secondly, the individual key figures tracked by each vehicle during its run through the system as described in 3.2.3 are collected. The data provided is displayed in Table 1.

With this data a first assessment of the evaluated infrastructure can be done. This contains evaluations concerning breakdown analysis of the vehicles with consideration of the given infrastructure or analysis of the effective charging time with regard to the amount of charging stations and the specified mobility patterns. The following subsection will give an example of an evaluation.

3.3.2 Infrastructure Assessment

This subsection describes an example of a scenario evaluation. This scenario was defined for the region Bremen/Oldenburg as part of the "Personal Mobility Center" project, supported by the funding of the German Federal Ministry for Transport, Building and Urban Development (BMVBS).

The scenario covers the private domain for installing the charging infrastructure. The focus lays on the possibilities to charge at home and at work. Therefore Bremen's district 'Sebaldsbrück' was selected as the investigated destination zone due to its industrial and commercial character. The mobility patterns were

Table 1 Output data

Simulation-wide	Vehicle specific
Home fuel availability (yes/no)	Standing time (separated by each driving reason)
Number of vehicles	Driving time (separated by each driving reason)
Number of vehicles fueled	Charging time (separated by each driving reason)
Fueled percentage	Waiting time (separated by each driving reason)
Number of breakdowns	Driven distance (separated by each driving reason)
Breakdown percentage	End state
Aggregated fuel time	Starting time
Effective fuel time per day	Vehicle type

characterized by vehicles from all over the model region to 'Sebaldsbrück' with 'work' as driving reason. Also a daily curve was provided by the project partner CRIE. Following questions were targeted (not exhaustively):

- Which amount of charging stations are needed to cover the users' needs with consideration of different exchange rates?
- Which influence do different ranges of the vehicle have on the first question?

After configuring the planning model, several runs were made. Before each run the exchange rate of the conventional vehicles (1–5 %) (Oliver Wyman 2009) and the amount of charging stations at the destination zone was changed. For a first run the infrastructure type was set to a normal German power connection with 230 V/16 A. In this scenario, the users living in Bremen had a possibility of 43 % to have a private parking place with power access at home. People living outside of Bremen had a chance of 71 % (Wietschel et al. 2009).

The first analysis deals with the amount of charging stations. Figure 9 shows the effective charging time dependent on the exchange rate and the amount of charging stations.

In this scenario 80 charging stations will be sufficient to cope with all vehicles with an exchange rate of 1 % (see Fig. 8). The effective charging time per charging station is quite low though (1.6 h per station and day). Even changing the infrastructure type to a fast charging station (400 V/32 A) does not have an effect on these figures. This can be explained by the missing release of the vehicles during their activities. To reduce the amount of needed charging stations or to raise the effective charging time, strategies to unblock the charging stations when the vehicle is sufficiently charged must be implemented.

The second analysis deals with the influence of different ranges of the vehicle (see Fig. 9). The comparison is done with two estimated ranges—180[2] and 100 km. Again, the possibility of breakdowns is analyzed with regard to different amounts of charging stations and the exchange rate of the vehicles. As one can see in Fig. 9, the chance of a breakdown rises especially with a small amount of charging stations. Up to 19 % more breakdowns are expected. Even with an amount of 250 charging stations the danger of a breakdown is 4 % higher.

This example gives only a small insight into the possible evaluations the whole planning and simulation tool can offer. With this evaluation the user of the system can easily compare different sets of input parameters in an easy and convenient way.

---

[2] The data of the reference vehicle (Think City) was provided by the project partner Deutsches Forschungszentrum für Künstliche Intelligenz GmbH (DFKI).

Fig. 8 Effective charging time considering different exchange rates (1–5 %)

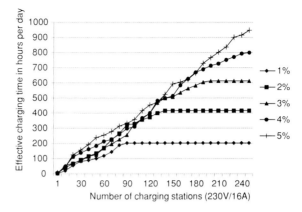

Fig. 9 Breakdown possibility considering different exchange rates (1–5 %)

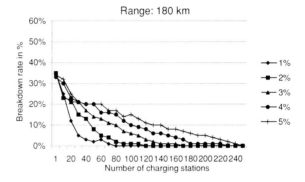

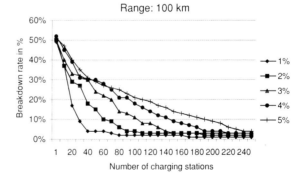

4 Conclusions

In this paper, a methodology to enable analyses of the infrastructural design to support traffic for electric vehicles within a given region is presented. A planning and simulation model is presented including an overview of the many possibilities to configure and calibrating the model which enables the user to create different scenarios for infrastructure layout assessment and optimization. Multiple simulation runs can be aggregated to evaluate different sets of input parameter. An example of a scenario evaluation was described to show a set of possible questions concerning the infrastructure layout and the corresponding analysis dimensions.

During the explanation it was shown that many factors have an influence on the outcome of the infrastructure layout. This input data must be of high quality and need a high level of detail. Thus creating a holistic scenario either requires a high effort to collect and aggregate the data or forces the user to make assumptions of the underlying system. Since second option may decrease the correctness of the results, the future work will focus on defining a specific and closed scenario with a high availability of data. This could be for example a large scale enterprise that wants to provide charging stations for its employees.

Next to finding a closed scenario, the extension of the simulation tool is planned to integrate important features of the electromobility. This includes the dynamic release of charging infrastructure by the vehicles with regard to the actual charging state and capacity needs. The integration of intelligent routing by the vehicle dependent on traffic situation and energy effectiveness is also intended.

All in all, the planning and simulation model is easily extensible so that future additions can be done with a low amount of time and effort.

References

Arnold H, Kuhnert F, Kurtz R, Bauer W (2010) Elektromobilität: Herausforderungen für Industrie und öffentliche Hand. Accessed 23 Nov 2011
Azarmand Z (2009) Location allocation problem. In Farahani RZ (ed.) Facility location: concepts, models, algorithms and case studies. Physica-Verlag, Berlin, pp 93–109
Balmer M, Axhausen K, Nagel K (2006) Agent-based demand-modeling framework for large-scale microsimulations. J Transport Res Rec 1985:125
Choy MC, Srinivasan D, Cheu RL (2003) Cooperative, hybrid agent architecture for real-time traffic signal control. IEEE Trans Syst Man Cybern Part A Syst Humans 33:597
Daneshzand F (2009) Multifacility location problem. In Farahani RZ (ed) Facility location: concepts, models, algorithms and case studies. Physica-Verlag, Berlin, 69–92
France J, Ghorbani A (2003) A multiagent system for optimizing urban traffic. In IEEE/WIC international conference on intelligent agent technology IAT 2003, pp 411–414)
Gringmuth C, Liedtke G, Geweke S, Rothengatter W (2000) Impacts of intelligent information systems on transport and the economy—the micro based modelling system OVID. In: Advances in modeling, optimization and management of transportation processes and systems: theory and practice—10th meeting of the EURO working group transportation (EWGT)
Hernándes JZ, Ossowski S, García-Serrano A (2002) Multiagent architectures for intelligent traffic management systems. Transport Res Part C Emerg Technol 10:473
Hoberg P, Leimeister S, Jehle H, Krcmar H (2010) Elektromobilität 2010: Grundlagenstudie zu Voraussetzungen der Entwicklung von Elektromobilität in der Modellregion München. fortiss GmbH. http://www.fortiss.org/fileadmin/user_upload/FB3/Grundlagenstudie_Elektromobilitaet2010_fortiss_final.pdf. Accessed 23 Nov 2011

Hunecke M, Schubert S, Zinn F (2004) Mobilitätsbedürfnisse und Verkehrsmittelwahl im Nahverkehr. Ein einstellungsbasierter Zielgruppenansatz. Internationales Verkehrswesen 57(1/2):26–33

Infas, DLR (2010) Mobilität in Deutschland 2008: Ergebnisbericht. Struktur—Aufkommen—Emissionen—Trends. http://www.mobilitaet-in-deutschland.de/pdf/MiD2008_Abschlussbericht_I.pdf. Accessed 23 Nov 2011

Junges R, Bazzan ALC (2008) Evaluating the performance of DCOP algorithms in a real world, dynamic problem. In: AAMAS'08: proceedings of the 7th international joint conference on Autonomous agents and multiagent systems, Vol 2. International Foundation for Autonomous Agents and Multiagent Systems, Richland

Klügl F (2001) Multiagentensimulation: Konzepte, Werkzeuge, Anwendungen. Addison-Wesley, München

Kosonen I (2003) Multi-agent fuzzy signal control based on real-time simulation. Transport Res Part C Emerg Technol 11:389

Moradi E (2009) Single facility location problem. In Farahani RZ (ed) Facility location: concepts, models, algorithms and case studies. Physica-Verlag, Berlin, pp 37–68

Nagel K, Schreckenberg M (1992) A cellular automaton model for freeway traffic. J de Physique I 12:2221

Nunes L, Oliveira E (2004) Learning from multiple sources. In: AAMAS-2004—proceedings of the 3rd international joint conference on autonomous agents and multi agent systems, IEEE Computer Society, pp 1106–1113

de Oliveira D, Ferreira P, Bazzan A, Klügl F (2004) A swarm-based approach for selection of signal plans in urban scenarios. In Dorigo M, Birattari M, Blum C, Gambardella L, Mondada F, Stützle T (eds) Ant colony optimization and swarm intelligence, Vol 3172. Springer, Berlin, pp 143–156 (Lecture Notes in Computer Science)

Oliver Wyman (2009) Oliver Wyman-Studie „Elektromobilität 2025: Powerplay beim Elektrofahrzeug. http://www.oliverwyman.com/de/pdf-files/ManSum_E-Mobility_2025.pdf. Accessed 23 Nov 2011

OpenStreetMap Foundation (2011) http://www.openstreetmap.org/. Accessed 23 Nov 2011

Ossowski S, García-Serrano A (1999) Social structure in artificial agent societies: implications for autonomous problem-solving agents. In Proceedings of the 5th international workshop on intelligent agents V, agent theories, architectures, and languages ATAL'98. Springer, London, pp 133–148

Rochner F, Prothmann H, Branke J, Müller-Schloer C, Schmeck H (2006) An organic architecture for traffic light controllers. In Hochberger C, Liskowsky R (eds) Informatik 2006—Informatik für Menschen, Vol 93. Köllen Verlag, pp 120–127 (Lecture notes in informatics (LNI))

Salhi S, Gamal MDH (2003) A genetic algorithm based approach for the uncapacitated continuous location–allocation problem. Ann Oper Res 123:1

Spath D, Bauer W, Rohfuss F, Voigt S, Rath K (2010) Strukturstudie BWe mobil—Baden-Württemberg auf dem Weg in die Elektromobilität. Fraunhofer-Institut für Arbeitswirtschaft und Organisation IAO. http://www.iao.fraunhofer.de/images/studien/strukturstudie-bwe-mobil.pdf. Accessed 23 Nov 2011

Steierwald G (2005) Stadtverkehrsplanung: Grundlagen, Methoden, Ziele, 2nd edn. Springer, Berlin

Vasirani M (2009) Vehicle-centric coordination for urban road traffic management: a market-based multiagent approach. Diss. Universidad Rey Juan Carlos, Madrid. http://hdl.handle.net/10115/5146

VDE (2010) Smart energy 2020: Vom smart metering zum smart grid. Verband der Elektrotechnik Elektronik Informationstechnik e.V

Wiering M (2000) Multi-agent reinforcement learning for traffic light control. In: Proceedings of the 17th international conference on machine learning. Morgan Kaufmann Publishers, Burlington, pp 1151–1158

Wietschel M, Kley F, Dallinger D (2009) Eine Bewertung der Ladeinfrastruktur für Elektrofahrzeuge. ZfAW. Zeitschrift für die gesamte Wertschöpfungskette Automobilwirtschaft 2009(3):33–41

Zarinbal M (2009) Distance functions in location problems. In Farahani RZ (ed) Facility location: concepts, models, algorithms and case studies. Physica-Verlag, Berlin, pp 5–17

Author Biographies

Tim Hoerstebrock graduated 2006 from the University of Göttingen, Germany and holds a diploma in business engineering. After working as a SAP consultant for three years, he became a research associate at the Institute for Information Technology (OFFIS) in Oldenburg, Germany. Since 2009 he is involved in major projects concerning eco-efficient people and freight transport evaluating the effects of alternative fuel types. He has experience in information systems, data base systems, enterprise resource planning and simulation.

Axel Hahn has special experience in the domain of inter- company information exchange, simulation, transportations systems and field level control in manufacturing. After his PhD, Prof. Hahn worked five years as head of development for a SME software company dealing with technical product information management. He returned to academics in 2002 as a professor for business information systems at the University of Oldenburg. 2005 he hold a professorship at the Beuth Hochschule in Berlin for production informatics. Since 2006 he is professor for business engineering at the University of Oldenburg and he his member of the board for the research group Transportation at OFFIS. With a strong background in energy information technologies his team at OFFIS provides major parts of the energy aware modelling and computing infrastructure.

Part III
Practical Insights

Coping with a Growing Mobility Demand in a Growing City: Hamburg as Pilot Region for Electric Mobility

Sören Christian Trümper

Abstract Hamburg, Germany's second largest city and economic centre of Northern Germany, features a high level of cargo handling and port-originated surface transport leading to the respective burden for air quality. Both demographic growth as well as common changes in age structure increase and change the associated mobility needs. The resulting demand for new mobility systems, the vicinity to large offshore wind power sources as well as existing technology clusters constitute the ideal test bed for electric transport applications combined with innovative mobility solutions to cope with these challenges. The city was thus selected as a pilot region for the demonstration and market introduction of electric mobility systems from the German federal government, which provides a total of 500 million euros from the Second Economic Stimulus Package to fund the development and commercialization of electric mobility in eight selected German regions. Between 2009 and 2011 players from academia, industry and local authorities were co-operating closely in projects in order to collect real-life data and hands-on experience with the technology, promote the development of infrastructure and to identify links and barriers to the widespread take-up of electric mobility.

1 The German Federal Funding Programme

Electric mobility is seen as a potential means to decarbonise urban traffic and thus to reduce the detrimental effects of our today's largely fossil fuel based mobility needs on the climate. In order to cut the emission of greenhouse gases and sustain Germany's position as one of the world's largest car-manufacturer and technology

S. C. Trümper (✉)
Institute for Transport Planning and Logistics, Hamburg University of Technology, Schwarzenbergstrasse 95 (E) 21073 Hamburg, Germany
e-mail: truemper@tuhh.de

leader in a global competition, the German federal government initiated a national funding programme for electric mobility in the year 2009. With large parts being used for the build-up of a battery test centre, a still substantial amount of funding (115 million Euro) was directed to eight selected regions (Fig. 1).

These regions should both apply the technology of battery electric vehicles (BEV) and the accompanying recharging infrastructure, as well as further develop and improve the system 'electric mobility' in all its facets. In detail, this includes to

- evaluate the technical maturity and estimate a future market potential
- identify further optimization needs and market entry barriers in legislation or with the users
- trial use cases and possible business cases
- create local added value through build-up of specific knowledge or a first domestic demand.

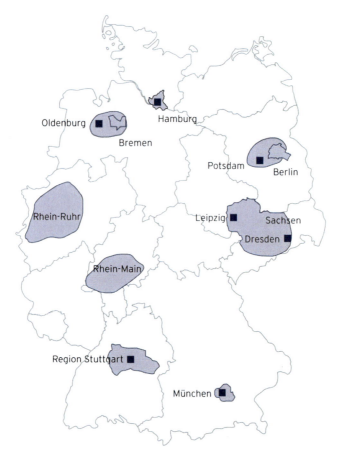

Fig. 1 Eight German regions were selected as pilot regions for the application and further development of electric mobility (Tenkhoff et al. 2011)

2 Pilot Region Hamburg

Hamburg was chosen as pilot region for a number of reasons, first of all its representative size and setting as a large urban cluster. With a population of 1.8 m, Hamburg is Germany's second largest city and an economic leader on many fields. Its gross value originates mainly from cargo handling, service industry, and finances (Statistisches Amt für Hamburg und Schleswig–Holstein 2010a). Important technology clusters are renewables, aeronautics, shipbuilding and logistics. Hamburg is also home to Europe's second largest container port and thus acts as a central logistic hub for the entire north of Germany with goods even delivered as far as Northern Italy. This results in a large share of port-originated surface transport in and around the city, leading to a heavy burden for the existing infrastructure and air quality.

Air quality, and more importantly, CO_2 emissions are in the focus of Hamburg's climate protection program. It aims at a reduction of CO_2 emissions of 40 % by the year 2020 (based on 1990 levels), which is so far in line with the national government's goals. Currently, 27 % of the city's local CO_2-emissions derive from transport (Statistisches Amt für Hamburg und Schleswig–Holstein 2010b) and emission-free electric drives can make a valuable contribution to lower this figure.

However, while national targets for the introduction of battery electric drivetrain technologies mainly focus on reducing CO_2 emissions and technology leadership, there are several other beneficial factors of electric vehicles from a municipal perspective. Battery electric cars neither emit CO_2 nor any other harmful substances, like NO_X, PAH or soot. With the currently limited driving ranges, they are also most likely to be seen in inner-city use rather than for inter-regio traffic, thus improving air quality where it is most needed.

While this might be true for the point of use, it bears the danger of relocating tailpipe emissions to more distant places where the energy is generated in fossil fuel fired powerplants. The use of renewable energy is, therefore, mandatory. While a higher demand for renewable energies is often associated with challenges in grid-balancing, in fact, battery cars can even be an opportunity for a better load balancing of the electricity grid in areas with a high percentage of intermittent renewable sources when flexible charging technologies, like delayed charging, are applied. Hamburg with its vicinity to large offshore wind parks does and will have times of energy surplus in its control area at times of low electricity demand since the transmission grids are not yet fully extended to cope with the projected incoming offshore capacities. Table 1 shows the increase of overload events due to excessive wind energy generation in the control area surrounding Hamburg (including the northern coast line) during the years 2005–2011. An event is here defined as an occurrence where too much wind generation in the control area leads to a grid frequency instability[1] and where, in order to protect the grid, some form

---

[1] According to German law renewable energy must be fed into the grid with higher priority than conventionally generated power.

Table 1 Overload events due to excessive wind energy feed-in in the control area surrounding Hamburg (TenneT TSO 2012)

Years	2005	2006	2007	2008	2009	2010	2011
Events	51	172	387	228	312	290	1024

of immediate countermeasures need to be performed. This could either be technical measures like switching operations and use of grid bypasses or market measures like redispatch and countertrading. Albeit these problems, only 90 km of the required 850 km additional transmission grid to feed-in the expected amount of renewables (i.e. offshore wind energy in Northern Germany) were built up to date (Kohler et al. 2010).

3 Addressing Personal Mobility Needs

Hamburg is one of the few large German cities to show a net demographic growth. The city with its large service industry, high-tech and media companies attracts young professionals from the rest of the country, who naturally have a strong demand for mobility. While market research have sensed a new "demotorisation" trend, coming from Japan and meaning that cars are slowly being superseded by products like mobile phones etc. as status symbols, such a new paradigm of "using rather than owning" mobility only works with a sufficient and highly individualised alternative. Mass transport as known today, with decreasing frequencies in the outskirts or at night, sometimes not fully cater for those needs. It does fulfil the needs of the elderly, though, and will need to be extended since even Hamburg cannot pull away from the national trend of an ageing society (Statistisches Amt für Hamburg und Schleswig–Holstein 2010c), meaning more citizens being dependent on public transport systems. The main requirements for an urban future mobility for all individuals in the coming decades are thus a strong public transport system with intermodal, i.e. individual, options which cater for all mobility needs—regardless of location, time of the day, or distance.

An intensified use of public transportation will automatically reduce congestion, emissions and noise in the city, helping to reach the municipality's emission reduction targets. Emissions and noise can even be further reduced by using electric vehicles in public transport, which is why Hamburg is deploying five Evobus Citaro diesel-electric hybrid buses. These buses feature a serial hybrid, which allows to drive fully electric for 2–5 km, meaning that sensitive urban areas can be completely spared from any exhaust fumes and engine noise or, the other way round, new areas formerly excluded from traffic can be developed to improve the accessibility to public transport. Since the system 'bus' has been entirely functional for decades the ongoing bus trial is mainly focussing on technical aspects like flawless operation and availability, infrastructure needs and possible fuel savings through hybridisation.

There are also six battery-electric cars (Daimler Smart Fortwo Electric Drive) deployed in carsharing for the "once in a while shopping tour" or just the need to be independent from any scheduled system. Studies have shown that a shared car replaces on average eight individually owned cars, helping to minimise both emissions as well as traffic congestion (Loose 2010). Battery cars can make carsharing even more environmentally friendly by adding the zero-emission bonus to it. The cars are located at Hamburg's major railway stations and selected spots in the city-centre. The trial will help to prove the general capability of battery vehicles for this purpose (e.g. required driving range) and evaluate user acceptance of the new technology (e.g. charging).

Cars for single individuals are not provided in the trial for various reasons. First, privately owned electric cars still claim the same share of limited inner city space as vehicles with an internal combustion engine (ICE). The large scale replacement of these with zero-emission vehicles does reduce CO_2-emissions, but therewith only tackling one of the problems associated with urban traffic (cf. sect. 2). Second, the still limited driving range combined with long recharging times often relegates the battery car to be the second or third car, since according to market research the expected driving range for a family car is about 500 km. A recent sur-vey by the Technical University of Munich (von Wangenheim 2011) showed that even for electric cars this value is still near 300 km and far off from what today's battery cars can deliver, therefore a conventional car will always be the base transportation device for the time being.

4 Adressing Commercial Transport Requirements

As shown, private individuals can use public transport for most of their journeys while businesses are usually dependent on their own transportation which they need to

- deliver goods to customers
- carry equipment to distant work places or construction sites
- use customised accessories, like shelves or pressure tanks in their cars
- link different routes and cover large areas of the city on a tight schedule.

In addition to these economic reasons there is also an environmental rationale for fleets being important addressees of a change in drivetrains. Figure 2 shows an analysis of the entire German car fleet and their emission standards. Data were drawn from the national car register. Since the mere number of emission reduction standards in the registered vehicle stock does not necessarily represent their percentage in the actual moving traffic, the data were modified and take into account data from traffic counts to better represent the share of emission standards in Hamburg's moving traffic.

In the field of light duty vehicles, which include vehicles up to 7.5t and which can mostly be seen in commercial fleets, a large proportion (>40 %) of the fleet still

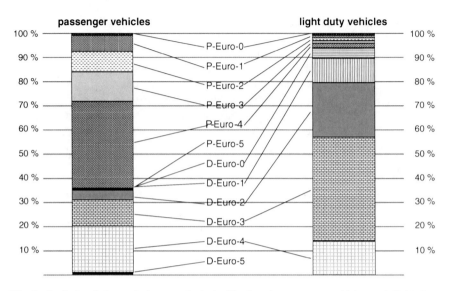

Fig. 2 Analysis of the emission standards in Hamburg's passenger vehicles and light duty vehicles fleet according to their percentage of the actual moving traffic (alteration of Lorentz et al. 2010); *P* stands for petrol vehicle emissions standards and *D* for diesel vehicle emission standards

has Euro-3 standard, more than 35 % even below that. For comparison: Euro-4 was introduced in the year 2005, Euro 5 in the year 2009. Substituting this vehicle sector with emission-free alternatives will tackle a large and relevant emission source in urban transport.

It is beyond the scope of this paper to evaluate the benefit of a longer use of an existing vehicle with low emission standards versus the purchase of a newly manufactured car with higher emission standards. However, recent lifecycle assessments for lithium-ion batteries—previously thought to have a large negative impact on the overall carbon footprint of BEVs due to their energy-intense production—indicate that emissions associated with their manufacture are negligible compared to the emissions of a car from propulsion energy ("fuel") (Notter et al. 2010). Also, following the merit-order effect BEVs will substitute those cars in the fleet with the highest cost of operation, which will be older cars with a higher fuel consumption and lower emission standards.[2] A sensible way to reduce emissions and noise in the commercial transport sector is, therefore, the introduction of technologies that offer zero-emission to one of the most polluting vehicle sectors, while maintaining the required degree of individuality. Battery electric vehicles will suit both needs.

---

[2] Emission standards correspond to the initial registration date of a vehicle and, for instance, a car first registered in the year 2011 (and likely to be manufactured about the same time) has to fulfill EURO-5 emission standard.

In Hamburg, 318 BEVs were brought into a 24-month field trial in November 2009. In detail these are 50 Daimler Smart Fortwo Electric Drive, 20 Karabag E-Fiorino, 200 Karabag New500E, 15 Karabag E-Ducato, 18 Mercedes-Benz A-Class E-Cell, and 15 Renault Kangoo Z.E. The different car models cater for different needs amongst the corporate fleet users (Table 2).

All vehicles are given to users on a rental or leasing contract with a duration of one to four years. Each car is also provided a charging spot installed on company grounds. A commercial fleet environment for the demonstration of these vehicles was chosen since corporate fleets have, in addition to the above identified economic and environmental reasons, a number of practical advantages for the testing and the introduction of this new technology.

4.1 Fall Back Options with ICE-Cars in the Fleet

When a company starts introducing battery cars into their fleet, they will still have the majority of cars equipped with a combustion engine. This backup allows for a certain failure of the new technology, e.g. in terms of availibility in cold weather or the driving range.

4.2 Regular or Predictive Use Patterns

Most fleets will have a very good understanding of their daily routes. Either they drive the same ways every day or, if that varies, at least know their trips in advance. Especially couriers, parcel and mail services or maintenance and metering companies plan their routes every day anew and with great detail. This gives the opportunity to use the electric cars on those routes that are most suitable in terms of distance or recharging spots available en route. Through their daily operations the fleet operators can slowly approximate the best use case, e.g. in respect to the maximum distance with a given state of charge, and gain statistically sound values on how the battery cars can be used in commercial fleets.

4.3 Short Cruising Ranges

Albeit the fact that fleet drivers use their vehicles all day long, they do not necessarily drive many kilometers per day. Inner-city traffic and many stops limit the range to an acceptable level for battery electric vehicles. In addition, most fleets that participated in the trial had the choice between a variety of route lenghts within their daily operations and could thus ensure that the vehicles are only used according to their (yet limited) technical ability. For instance, the parcel delivery

Table 2 Overview of battery electric vehicles in the pilot region Hamburg (buses not included) and a selection of technical details that are relevant for corporate fleet users

Make Model	Smart Fortwo electric drive	Karabag E-Fiorino	Karabag New 500E	Karabag E-Ducato	Mercedes-Benz A-Class E-Cell	Renault Kangoo Z.E.
Numbers in use in Hamburg	50	20	200	15	18	15
Seats	2	5	4	2	5	5
Anticipated range [km]	135	100	100	90	255	170
Battery capacity [kWh]	16	21	11	42	36	22
Allowed payload [kg]	230	535	325	1,263	350	650
Gross vehicle weight rating [kg]	1,150	1,755	1,420	3,500	1,960	2,078
Difference in gross veh. weight to non-electric version	Yes	Yes	Yes	Yes	Yes	No
Difference in loadspace volume to non-electric version	No	No	No	No	No	No

company Hermes has a daily mileage on their routes of up to 200 km, but also routes with 70 km which were mainly chosen in the trial (Hermes Logistik Gruppe Deutschland GmbH 2012).

4.4 Addressable by Regulatory Instruments

Since commercial fleets are a well-defined group of vehicles, they are easily addressable by regulatory instruments, which could be used to help the market ramp-up of electric vehicles. Measures include tax exemptions, free parking or congestion zone entry. While these also apply to individual car owners, fleet operators can also benefit from e.g. extended delivery times in housing areas. If a noiseless electric drive allows the delivery of a supermarket 1 h earlier in the morning or 1 h later in the evening it will have a substantial operational and thus economic effect for the delivering company.

4.5 Company Cost Accounting

Extended operation hours, tax credits or just the green image a company gains from using electric vehicles in their fleet are all monetizable components for companies—extra profits that can be credited against the higher purchase costs of battery cars, making their deployment reasonable if not profitable much earlier than for private car owners.

4.6 Early Markets and Potential Replication

In Germany, there are more than 570 cars per 1,000 inhabitants, meaning that each household has statistically between one and two cars (Shell Deutschland Oil GmbH 2009). While this is a large potential for future markets, it can also be a limit in early markets since a great deal of individuals have to take the decision to purchase an electric car before sales figures are high enough to have a relevant effect on manufacturing costs. With fleets, on the other hand, a small number of relevant players need to be addressed in the first place. If those "early adopters" can see a benefit in using electric cars in their vehicle stock, either for environmental or marketing or operational reasons, they are likely to purchase more cars. It is those early sales that the industry needs to trigger economies of scale, ultimately bringing down the manufacturing costs and thus preparing a mass market which also includes private car owners.

4.7 Centralised Charging Infrastructure

Most of the fleet operated cars, especially in multi-user environments or shift operation return to their base on a daily basis. For battery electric cars this means they can be easily recharged on corporate grounds once a day, assuring their maximum driving range for the users during worktime. That is, however, only possible for individual car owners if they have access to an exclusive parking lot or a garage, which applies to only 20 %[3] of the private car-owners in Hamburg. In addition, a centralised charging infrastructure on corporate grounds is cheaper and allows for innovative charging technologies, as will be explained in Sect. 5. Public charging is no alternative as long as the geographical density of charging points is still low and fast-charging is not yet widely available.

5 Charging Infrastructure

In the pilot region two different systems of charging infrastructure were set up, a corporate charging infrastructure which resides on the private grounds of a company, and a public charging infrastructure which is installed in the public space and can be used by anyone.

5.1 Corporate Charging Infrastructure

A centralised, corporate charging infrastructure has many advantages, such as lower installation costs and simpler maintenance. In contrast to public installations there are no administrative hurdles during the build-up and no permissions to gain. Also, the devices are cheaper, because no vandalism-proof housing is required and access could be granted through a low-tech system without identification or billing functions. When cars are charged in turn, fewer devices are needed, further lowering the cost of infrastructure per car.

In Hamburg, each car operated in a fleet is assigned its own charging spot, a so called "wall box" in contrast to the pillar-type devices anchored in the ground in public spaces. Although not necessary, all wall boxes have an identification module and are linked via a mobile phone connection to a central computer centre backend located at one of the utilities participating in the trial. This allows to gather usage data during the field test, for instance charging time, energy consumption or charging duration. Also for practical reasons both corporate as well as

---

[3] This value is only an estimate from experts and the local transportation office (Amt für Verkehr), taking into account housing types (detached, semi-detached, terraced), home ownership rate, and representative figures from small scale census or similar surveys.

public infrastructure have the same electrical specifications and host the same technical features (cf. Sect. 5.3).

5.2 Delayed Charging

All wall boxes in Hamburg are equipped with a remote control unit to perform "delayed charging". This technology basically allows the utility to start and stop the charging process of a battery car without interaction from the car owner. The necessary charging signals are transmitted via GSM (mobile phone connection) to the wall boxes. The car user only needs to define a time frame in which he allows the utility to perform delayed charging. This can be done via the internet or through a smartphone application. The car user can also specify further parameters in regard to the charging process, e.g. the required state of charge of the battery at a certain time etc. Within this given time frame, the utility can use the car as a flexible load to balance fluctuations in the grid.

Such fluctuations occur when there is a high availability of renewable electricity in the grid but a low electricity demand, usually at night time. When the electricity cannot be transported to adjacent control areas (national or international) or be used in pumped-storage facilities the generation needs to be reduced (cf. Fig. 3).

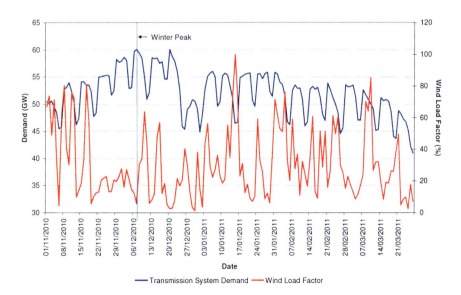

Fig. 3 Daily percentage of wind energy in the grid (*bottom graph*) against daily electricity demand (*top graph*) in the UK. Also indicated the winter peak as the day with the highest electricity demand in winter (National Grid plc. 2011)

In this case renewable production plants like wind farms will be shut-off to protect the grid from instabilities in its frequency, meaning that not only valuable green energy is lost, but also valuable profit for the wind farm operators cannot be gained. This is and will be particularly true for Hamburg with its vicinity to large offshore wind farms in the North Sea (cf. Sect. 2). Delayed charging allows to utilise this energy and hence to "green the grid".

5.3 Public Charging Infrastructure

In Hamburg, 50 publicly accessible charging spots were built between the years 2009 and 2011: 47 on public grounds, another three on semi-public Park & Ride areas. Each spot has two independent sockets, allowing two cars to recharge at the same time. In front of the devices there is a reserved parking space with a road marking on the ground and a signpost, indicating that only electric cars are allowed to park. Once there is a larger market penetration of BEVs and a more intense use of the public charging spots, a second parking space can be marked and reserved. The number of the charging spots was evenly distributed over the city's boroughs, locations are usually places of good visibility and close to points of interest. In order to install the devices, permits needed to be obtained from the borough council, which again checks with the local police if the new charging spots and their use do not pose a risk for pedestrians or cyclists.

Access is granted through an RFID-card (radio-frequency identification), that identifies a user (technically a billing account) to the system. Charging is then carried out in "mode 3" (IEC 61851) with up to 3×16 A and 400 V, although most cars still charge with single phase 16 A and 220 V (all AC). The charging spots can be easily upgraded to 3×32 A in the future through the replacement of a fuse. Car owners need a "type 2"-plug (VDE-AR-E-2623-2-2) to use the installation. Plugs always have to be provided by the vehicle user. In case a car with a different outlet or some other electric vehicle (pedelec, moped etc.) wants to use the charging spot, an additional outlet with a common German household socket ("Schutzkontaktstecker/Schuko", CEE 7/4, 13 A/220 V) is provided underneath the car plug socket.

5.4 Third-Party Access

One of the strong distinctions of the pilot project in Hamburg to other German pilot regions is the mandatory third-party access to the public infrastructure. In contrast to roaming models, where the infrastructure provider distributes his "own" electricity and charges third-party customers for "his" power plus a roaming fee, in Hamburg third-parties have direct access to the public charging spots—as long as they distribute green electricity. The disadvantage of roaming

Coping with a Growing Mobility Demand in a Growing City

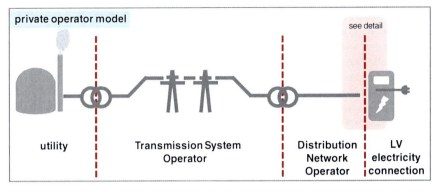

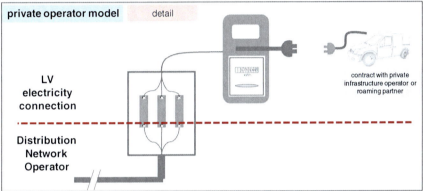

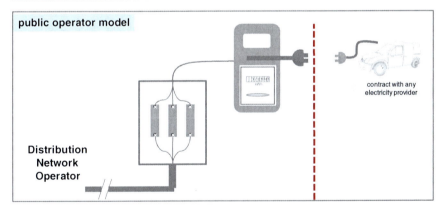

Fig. 4 Private operator model versus public operator model of charging spots in public space, showing the generation and the path of electricity from its point of origin to the single electricity outlet (*top section*), the legal boundaries at the interconnection point (*middle section*), and the change of the interconnection point in the public operator model (*bottom section*) (Trümper 2012)

models is the variable price the infrastructure owner can charge third-parties for his electricity, possibly discriminating smaller electricity providers with fewer customers and a lower energy consumption. Also, third-parties have less influence on the quality of the distributed electricity and generally have to accept whatever the infrastructure owner provides when they do not want to disappoint their customers in terms of available infrastructure in public space.

The advantage of the model in use in Hamburg is that each electricity provider can dispense "his" electricity at his price, paying only a standard transaction fee ("wheeling fee") to the owner of the charging spot (infrastructure operator). This gives the provider the freedom, for instance, to sell very specific qualities of green electricity to customers who pay an extra price for this extra bit of "green" (e.g. only solar power rather than hydro-power). Although third-party access is nothing special since the introduction of the common rules for the internal EU electricity market, it does represent a novelty when used for charging infrastructure in public space.

Figure 4 depicts the differences between the two different business models, which are roaming, also called the private operator model, and a third-party access model, also called the public operator model. It starts with showing the entire path from the power station to the outlet, from where the utility feeds its generated electricity into the transmission grid (220–380 kV) and the subsequent transformation down to the distribution network (110 kV) (top section).

The distribution network operator (also known as distribution system operator) supplies both industry and households with low voltage electricity up to their property boundary. Behind that boundary, owners or operators of a private charging infrastructure can use and sell electricity at their own discretion, including defining roaming agreements and pricing, thus called a private operator model (see middle section in Fig. 4).

A public operator model, in contrast, would need the charging infrastructure to be owned by a neutral body that allows every electricity provider access to the infrastructure. Access would be granted on the basis of wheeling agreements, a regulated and every-day process in the internal market, meaning that electricity is just "passed" through the grid. The neutral body, preferably the distribution network operator (since it is a already defined as neutral by law), will assign the single meter readings of the charging spot to varying electricity providers, depending on the car user's choice of provider (see bottom section of Fig. 5). For this task, the current metering systems, which assign one customer to one meter, will not suffice. Instead, the meter needs to log all single charging operations with time and energy consumption and assign them to electricity contracts with the respective providers. By the customer authenticating himself to the charging spot with some form of identification device (e.g. chip cards or RFID devices) the charging spot operator knows who to assign the charging data to. On an agreed timely basis, the charging spot operator will then send the collected charging data of each electricity provider's customers to the provider who will then use these data for billing his customers (Fig. 5). Obviously, charging would also work in a modified way with cash payment on the spot.

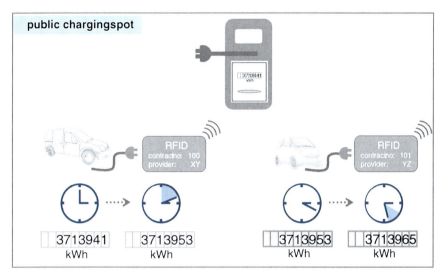

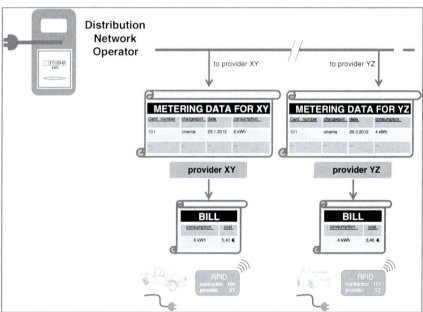

Fig. 5 Assignment of charging data to multiple users of a public charging spot with a public operator model (*top section*) and subsequent billing of the customer (*bottom section*) (Trümper 2012)

In the public operator model the distribution network operator qualifies as the sole charging infrastructure owner and operator for a number of reasons:

1. the distribution network operator is bound by law to grant third-party access to his infrastructure and transporting other companies' electricity is his daily business
2. since the distribution network operator works on the basis of a concession of the municipality, the city can regulate both roll-out as well as operation of public charging infrastructure; it can influence in which areas infrastructure will be build up, limit or increase the overall density in the city or, for instance, make green energy mandatory for charging vehicles
3. the distribution network operator can pass on his costs for the build-up and operation of the infrastructure not only to its users, but to all electricity customers in his distribution grid, lowering the specific costs "at the pump"; the reason is that in his position as distribution network operator his wheeling fees are calculated from his overall cost for infrastructure provision and operation in the respective distribution network.

While the first two points could be carried out by any private company providing an appropriate concession agreement, the third reason only applies to the distribution network operator. It needs to be debated, however, to what extent these costs could or should be socialized, making the wider public pay for a technology currently only fractions of the population use.

5.5 Green Electricity

When the environmental benefits of electric mobility are only assessed from a local perspective, it bears the danger of relocating tailpipe emissions to more distant places where the energy is generated in fossil-fuel fired powerplants. Only electric mobility driven by green energy is a plus for the environment as the following calculations show.

Equation 1 illustrates the CO_2 emissions of a BEV per kilometer when using the average German electricity mix

$$0.13 \text{ kWh/km} * 565 \text{ g } CO_2/\text{kWh} = 73.45 \text{ g } CO_2/\text{km} \tag{1}$$

with 565 g CO_2/kWh as the average CO_2 emission from a German powerplant in the year 2009 producing 1 kWh of electricity (Umweltbundesamt 2011) and 13 kWh as the average electricity consumption per 100 km of a Renault Kangoo (2011a).

Assuming in cold conditions or after some battery tear, the energy consumption increases to 0.22 kWh/km this would result in a CO_2-emission of 124 g/km. For comparison, an equivalent model with the smallest available diesel combustion engine emits 140 g CO_2/km (Renault 2011b).

Equation 2 illustrates the CO_2 emissions of a BEV per kilometer when using wind energy

$$0.13 \text{ kWh/km} * 22.5 \text{ g } CO_2/\text{kWh} = 2.93 \text{ g } CO_2/\text{km} \qquad (2)$$

with 22.5 g CO_2/kWh as the average CO_2 emission from wind energy (onshore and offshore) (Fritsche et al. 2007) and 13 kWh as the average electricity consumption per 100 km of a Renault Kangoo (2011a).

This clearly shows that the vision of zero-emission transport can only be realised if green electricity is used to charge the vehicle's batteries, which is why solely green electricity is currently dispensed in Hamburg at all charging devices (public and private).

6 Economic Benefits

One of the main goals of the national funding programme was to give a stimulus to the at that time struggling economy (2009) and to cushion the negative effects of the worldwide recession by sustaining and extending Germany's position as global technology leader. Much of this leadership is formed at local level with small and medium enterprises building up knowledge and competence and contributing with developments and solutions to the introduction of the new technology. In Hamburg, three types of local involvement are taking place:

6.1 Training and Preparation of Local Car Dealers

All three car makers providing BEVs for the field test have trained their local staff or representatives to carry out maintenance and small repairs. Since electric cars use high-voltage equipment a special qualification is required by those working on the car. In addition, the workshops were prepared for the maintenance of electric cars, e.g. with the installation of battery trays, mains outlets, ground closure etc. Both the building modifications as well as staff qualification are a prerequisite for the evolving market of BEVs. Also, sales force and other employees get familiar with the new products and learn about customer demands or acceptance issues, which they can report back to their company's headquarters. With all those exclusive early information and knowledge build-up the manufacturers get a significant head start in the upcoming market while their staff prepare for future customer requests and obtain technical qualifications required in little time.

6.2 Users Collect First-Hand Experience

318 vehicles give 318 test reports on the new technology. The field trial gives the owners of the electric cars in use an early insight into the pros and cons of electric mobility. While enjoying the amenities of electric driving users are getting a realistic opinion about driving range, handling, charging times or costs at the same time. This helps them to assess the maturity of the technology in general and, more important, the applicability ("use case") for their own service or product portfolio—giving industry and government a more realistic view on the size of the future market and potential show stoppers or market barriers in the foreseen market ramp-up.

6.3 Entrepreneurship

The Hamburg based company Karabag is a local Fiat-dealer supplying 20 E-Fiorino, 15 E-Ducatos and 200 transformed Fiat 500 (now Karabag New500E) to the pilot region. In addition to the supply with vehicles and extensive training of workshop staff, the company has gone a step further and is now actively shaping the electric future of cars: after importing transformed[4] electric fiats from Italy it has started to transform the cars itself in Hamburg for which additional staff was hired. Seeing the needs of its clients the company is currently developing a system for inductive charging together with the local university and is setting up a project in which single family houses are equipped with solar-panel carports and battery storage to supply electric cars.

7 Lessons Learned

During the past two years, a great deal of experience has been gained and several lessons were learned—positive as well as negative—owing to the novelty of the technology and the size of the demonstration. These can be divided into fleet operation and the use of the vehicles, and the infrastructure.

---

[4] The conventional drivetrain from an ICE car is replaced by an electric motor, battery and the respective energy management system, which are all fitted into the existing space in the conventional cars; strictly speaking, all BEVs on the market are transformed cars since none of them were designed on purpose for a full electric drivetrain.

7.1 Fleet Operation and Vehicle Use

The following experiences were gained, advantages and challenges identified and solutions or recommendations developed.

7.1.1 Purchase Price

High purchase prices for battery electric cars are still a constraint for fleet operators to switch large parts of their fleet to zero-emission cars. This applies especially to small and medium enterprises (SME), like craftsmen etc. While leasing models, as practiced in Hamburg, spread the financial burden to a monthly payment, it still remains high in comparison to the usual leasing rates for conventional vehicles. With increasing numbers of vehicles on the streets, vehicle prices must and hopefully will drop significantly. Future smart grid applications might help to partly refinance the vehicles or at least lower the operation cost further, either by charging at times of low electricity prices or actively using the vehicles as storage in times of surplus-energy in the grid.

7.1.2 Availability of Vehicles

The general availability of BEVs is (currently) low. Although the media often give the impression of readily available products off the shelf, it is in fact very hard to secure sufficient numbers of vehicles for demonstration purposes. The pilot region Hamburg, for instance, could secure itself a share of 50 Electric Smarts from Daimler's first and limited production batch of 1,000 vehicles in 2009 (Daimler 2009). Renault contributed 45 electric vehicles in two German pilot regions: 30 in Rhein-Ruhr and 15 in Hamburg (Renault 2011c, d). This general "scarcity" is especially true for larger vehicles ($\geq 2.8t$), where there are hardly any models for sale or leasing. The company Karabag specifically transformed the 15 E-Ducato vans for the purpose of this pilot project. However, these are the vehicles much needed by many commercial fleets, e.g. parcel deliveries etc. Manufacturers also need to address this market segment and realise that commercial users have a large potential in the deployment of cars in the short-term.

7.1.3 Technical Optimisation and Concepts

Although sound and mature in general, a few cars still showed various teething troubles, like software problems. Most of the technical faults, though, had their origin in a wrong operation by the car users or were faults in the non-electrical part of the car and thus could had happened in any other conventional car as well. For instance, some of the Renault Kangoo Z.E. had a leakage in the petrol pipe that feeds the auxiliary heating, and the cars had to return to the dealer for an exchange of some parts. Like with this problem, all faults were addressed appropriately and are seen as an important outcome of this publicly funded R&D-programme.

Apart from these single occurences the bigger challenge will be to find answers on how future vehicles and users deal with the inherent limitations of the incumbent technology (lithium-ion batteries etc.), for example in regard to driving range or the reliability in cold conditions. Different drivetrain concepts, such as range extenders, are worth taking into account in future scenarios since they would allow for a larger replacement of fleets with electric cars without loosing fallback options in regard to distance and availability. Going beyond technical solutions, alternatives would be an adapted use case, deploying auxiliary technologies like vehicle telematics or even combined urban delivery services.

7.2 Public Infrastructure

The build-up of a public car charging infrastructure was very much of a novelty for all participants of the field test and the following experiences show how much progress was made during the 2 years in the trial, but also what questions were identified for future installations.

7.2.1 Build-up

Before the year 2009, when the German pilot regions for electric mobility started their activities, there was no such thing like a public vehicle charging infrastructure in Germany. In fact, planning regulations do not know an electricity outlet on public grounds for the described purpose. As a result, a great deal of preparation needed to be accomplished as well as a number of administrative barriers to be broken down, lacking any comparisons or previous experiences. First, the locations needed to be chosen, for which subsequently building permits had to be obtained from the borough which in turn needed to consult the local police in regard to the safety of the remaining traffic. Future installations will most likely be less time-consuming on the administrative side due to the gained experience, but will remain complex. Further investigations are also needed into how much public charging infrastructure is needed in terms of technical functionality, legal status (public, semi-public, or private grounds), or spatial distribution.

7.2.2 Purchase Price and Return of Investment (Business Model)

The planned large scale roll-out of public charging spots during the future market ramp-up will face great challenges in respect to the return of investment for the infrastructure. Costs per double-outlet charging column were about 16,000 € per installation including ground work during the project. Although there are cheaper makes available on the market, those often lack the necessary functionality for smart charging. Smart grid applications, on the other hand, are often seen as a possible means to generate the necessary profit from the cost intensive devices with prices that vary with market demand over the course of a day, while today an electric Kangoo Z.E. with 3.2 kW/h charging speed and an electricity price of 0.24 €/kWh (incl. tax) will make this goal seem difficult to reach. Alternative policies suggest the ownership of the infrastructure to lie with the local grid operator. The advantage is that, in contrast to single entities on the electricity market (providers), the grid operator can pass on his fixed costs—including those for charging spots—onto all electricity customers, analogous to the feed-in tariffs for renewables in Germany. This, however, will require an amendment of national energy legislation.

7.2.3 Billing

Billing still remains more complicated than with normal household contracts as explained in Sect. 5.4. This does not only apply to the electricity providers who sell their electricity through someone's infrastructure, but in particular to end users and, more over, to external users (i.e. not part of the demonstration projects) who wish to use the public infrastructure in Hamburg. They need an additional electricity contract with a company that has access to the local public infrastructure and obtain an electronic card to get access to the charging spots prior to their use. Future systems need to be more comfortable, at least for car owners, and easy to use without prior registration. It is also necessary to address concerns about data protection, a discussion that has only started in relation with smart grids. Alternatives could be pay-as-you-go charging spots with micropayment or pre-paid systems.

7.2.4 Low Use

Analyses of the infrastructure monitoring data showed that hardly any public charging spots were used during the demonstration phase. There are two explanations for this: firstly, the cars to use those spots arrived at a rather late stage of the project, limiting the time for using the infrastructure. Secondly, all vehicles are deployed in commercial fleets which usually charge their cars over night on their own grounds. A forthcoming paper will look far deeper into the use of the infrastructure, analysing both temporal as well as spatial use, migration over time etc.

The usage rate, however, which is not likely to change dramatically in the near future even with a sharp market ramp-up, raises questions to how quick the return of investment can be achieved.

7.2.5 Blockages from Parking Offenders

The previous months of operation showed a problem common to all German pilot regions and their public charging spots: a great number of them is partly or permanently blocked by other (conventional) vehicles (Tenkhoff et al. 2011). Shortage of inner city space and heavy traffic make it difficult to keep the reserved parking lots free of the conventional traffic, especially since due to low usage they appear to be unoccupied and available for the common driver. Although no-parking signposts had been put up in Hamburg as a trial and for psychological reasons, national transport law does not yet give sufficient legal certainty to tow away non-electric vehicles. Hamburg, though, is the leading region ("Bundesland") in an initiative to change the respective laws and is looking into trialling alternative ways to apply legal sanctions (Bürgerschaft der Freien und Hansestadt Hamburg 2011, Bundesregierung 2010).

7.3 Private Infrastructure

The private infrastructure on corporate grounds proved to work reliably and flawlessly. Handling was easy for the employees and technical faults were not reported. Since companies can park all their electric cars next to each other on their parking decks it is possible to install master–slave systems which are easier to install and cheaper. The same goes for the device itself which needs to be less "armoured" against vandalism in the public space.

8 Summary and Outlook

Between the years 2009 and 2011 more than 300 battery electric vehicles were operated in a large-scale trial in Hamburg as part of a national R&D programme. Commercial fleets were identified as relevant lead users and future markets for both economic, technical and environmental reasons since fleets usually have a fallback option with conventional cars in their vehicle stock, drive short and regular routes, and benefit from a positive image towards customers. Fleets can also use a simple charging infrastructure on corporate grounds. In public space, however, charging spots are more expensive to build due to a different and more sturdy construction. Yet, it is important to have a visible number of such charging facilities in the city to give owners of BEVs the confidence to be able to recharge their car away from their own garage. Therefore, it is recommended to make public charging spots part of the distribution grid to distribute the costs for installation among all electricity customers and thus have an incentive to a build-up of a public charging network. It needs to be ensured, though, that those parking spots with a charging device are not blocked by parking offenders.

For the next years the electric mobility coordination office for Hamburg will prepare new projects for the region. These will focus on increasing the overall number of BEVs in Hamburg, but also on the inclusion of more small and medium enterprises which are regarded as having a large potential in the utilization of the new technology. Another focus lies on the build-up of larger fleets in geographically or operationally defined areas such as the port which gives better insights into the system electric mobility. Also, carsharing in neighbourhoods or the inclusion of electric mobility services in residential building projects will widen the application to more private car owners, slowly tapping this potential as well. In this course, a better integration of public transport and mobility services will be introduced as well as new innovative charging systems such as inductive charging. Further initiatives and projects will be derived from a master plan for the region, which will be prepared based on the experiences of the current demonstration serving as a strategic guideline, identifying the largest potentials for Hamburg in the field of electric mobility.

References

Bürgerschaft der Freien und Hansestadt Hamburg (2011) Drucksache 19/4906, 19. Wahlperiode. Freie und Hansestadt Hamburg, Hamburg

Bundesregierung der Bundesrepublik Deutschland, Bund-Länder-Fachausschuss Straßenverkehrsordnung/Straßenverkehrsordnungswidrigkeiten (BLFA StVO/Owi) (2010) Anlage 3 zur Tagesordnung I/2010. Bundesregierung der Bundesrepublik Deutschland, Berlin

Daimler (2009) Produktionsstart für den smart fortwo electric drive im Werk Hambach. Press release (19 Nov 2009)

Fritsche UR, Rausch L, Schmidt K (2007) Treibhausgasemissionen und Vermeidungskosten der nuklearen, fossilen und erneuerbaren Strombereitstellung—Arbeitspapier. Öko-Institut e.V, Darmstadt

Hermes Logistik Gruppe Deutschland GmbH (2012) Personal communication with Mr. Stefan Hinz

Kohler S, Agricola A-C, Seidl H (2010) Dena grid study II. Deutsche Energie-Agentur GmbH, Berlin

Loose W (2010) Aktueller Stand des Car-Sharing in Europa. Endbericht D 2.4 Arbeitspaket 2. (Final Report Workpackage 2, Momo Car-Sharing (www.momo-cs.eu), EU-funded project within Intelligent Energy Europe (IEE), contract no. IEE/07/696/SI2.499387). Bundesverband CarSharing e.V, Freiburg

Lorentz H, Schmidt W, Düring I (2010) Berechnung Kfz-bedingter Schadstoffemissionen und Imissionen in Hamburg. Gutachten für die Stadt Hamburg. Behörde für Stadtentwicklung und Umwelt, Hamburg

National Grid plc. (2011) Winter consultation 2011/2012. In: Winter outlook report. National Grid plc, London

Notter D, Gauch M, Widmer R, Wäger P, Stamp A, Zah R, Althaus H-J (2010) Contribution of Li-Ion batteries to the environmental impact of electric vehicles. Environ Sci Technol 44:6550–6556

Renault (2011a) Premiere für Elektrotransporter Kangoo Maxi Z.E. Press release (Renault Österreich GmbH - 23.02.2011)

Renault (2011b) Renault Kangoo Rapid, Renault Kangoo Rapid Compact, Renault Kangoo Rapid Maxi. Product brochure. Renault s.a.s.. http://renault-preislisten.de/fileadmin/user_upload/

Broschuere_Kangoo_Rapid_NFZ_111215.pdf. Accessed 26 Jan 2012 (based on NEDC combined cycle, model Kangoo Rapid dCi 70, 50 kW)
Renault (2011c) Hamburger Wirtschaft setzt auf Elektromobilität. Press release (PRW 38/11 - 04.05.2011)
Renault (2011d) Bergische Energie- und Wasser GmbH übernehmen Kangoo Z.E. Press release. (PRW 32/11 - 27.04.2011)
Shell Deutschland Oil GmbH (2009) Shell Pkw-Szenarien bis 2030. Hamburg: Shell Deutschland Oil GmbH
Statistisches Amt für Hamburg und Schleswig-Holstein AöR (2010a) Statistisches Jahrbuch Hamburg 2010/2011. Statistisches Amt für Hamburg und Schleswig-Holstein AöR, Hamburg
Statistisches Amt für Hamburg und Schleswig-Holstein AöR (2010b) Statistische Berichte (P V 2 – j/07 H)-Umweltökonomische Gesamtrechnungen- Treibhausgasemissionen in Hamburg 2007-3. Kohlendioxid (CO_2)-Emissionen- Quellenbilanz-3.1.3 Kohlendioxid (CO_2)-Emissionen und Energieeinsatz 2007. Statistisches Amt für Hamburg und Schleswig-Holstein AöR, Hamburg
Statistisches Amt für Hamburg und Schleswig-Holstein AöR (2010c) Statistisches Jahrbuch Hamburg 2010/2011. Therein 12. Koordinierte Bevölkerungsvorausberechnung (Variante 1- W1). Statistisches Amt für Hamburg und Schleswig-Holstein AöR, Hamburg
Tenkhoff C, Braune O, Wilhelm S (2011) Ergebnisbericht 2011 der Modellregionen Elektromobilität. Berlin: Bundesministerium für Verkehr, Bau und Stadtentwicklung (ISSN-16148045)
TenneT TSO GmbH (2012) Personal communication with Ms. Ulrike Hörchens
Trümper SC (2012) Author's own drawing
Umweltbundesamt (2011) Entwicklung der spezifischen Kohlendioxid-Emissionen des deutschen Strommix 1990–2009 und erste Schätzung 2010 im Vergleich zu CO_2-Emissionen der Stromerzeugung. Umweltbundesamt, FG I 2.5. http://www.umweltbundesamt.de/energie/archiv/CO_2-strommix.pdf. Accessed 05 Aug 2011
Wangenheim von F et al (2011) Auswertung eFlott. Fragebogen 1-3. München: Technische Universität München, Lehrstuhl für Dienstleistungs- und Technologiemarketing

Author Biography

Sören Christian Trümper is an Environmental Scientist by training. After working several years in the field of hydrogen and fuel cells as a project manager and consultant to the European Commission's research policies division he became engaged in electric mobility. As part of a public–private partnership he co-ordinated the vehicle based R&D projects on behalf of the city of Hamburg (Germany) and supported the strategy development together with the local authorities. Recently, he has returned to academic research where he focusses on the potential of electric mobility to reduce the carbon footprint of the city's commercial fleets as well as the energy consumption of battery electric cars in dependence of their driving patterns.

The City of Bremen and Its Approach Towards E-Mobility

Michael Glotz-Richter

Abstract Transport is a politically very sensitive area in all cities. There are high expectations related to electric mobility. From the viewpoint of a municipality, a change of propulsion technologies does not solve the transport problems. The potential of alternatives to petrol and diesel needs to be integrated into wider strategies—especially a modal shift to the sustainable modes. Bremen is an interesting showcase in the field of transport—with challenges of a traditional harbour city and being at the same time a "cycling city". Bremen presents a wider approach of sustainable mobility—including the benefits of further electrification.

1 Introduction

There is an intense debate about electric mobility—with calls for millions in subsidies [electrive.net (Electrive.net: daily electronic newsletter with information about electric mobility (in German) [www.electrive.net/newsletter@electrive.net]), FAZ 2012]. This contribution wants to highlight the expectation from the viewpoint of a municipality that is ambitiously active in the field of sustainable transport strategies—and involved as part of a model region for electric mobility (Fig. 1).

The City of Bremen is located in the north of Germany and has about 550,000 inhabitants. As a traditional harbour city, transport has a major function. For some decades already, Bremen's transport policy has focused on extending the public transport network and improving the conditions for cycling rather than on building inner city highways. There is a visible pay-off of that policy in Bremen's modal split: 25.8 % of all journeys are taken by bicycle and about 20 % on foot.

M. Glotz-Richter (✉)
Freie Hansestadt Bremen Der Senator für Umwelt, Bau und Verkehr, Referent 'nachhaltige Mobilität', Ansgaritorstr. 2, 28915 Bremen, Germany
e-mail: michael.glotz-richter@umwelt.bremen.de

Fig. 1 Logo of the "Modellregion Elektromobilität Bremen/Oldenburg"

Fig. 2 Modal split city of Bremen (all trips of all citizens in 2008), (Dresden 2010)

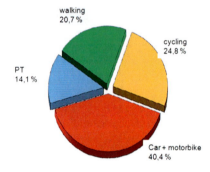

These carbon neutral modes account for more than motorised individual transport, which has about 40 % (MiD 2008) (Fig. 2).

Public transport has a split of about 15 %, but shows some increase. Part of it is related to the extension of the Bremen tram network—where today the largest extension ever is planned or under construction. Major business areas will be linked and residential areas in Bremen's periphery and surrounding communities will have a direct tram connection to the city centre. The annual number of public transport trips per capita in Bremen increased from 176 in 1995 to 187 in 2010 (BSAG 2011). It is worth mentioning that the tram has represented electric mobility for more than 100 years in Bremen. It is important to understand the attractiveness of the tram as a transport mode as well as the openness of the public transport operator to buses with electric components (as in hybrid buses) (Fig. 3).

The introduction of surface-based trams in areas which were previously served by buses has shown increases in ridership of 40–60 %. As trams have a higher capacity (especially more seating), fewer drivers are necessary to provide the same transport capacity—reducing operating costs. From the viewpoint of riders, the high share of seats and the convenience of smooth rail-based electric operation make the tram very attractive.

It should be mentioned that most of the rail tracks in Bremen and the region are operated electrically. The link into the harbour district with its freight village west of Bremen was also recently electrified. A rail track in the north was closed to passenger transport in the 60s and has now been re-opened and operated first with

Fig. 3 Electric mobility for more than 100 years in Bremen: the tram as backbone of public transport

self-propelled diesel units—and now been fully electrified—allowing the operation of electric through-trains from Bremen's central station to the very northern edge of Bremen.

I present this background to help you understand Bremen's position related to electric mobility and its options for the future.

2 Driving Forces for Electric Mobility

The transport sector is facing a discussion about environmental impacts—especially local emission, greenhouse gas and noise—as well as about the dependence on mineral oil. Today, road transport depends by more than 90 % on mineral oil (European Commission 2011). As most of the oil consumed in the EU is imported—and the share is increasing further—we face an increasing (and often unwelcome) geo-political dependence on oil-exporting countries. The more passenger transport depends on the car and the more freight transport depends on the truck, the more vulnerable our economy is. Is electrification of vehicles the road out of these problems?

Another theme that is very visible in Bremen is the impact of climate change. The city is located in the lowlands; most of the city is just above sea level. Millions are currently spent to improve protection by raising the level of dikes. We have to expect more extreme weather events which will also have impacts on business activities. Besides strategies of climate adaptation, Bremen is committed to reducing its CO_2 emission by 40 % (in comparison to 1990) by the year 2020. All sectors have to contribute if we want to achieve this target—including transport. The analysis and modelling (by the Wuppertal Institute for Climate, Environment and Energy) also include electric mobility (BET et al. 2009; Freie Hansestadt Bremen 2010).

Noise and local pollution are themes of high public perception. Seeing that electric vehicles can be operated with low noise and without local exhaust, what is the potential for air quality management and noise abatement plans?

As one of the hottest issues in transport planning is the parking question, what are potential and risks of electric mobility in that field?

3 The 2020 Target of One Million Electric Vehicles

The German federal government has defined a target of one million electric vehicles on the road by 2020 (Bundesregierung 2009; der Bundesregierung 2012). The target refers to cars and trucks as well as plug-in hybrids and hybrid buses and trucks, but does not include electrically-powered or supported two-wheelers. As well as become a leading market, Germany also wants to be a leading supplier in the field of electric vehicles. Research and development of components and vehicles for electric vehicles steer the development, indicating that the target is technology driven, not transport policy.

The target is ambitious in light of the number of electric cars today, but it only represents about 2 % of the entire fleet of motorised registered vehicles (cars, trucks, buses). This means that 98 % of the vehicles in operation will still be using combustion engines—albeit with much higher emission standards (majority Euro VI). The limited range of battery-electric cars and smaller vehicle types lead to one thesis that the mileage driven in electric vehicles may be lower than the average of vehicles with combustion engines. As the majority of electric vehicles are expected to be plug-in hybrids—using their combustion engines at some times, the impacts of the change might be in the range of 1–1.5 % of the mileage driven.

The Climate and Energy Programme of the Free Hanseatic City of Bremen analysed the climate protection potential of electric mobility (Freie Hansestadt Bremen 2010). The above-described limited shift to electrically driven kilometres also limits the impact. Only a shift in the generation of electricity from fossil generation to renewables can reduce the greenhouse gas emission. A shift on the demand side will not accomplish this.

3.1 Trucks and Buses

Today, heavy duty vehicles with their large diesel engines cause a disproportionate share of local emissions and noise problems. The progress in terms of emission standards, especially with the implementation of Euro V, and even more with Euro VI, can mitigate the problems of particulates and nitro-oxides. Also we can expect further reduction in noise levels.

Mainly due to the limitations of energy storage in batteries, we cannot expect battery-powered heavy duty vehicles in large numbers. Some special vehicles

Fig. 4 Full battery electric pushback truck in test at Bremen airport (2010)

(such as small urban buses or airport pushback vehicles) that do not have a long range for operation may benefit from the special performance of electric power trains (Fig. 4).

But electric power trains in combination with a combustion engine may increase vehicle efficiency—especially in combination with the recuperation of brake energy. These hybrid vehicles are especially interesting for operation in stop-and-go situations. Urban buses, delivery and courier vehicles and waste collecting trucks are the focus.

Today, these hybrid heavy duty vehicles are still much more expensive than diesel vehicles but increasing diesel prices, higher investment costs for Euro VI trucks and further developments with hybrid trucks and buses can bring the operational costs of hybrids close to or even below those of diesel powered buses and trucks. Noise reduction is another point that favours electric drivelines (e.g. when accelerating from a bus stop).

Bremen, as a tram city, started the operation of hybrid buses as early as 1998. Two very advanced buses came into operation: an 18 m articulated bus with wheel-motors and flywheel storage and a 12 m standard bus using batteries. The buses did not fulfil expectations. The performance was low and drivers did not like the buses. Reliability was also poor. The fate of technology that was too advanced?

But the idea of electric propulsion in buses was not abandoned. Two buses of the next generation came into test in 2006 and 2007–2008. One bus failed completely, the other brought a reduction in fuel consumption of about 10 %, not enough to justify additional costs of 120–160,000 €. In 2011, two hybrid buses came into operation as part of Bremen's activities in the model region for electric mobility. 50 % of the additional costs of more than 400,000 € (per bus!) came from the federal government—but the remaining additional costs will not be made up by reduced fuel consumption.

A hybrid waste collecting truck has been in operation since May 2011. The technology looks promising. The large diesel engine of the truck is only used for fast movement from and to the depot. Once the truck has reached the neighbourhood, the diesel engine is switched off and a fully noise capsuled small diesel

Fig. 5 Hybrid waste collecting truck in Bremen/cockpit display for electric components

engine (from Volkswagen Passat) takes over in a serial hybrid drive train configuration. Supercaps store energy which is used for the components of waste collection and for the movement of the vehicle (max about 30 km/h). The operator expects a remarkable reduction in fuel consumption, but the most perceptible change is in the noise level: a reduction of 18 dB has been measured in collecting operation, which is more than 90 % of the noise! But this also includes the usually very noise hydraulic components for waste collection (Fig. 5).

In the field of hybrid commercial vehicles, we expect some development in the next years, leading to wider operation and coming closer to a market-based implementation that does not depend on subsidies as much as today.

3.2 Electric Cars and Recharging Infrastructure

Today, we see a high awareness of electric cars. Every motor show presents new models and has more new announcements. The expectations are high. At the same time, there is still quite a high price level of electric cars—combined with still comparatively limited performance.

Bremen's Senate Department for Environment, Construction and Transport has put an electric car into its own fleet, with the very visible label: "Others talk about it. We try it out." (together with the labels for the funding sources within the model region for electric mobility). This label represents the open approach of the Senate Department—not raising too high expectations, enabling testing and setting a framework of justifiable conditions (Fig. 6).

The aspects of target groups for electric vehicles are of great relevance when it comes to recharging infrastructure and the involvement of municipalities—especially related to recharging infrastructure in public street space.

The provision of these recharging points and of electric energy is a task for private actors that raise the questions of how to justify a privileged use of

Fig. 6 Battery-electric car as part of the fleet of the senator for environment, construction and transport (May 2011)

dedicated public street space and what conditions should exist for operators of recharging points.

A privileged use of dedicated public street space can be justified by creating a general benefit. This could be seen in the recharging process—but when it comes to a privileged parking space, there is a problems of justification (as no electricity can be sold when it is pure parking without recharging).

In terms of conditions for operators, they will be required to supply electricity only from renewable sources. A further task is access to any electricity provider (discrimination free access) which will require some common standards e.g. for access and billing processes. The National Development Plan for Electric Mobility has put some emphasis on that point—but in the implementation we do not yet see much progress.

The City of Bremen has prepared a regulation for recharging points on dedicated public street space requiring proper monitoring of the use and requiring discrimination-free access (Freie Hansestadt Bremen 2011). In order to give some time for this standardisation process, there is a general exemption until 1.7.2014. Industrial suppliers of recharging stations and electricity providers are now aware that the discrimination-free access needs to be implemented.

A general question relates to the demand of the market. We expect the use of electric vehicles mainly in fleets operated by companies. The number of private users will be rather limited. The experience with recharging infrastructure in pilots in Berlin and Munich also leads to the assumption that these vehicles will rather be recharged at the company or at home carports. The demand for 'lantern' recharging in public inner city streets is expected to be limited at least in the next years.

As those areas suffer from high parking problems, it is not desired to reserve parking space for electric recharging stations when there is not sufficient demand. It will rather undermine the acceptance and reputation of electric mobility if there is underused privileged space in such areas.

Fig. 7 Inauguration of recharging station in a public parking garage in Bremen (30.09.2010)

In addition, legal questions of liability have not yet been sufficiently addressed. Who will be responsible if someone stumbles over a recharging cable and gets hurt? Is it the user or the operator of the recharging point? Is there any risk for the municipality?

In a pragmatic approach to circumnavigate, these problems for recharging points on dedicated public street space, the city of Bremen promotes recharging stations in inner-city parking garages with public access. All risks are reduced; there is a charge for the parking time, which will limit the duration of using that space. On the other hand, the locations are attractive for combinations with inner-city business or shopping activities (Fig. 7).

The City of Bremen will run a campaign *Zum Laden in die Innenstadt* (a play on words meaning both to recharge and to shop in the city centre). Free parking vouchers will be offered to owners of electric cars in the model region electric mobility (within the range of electric cars). If there is a change in the real demand for recharging infrastructure, the city can react quickly.

3.3 Car-Sharing

Bremen is known world-wide as a leading city in the field of Car-Sharing. The collaboration with the local Car-Sharing operator and the integration with public transport and into urban development strategies led not only to some strong growth in numbers of users but also to recognition as model case. Bremen was selected in an international competition with Car-Sharing as one of 45 urban best practice examples at the 2010 World Exposition in Shanghai under the theme "Better City—Better Life."

The focus is on the impact of Car-Sharing to replace private car use. A variety of cars at decentralised stations is at the disposal of Car-Sharing users. Reservation and access is possible as a 24/7 service. The Bremen Car-Sharing service (operated by the private operator cambio) today has about 7,000 users and has removed more than 1,500 cars from the streets of Bremen (Fig. 8).

Fig. 8 Mobil.punkt car-sharing stations in Bremen: a symbol for integrated mobility services

At first glance, this seems to be ideal for electric cars—you have electric cars available for those trips where neither a longer range nor luggage space is necessary. And for the other trips you have 'conventional' cars at your disposal.

There are a number of problems related to electric cars in Car-Sharing fleets. As the Car-Sharing operator cambio is completely market-based (without any subsidies for operation from the public sector or the motor industry), the procurement and operating costs need to be covered by user fees. Today, electric cars are much more expensive (rather twice the price) and can not be given to a next user directly after return. The share of time when the car is available to users is lower. So at the end of the day, the fees for users under pure market conditions will be significantly higher—and the question remains: what will happen after the phase of curiosity, after the first test of the electric cars that we hear so much about in the media?

But there is an important role of the existing Car-Sharing services for users of electric cars—it offers a kind of mobility insurance for all cases when the electric car is not sufficient. It is recommended to integrate Car-Sharing access into the promotion of electric cars.

With such a combination, the fear of simply adding electric cars on top of the existing cars in our cities can be lowered. It would be a transport policy disaster to have one million electric vehicles on top of the existing car fleet. This aspect is not yet sufficiently addressed in the German strategies for electric mobility, but hits a very sensitive point of urban transport policy.

3.4 Pedelecs

The fastest development of electric vehicles in going on in the area of two-wheelers, especially electrically-supported bicycles (pedelecs). These vehicles are supportive not only to have elderly people on bikes but also in conjunction with trailers or as range extenders.

Fig. 9 Promotional website and blog for pedelecs in Bremen: www.pedelec-bremen.de (2011)

Bike parking is an issue as good pedelecs cost 1,500 € or more. The range is quite sufficient—and many models offer an easy disconnection of the battery for recharging at the workplace or at home. The question of recharging infrastructure is not so much a problem as the question of secure bike parking. In the mid-term, we see the need of providing more infrastructure for cyclists as the speed of electrically-supported bikes and even more of electric bikes is above the usual biking speed.

We will need more space for cyclists to organise the different speed levels e.g. for safe overtaking, etc. Already now, the number of pedelecs sold is in the hundreds of thousands and will further increase (Fig. 9).

4 Conclusion

Bremen is committed to sustainable urban development—including transport. In its tradition of open-mindness, Bremen is interested in the potential of alternative fuels and propulsion technologies.

Along with the necessity for reducing CO_2 emission and dependence on limited and more and more expensive mineral oil, there is the need to develop post-fossil transport options. These must not be limited to propulsion technologies but need to be seen in a wider context of mobility and accessibility.

Electric cars will not substantially contribute in the next ten years to reducing local emission or noise level. Hybrid solutions can mitigate noise problems especially near bus stops and in relation to waste collection.

The limitation in performance of electric cars is also an opportunity to foster the development of using cars instead of owning cars—as already well provided in the Bremen Car-Sharing operation. With such an approach, we can also solve problems around parking in inner city neighbourhoods.

But the electric car will need more development to become a real market option. Price and performance are not yet in a range to make them attractive options. We should be careful with raising high expectations, as was done before related to hydrogen or biofuels.

As a city, we have to extract the transport dimension of electric mobility—and keep a proper balance. In a longer view, the well developed Car-Sharing background in Bremen can be supportive to develop a new mobility culture, where fleets of various vehicles are at the disposal of users.

Today, we see a lot of technology developments mixed up with transport and mobility strategies in the field of electric mobility. Electric mobility definitely has potential for post-fossil mobility but electric mobility is not post-carbon mobility per se as we still depend too much on fossil sources for our electricity.

Most of the talk about electric mobility is in fact about electric transport. Replacing internal combusion cars with electric drives does not make electric mobility. Mobility concepts need to cover more than propulsion technology. We need to talk about modal shifts to cycling and collective modes, and we need to look more closely to the potential of replacing personal cars—such as with Car-Sharing concepts. And in both fields we need to closely analyse the potential contribution of electrification. Only in this way can we can develop sustainable mobility concepts.

Acknowledgments All photographs were taken by the author.

References

BET et al (2009) Energie. Energie- und Klimaschutzszenarien für das Land Bremen. Aachen, Bremen und Wupperta

BSAG (2011) Geschäftsbericht 2010. http://www.bsag.de/pdf/BSAG_GB_2010_www_ss.pdf. Accessed 29 Februar 2012

Bundesregierung (2009) Nationaler Entwicklungsplan Elektromobilität

der Bundesregierung (2012) http://www.bmbf.de/pubRD/nationaler_entwicklungsplan_elektromobilitaet.pdf. Accessed 29 Februar 2012

Dresden TU (2010) Mobilität in Städten Haushaltsbefragung SrV 2008. http://daten.clearingstelle-verkehr.de/224/01/Staedtepegel_SrV2008.pdf. Accessed 29 Februar 2012

European Commission (2011) White paper roadmap to a single European transport area. Facts and figures, Brussels

FAZ: Frankfurter Allgemeine Zeitung (2012) Industrie fordert Milliardenhilfe für Elektroautos— Bundesregierung soll nun auch noch die stromtankstellen finanzieren (Indsutry calls for support in billions for electric cars—Federal government shall finance also the recharging stations). Report 14 May 2012

Freie Hansestadt Bremen (2010) Klimaschutz- und Energieprogramm 2020. http://www.umwelt.bremen.de/sixcms/media.php/13/KEP-Brosch%FCre_Endfassung%20komplett.pdf. Accessed 29 Februar 2012

Freie Hansestadt Bremen (2011) Erlass über die Errichtung und den Betrieb von Ladestationen für Elektrofahrzeuge im öffentlichen Straßenraum in den Gemeinden Bremen und Bremerhaven, as from 20.07.2011

MiD (2008) Mobilität in Deutschland 2008 (Mobility in Germany 2008) infas Institut für angewandte Sozialwissenschaft GmbH and Institut für Verkehrsforschung am Deutschen Zentrum für Luft- und Raumfahrt e.V. (DLR)—on behalf of BMVBS (German federal ministry for transport, construction and urban development). http://www.mobilitaet-in-deutschland.de/engl%202008/index.htm

Author Biography

Michael Glotz-Richter is Senior Project Manager for Sustainable Mobility for the German city-state of Bremen in the Senate Department for Environment, Construction and Transport. He holds a diploma in Urban and Regional Planning from the Technical University in Berlin (1984). In his professional career in Berlin, Cologne and Bremen, Michael has always worked on the link between urban development, transport strategies, environmental protection and urban lifestyle. Since 1994, he has been responsible for sustainable mobility and for many internationally recognised model projects on sustainable transport and environmentally-friendly mobility. Michael is a representative of the City of Bremen in the model region for electric mobility.

Acceptance of Electric Vehicles and New Mobility Behavior: The Example of Rhine-Main Region

Petra K. Schaefer, Kathrin Schmidt and Dennis Knese

Abstract To sustainably introduce electromobility in Germany research concerning the acceptance of this new form of mobility within the general public is a necessity. Thus, the Department New Mobility of the Frankfurt University of Applied Sciences has conducted two different research projects on that matter: "Elektrolöwe 2010" determined the potential electric car user in Hesse, using available mobility data on traffic behaviour. The studies were conducted for three different areas which are typical for the state of Hesse: a monocentric (Kassel), a polycentric (Frankfurt), and a rural region (Lauterbach). One main result was that less than 10 % of all respondents in all areas stated that they travel 80 km or less per day. Considering that an average electric car has a range of at least 100 km before the battery needs to be recharged, the often discussed range issue is not a problem for most people on an average day. From this point of view, there is a great potential for electric vehicles in Hesse. Within the Model Region Electromobility Rhine-Main social-scientific accompanying research has been conducted. Through quantitative surveys it was possible to gain detailed information about the users' acceptance and the mobility behaviour before and after the introduction of electric vehicles.Survey results show that, in general, users were highly motivated. Even though the respondents have a similar background and form a rather homogenous group as far as age and gender are concerned, it is still possible to draw overall conclusions: electric cars are especially suitable for thedaily commute to work. Only 2 % of all users within the survey have journeys of more than 100 km to get to work, and could easily use an electric car for their daily commute.

P. K. Schaefer (✉) · K. Schmidt · D. Knese
Fachhochschule Frankfurt am Main FB 1, Architektur, Bauingenieurwesen, Geomatik
Nibelungenplatz 1, 60318 Frankfurt am Main, Germany
e-mail: petra.schaefer@fb1.fh-frankfurt.de

K. Schmidt
e-mail: kathrin.schmidt@fb1.fh-frankfurt.de

D. Knese
e-mail: dennis.knese@fb1.fh-frankfurt.de

Electric vehicles do not only replace regular cars but also bikes and pedestrian traffic. Public transport is not affected by electric vehicles, especially not for commuting. For short distances of 10 km or less Pedelecs are an alternative to non electric motorized vehicles. Even though most respondents have an overall positive attitude towards electromobility, only a few would consider buying an electric car due to the high costs of purchase. However, certain incentives, such as tax reduction and free parking, would change peoples' minds.

1 Introduction

Electromobility is not only a matter of technology but also of acceptance within the population. When the 'National Development Plan Electromobility' came into effect in 2009, there was no or hardly any data available on that subject. Thus, research was necessary to create a data base and analyze the results to figure out the potential for electric vehicles. The findings are supposed to create guidelines for the Federal Government of Germany as well as the State of Hesse for a successful and sustainable introduction of electromobility. The research projects "Elektrolöwe 2010" and "Social-Scientific Research in the Model-Region Electromobility Rhine-Main" contribute to interpret the users' acceptance of electromobility in Hesse and especially the Rhine-Main area.

This section gives an overview of the contents of these projects. Further, the activities within the Model Region Electromobility Rhine-Main are introduced.

1.1 Research Goal

In two different studies the Frankfurt University of Applied Sciences determined the potential of electromobility in different regions and identify chances and barriers for the use of electric vehicles. Both projects aimed to show the needs for sustainable future transport systems where electromobility can play a crucial role. Only by switching to more eco-friendly means of transport climate protection targets can be reached (Federal Government of Germany 2009).

It was the leading member of the **Social-Scientific Accompanying Research of the Model Region Electromobility Rhine-Main**. Together with partners from the Goethe University Frankfurt and the environmental consulting company e-hoch-3 GbR, the mobility behaviour, and in particular the acceptance of electromobility in the model region, have been determined and evaluated in order to optimize electromobility and sustainably implement it in the State of Hesse as well as in Germany. The social-scientific research included all demonstration projects implemented within the Model Region Rhine-Main. Thus, a large diversity of electric vehicles was provided. The variety of electric vehicles ranged from electric cars, electric bikes (in the following referred to as Pedelecs) and scooters to busses, delivery vehicles, and commercial vehicles. The users were asked to take part in extensive surveys, and to keep travel diaries. The results are now available and will be presented in this paper.

The studies within the model region profit from another project within the context of the Hessian Sustainability Strategy, also conducted at the Frankfurt University of Applied Sciences: **"Elektrolöwe 2010"—The Electric Driver in Hesse**. This project was to develop new, specific and short-term findings about the mobility behaviour of Hessian citizens regarding the substitution potential and use of electromobility. The "typical user" in three representative areas in Hesse was to be determined by using available data on traffic behaviour and additional surveys. These three areas are a monocentric, a polycentric, and a rural region. This project was one of the first detailed mobility analyzations in different cities in order to determine the potential for electric vehicles in these cities. The results were turned into guidances for federal, state and municipal authorities as well as for potential user.

1.2 Model Region Rhine-Main

Research activities within the Model Region Rhine-Main were summarized under the umbrella brand ZEBRA—Zero Emission: Best Practice in Regional Applications (Regionalmanagement Nordhessen GmbH 2011). This includes the Sustainability Strategy as well as the numerous demonstration projects within the Model Region Rhine-Main. All of these demonstration projects have been accompanied by social-scientific and technical research. There was also a task group taking care of the needs of cities and communities. The project structure in Hesse is shown in Fig. 1. The ZEBRA activities are intertwined with the nationwide BMVBS platforms.

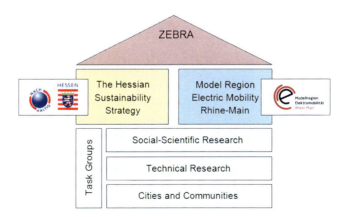

Fig. 1 The project structure in Hesse/Germany

The demonstration projects in the Model Region Rhine-Main include electric cars, transporters, Pedelecs, E-scooters, hybrid busses and commercial/utility vehicles. The social-scientific research in Rhine-Main is unique compared to all other model regions, as each demonstration project within the model region is included in the study to provide a comprehensive evaluation.

2 "Elektrolöwe 2010": The Electric Driver in Hesse

This section introduces the research project "Elektrolöwe 2010" and describes methods and goals as well as the research results in detail.

2.1 "Elektrolöwe 2010": Methods and Goals

"Elektrolöwe 2010", as part of the Hessian Sustainability Strategy, was funded by the HMUELV (Hessian Ministry of Environment) and accompanied by the HMWK (Hessian Ministry of Science). The project developed new findings about the mobility behaviour and usage patterns of Hessian citizens regarding the potential for the use of electric vehicles. The project was subdivided into three modules:

1. demand/user requirements,
2. supply/technical possibilities,
3. correlation of demand and supply.

Module 1: The mobility behaviour of the people in Hesse was analyzed, taking into account the different geographical and spatial structures of Hesse. Exemplarily, the city of Frankfurt as a polycentric area and the city of Kassel as a monocentric area were regarded. Mobility data was available for both cities from two different sources: 'Mobilität in Deutschland—MiD 2008' (Infas, DLR 2008) and 'System repräsentativer Verkehrs befragungen—SrV 2008 (TU Dresden 2008). Lauterbach was chosen as an example for a city in the rural area. As no data was available for that region an own survey was conducted among the citizens of Lauterbach (Fig. 2).

Module 2: This module was accomplished by Akasol Engineering, a project partner from Darmstadt. Akasol analyzed and evaluated the available and soon to be available vehicles and technologies on the market.

Module 3: This module combined the modules of demand and supply, added by an analysis of urban transport systems and circumstances for the introduction of electromobility in German cities, compared to other cities and countries worldwide. The results will be used as a guideline for politics.

Fig. 2 The three example areas of Hesse

2.2 Results of "Elektrolöwe 2010"

While the social scientific research in the Model Region focused on users of electromobility, "Elektrolöwe 2010" gave an overview of the general mobility behaviour of the Hessian people (Schaefer and Knese 2012). Nevertheless the results of both projects can be compared and the outcomes of the social scientific research in the Model Region can be verified based on the findings of "Elektrolöwe 2010". Thus, the results of "Elektrolöwe 2010" are described in the following section. The data sample for each city was sufficient, as Figs. 3 and 4.

Motorized individual transport in all three cities make up the largest share of the modal split, which is typical for German cities, especially in rural areas (Fig. 5). The second largest share is pedestrian traffic. As expected, public transport plays a

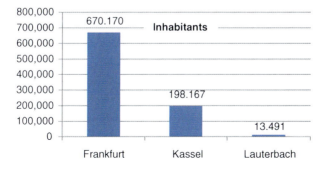

Fig. 3 Number of inhabitants in Hesse

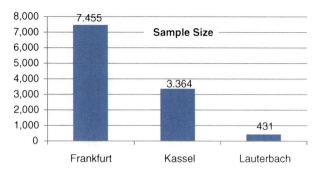

Fig. 4 Data sample size of the three typical areas in Hesse

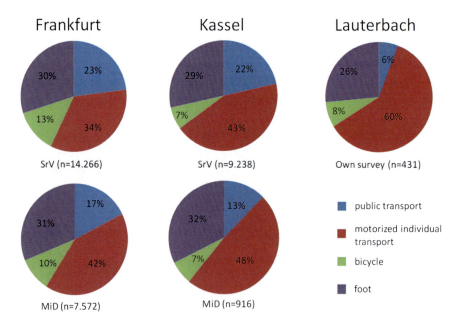

Fig. 5 Modal split

minor role for the population of Lauterbach and is more important for people who live either in Frankfurt or Kassel.

Looking at the daily kilometres travelled per person, it shows that only 5 % (SrV data) of the respondents in Frankfurt as well as in Kassel travel more than 80 km per day. According to MiD data this is 9 % for Frankfurt and 8 % for Kassel. The survey in Lauterbach indicates a similar tendency: 10 % of all respondents claim that they travel more than 80 km a day (Fig. 6). Thus, the often discussed range issue is not a problem for the average daily transport. From this point of view, there is a great potential for electric vehicles in Hesse.

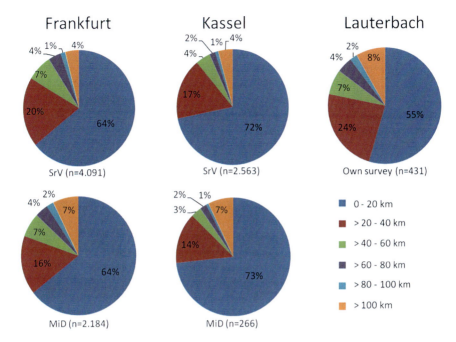

Fig. 6 Daily kilometres travelled per person

Even though most journeys could be accomplished by electric vehicles, there is still the question of how to get to places further away, possibly even with an overnight stay. According to the study people do not travel as much as expected. One third of the respondents in Frankfurt and Kassel did not go on any overnight journey within the past three months. In Lauterbach this is the case for almost half of the respondents. Nevertheless, for a successful implementation of electromobility in Hesse it is necessary to develop a concept for long distance journeys (e.g. by renting a car, going by train, or using a hybrid car).

In summary it can be said that the "range problem" is not as high as expected and exists only to a certain extend.

Further, the study "Elektrolöwe 2010" found out that, according to SrV and MiD, even in Frankfurt the majority of the respondents have a private parking spot (e.g. a garage or a carport) available at home. According to the MiD data, this is the case for 44 % of the respondents in Frankfurt and 57 % in Kassel. SrV shows even higher percentages for private parking space ownership (55 % in Frankfurt respectively 70 % in Kassel). The survey in Lauterbach found out that 88 % have access to a private parking spot. And only 6 % of the respondents have no possibility to park their car either at home or at work. This leads to the conclusion that the introduction of electromobility does not require extensive public charging infrastructure in public areas. Instead, selective charging spots should be established, e.g. at Park and Ride locations.

Taking all this into account it can be said that for a large part of the population, only from the view of mobility behaviour and today's transport system, electric vehicles are already an appropriate alternative in daily transport.

3 Acceptance of Electromobility in the Model Region Rhine-Main with the Help of Quantitative Surveys

Within the Model Region Electromobility Rhine-Main, quantitative surveys have been conducted to receive detailed information about the users' acceptance. In the following, the procedural methods of these surveys are described and the key results are pointed out.

3.1 Social-Scientific Research: Methods and Goals

The project's goals are to find out about the mobility behaviour of the participants of the demonstration projects in Rhine-Main area, thus determining the substitution potential for electromobility. Essential for a sustainable implementation of electromobility in Hesse as well as in Germany, is the acceptance of this new form of transportation among the population. The Department New Mobility focuses on determining the user acceptance in the model region electromobility Rhine Main.

The research team has been conducting extensive surveys among the users within the demonstration projects of the model region electromobility Rhine-Main. In general, these surveys consist of three parts—T0, T1 and T2. The contents of each survey are as follows:

Opening survey—T0:

- Socio-demographic data
- Mobility behaviour before the use of electric vehicles
- User's attitude towards electric vehicles
- Environmental orientation

  – The T0—questionnaire was handed out before the users first got into contact with electromobility to make before and after comparisons possible.

Second survey—T1:

- First impressions on the electric vehicles used so far

  – The second survey was conducted to show the user's first experiences with the electric vehicles.

Acceptance of Electric Vehicles and New Mobility Behavior

Final survey—T2

- Final evaluation of electric vehicles
- "Stated preferences"
- Infrastructure and charging needs

The final survey was conducted to give a conclusive evaluation of the acceptance of electromobility as well as to point out possible changes in the user's mobility behaviour.

In addition to surveys, users were asked to keep travel diaries—one before the start of the project, to get an idea of their conventional mobility behaviour without electric vehicles, and one during the time of the study, to see if their behaviour changes with the availability of electric vehicles. Also, by the help of travel diaries it is possible to get a direct before and after comparison.

This standard survey structure was used for all demonstration projects that include Pedelecs, e-scooters and e-cars, altogether seven different demonstration projects.Utility vehicle projects and hybrid-/electric-bus projects were treated differently, as described below.

- **Hybrid garbage collecting vehicle:** A survey was being conducted in which pedestrians were questioned about whether or not they noticed a difference between the hybrid garbage truck and a regular garbage collecting vehicle. The persons interviewed were randomly chosen among pedestrians in the immediate vicinity of the operating vehicle.
- **Hybrid-/electric busses:** So far these vehicles have not been used in regular line operation/route service. It is planned to conduct surveys at bus stations and on the bus. Due to delivery problems of the vehicles, results are presently not available.

3.2 Survey Results

3.2.1 User's Motivation and Representativeness

Depending on the demonstration project the return rate of the survey sheets was almost 100 %. With standardized surveys the data of 399 users could be collected. Altogether 790 survey sheets have been completed in an opening, a second and a final survey. Users in the Model Region Electromobility Rhine-Main are a homogenous group. The respondents are predominantly male (69 %). In comparison, 49 % of the population of Hesse is male and 51 % femmale (Hessisches Statistisches Landesamt 2011a). The age group is in between 42 and 46 years of age,which is partly due to the structure of the demonstration projects. In Hesse the age group between 38 and 55 is the strongest (Hessisches Statistisches Landesamt 2011b).

The respondents in the model region Rhine-Main are remarkably well educated in comparison to the Hessian standard. The users are to a great extend academics with high income. 58 % of the users have a general qualification for university entrance (German "Abitur"), approximately two third of the respondents have a university degree or are working towards it—9 % even with a PhD. According to thata large number of the respondents (34 %) have a relatively high household net income of more than 3.600 Euros per month. In Hesse only 22 % have a household net income of more than 3.200 Euros per month, according to the (Hessisches Statistisches Landesamt 2011b). In addition, the respondents of Rhine-Main have a pronounced environmental awareness. For instance, more than half of them already use electricity from regenerative sources.

Overall the sociodemographic results show that the respondents in the Model Region Electromobility Rhine-Main are not representative for the total population of the State of Hesse. The majority of the survey respondents are employees of the company in which the demonstration project is being accomplished. Therefore they are biased in a way that their company is environmentally friendly orientated, e.g. by using regenerative energy. A lot of these companies are also very open-minded towards technical innovations. A large amount of experiences in Rhine-Main is also based on Pedelec users due to a better availability of these vehicles compared to electric cars.

3.2.2 Electric Vehicles for Daily Mobility

Assuming that an average electric vehicle—even in bad weather conditions—has a range of 100 km, it can be concluded that electric vehicles are especially suitable for thedaily commute to work. Interesting enough, only 2 % of all users within the survey have journeys of more than 100 km to get to work, which means on the other hand that 98 % of the respondents could easily use an electric car for their daily commute—provided that they can recharge their vehicle at workand at home. But even if it was only possible to charge either at home or at work, electric cars present a possibility for over 89 % of all users that participated in the survey.

Figure 7 shows that the range problem for average work days is not as grave as generally expected. Even though most respondents cover distances of less than 100

Fig. 7 Distance from home to work (n=273)

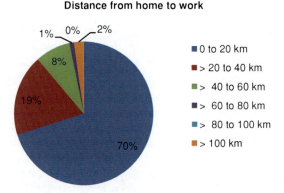

km a day, almost two fifths state that electric cars should cover distances of at least 200 km before they take into consideration to buy one.

The results shown above are congruent with the findings of "Elektrolöwe 2010" (in chapter Socio-economic aspects of electric vehicles: a literature review). Both studies taken into account it can be said that a change of mobility behaviour is necessary for a successful implementation of electromobility. Potential users still need to be made aware of this.

3.2.3 Pedelecs: An Alternative for Short Distances

It shows that Pedelecs are a real alternative to motorized individual traffic, especially for short distances. This is due to the fact that 36 % of the respondents have ways of 5 km or less to get to work. Of course this distance would also be suitable for regular bicycles. More difficult are larger distances of up to 10 km. Pedelecs are in contrast to normal bicycles also used for distances of more than 5 km so that it can be assumed that users travelling 10 km or less on their daily commute can do this using a Pedelec. It shows that a little more than half of the respondents fall under this category (Fig. 8).

Battery range is not critical since most Pedelecs can easily cover 40–50 km depending on weight, support und efficiency of the engine, head wind, ascending slope and temperature (ADFC 2010).

Travel diary results show that three out of four trips by Pedelec are shorter than 10 km and just over half of all trips by Pedelec are longer than 5 km (Fig. 9).

3.2.4 Electric Vehicle Usage

Only 34 % of the electric vehicles in the Model Region Rhine-Mine are used privately. A larger extend (49 %) is used as part of a company vehicle fleet, as a sharing vehicle or as a company car with exclusive use for one person. The latter

Fig. 8 Distance from home to work up to 10 km (n=273)

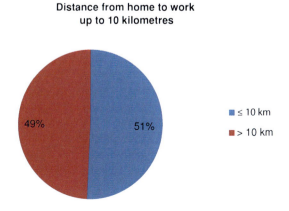

Fig. 9 Distances by pedelec (n=451)

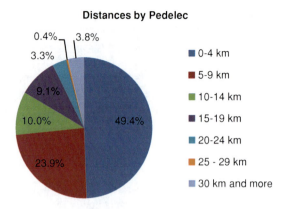

as well as privately used electric vehicles are used on a regular basis: Three quarters of the respondents use "their" electric vehicle at least one to three days per week. For sharing vehicles this only applies to half, for company vehicle fleet users to one thirdof the respondents.

All in all the users of Rhine-Main use the electric vehicles regularly but not as much as expected before the vehicles were implemented. This is, at least as far as Pedelecs are concerned, due to bad weather conditions (usage during the winter months). Among those users that took part of the opening as well as the second survey, 60 % claimed to use the electric vehicle on at least one to 3 days a week. Before 79 % expected that they would use the electric vehicle on at least one to 3 days a week.

3.2.5 Users' Expectations

Users' expectations towards electromobility were to a great extend positive which could be verified during the test phase with only a few exceptions. The majority of the users were fully convinced that electric vehicles would be useful in daily life, that there was hardly any noise emission, and that electric vehicles are environmentally friendly. These expectations were confirmed during usage.

Furthermore, users expected—though with few restrictions—an easy use of the vehicles as well as an easy handling for charging the battery. This could also be confirmed during usage. The same applies for maximum velocity, comfort, and safety of electric vehicles. The factors of cost savings and a sensible integration of electric vehicles in everyday life were rated better during usage than expected before. Only the availability of charging infrastructure and the loading space inside the vehicles was evaluated slightly worse than expected before.

3.2.6 Substitution Potential of Electric Vehicles

Travel diaries before and during the use of electric vehicles show differing results, depending on the respective demonstration project and trip purpose. The frequency of usage of the different usage types as well as the substitution of conventional vehicles by electric vehicles varies depending on the demonstration project, user structure, location and available electric vehicles. However, there is one aspect that all demonstration projects have in common: For the daily commute electric vehicles mainly replace conventional electric cars and bicycles, whereas in recreational traffic mainly pedestrian and bicycle traffic gets replaced. The amount of ways with conventional cars neither decreased in recreational traffic nor in traffic due to shopping/errands, whereas for shopping/errands the pedestrian traffic as well as bicycle and motorbike travel decreased.

Figures 10 and 11 show the share of the different means of transportation before and during the use of electric vehicles exemplarily for a company that had a large amount of Pedelecs and electric scooters as well as a smaller number of electric vehicles.

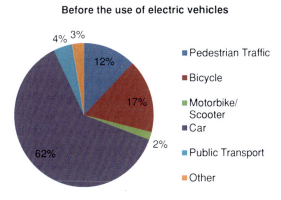

Fig. 10 Distribution of vehicle use before the use of electric vehicles (n=1,111)

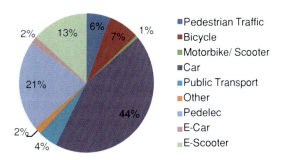

Fig. 11 Distribution of vehicle use during the use of electric vehicles (n=1,728)

3.2.7 Acceptance of Higher Prices for Electric Vehicles

Electric vehicles are—due to a low demand as well as a rather expensive technology, in particular for the batteries—associated with high costs of purchase compared to conventional vehicles. It is assumed that these additional costs are an obstacle for many otherwise enthusiastic users as far as willingness to purchase is concerned. Hence, of 146 respondents only one sixth is willing to accept additional costs of 20 % or more compared to conventional cars.

Thus, it was interesting to find out if certain incentives would increase the willingness to accept the higher purchase costs for electric vehicles: Monetary criteria, such as cheaper vehicle taxes and insurances, free of charge parking spaces for electric vehicles, and cheap electricity for rechargingare a great inducement to buy an electric vehicle. But also reserved parking for electric vehicles is seen as an effective incentive. The permission of using bus lanes was often discussed among politicians but proved less important than monetary criteria.

3.2.8 Public Transport is not Weakened by Electric Vehicle Use

Using the method "Stated Preferences" it was evaluated which means of transportation users would chose for **commuting to and from work** under certain boundary conditions. The respondents could choose between conventional cars, electric cars, public transport, and, depending on the distance, Pedelecs and conventional bicycles. Boundary conditions were defined as follows:

1. Distance: 20 km
   Metropolitan area
   Very good transport connection to public transport and road networks
2. Distance: 20 km
   Rural area
   Public transportation available hourly, access to federal road
3. Distance: less than 5 km
   Metropolitan area
   Very good transport connection to public transport and road networks
4. Distance: 60 km
   Provincial town
   Train station (long distance trains) relatively close to the motorway

It shows that except for alternative 3 most respondents would use public transport rather than other means of transportation. Compared to alternative 1 and 4 the use of public transport decreases slightly for alternative 2, but still stays the most used means of transport. On second place are electric cars for alternatives 1, 2, and 4. The decrease of public transport and electric car use for alternative 3 goes towards bicycles and Pedelecs. It shows that distances of less than 5 km are mainly

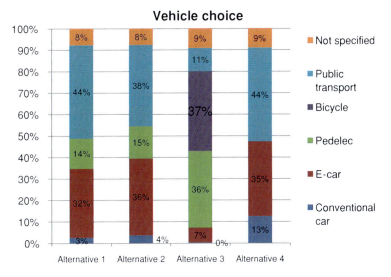

Fig. 12 Distribution of vehicle choice under certain boundary conditions (n=238)

accomplished by two-wheeled vehicles (bicycles and Pedelecs) and not a single respondent chose conventional cars for short distances (Fig 12).

4 Conclusion

The studies accomplished at Frankfurt University of Applied Sciences show that the much discussed range problem only exists to a certain extend and that an extensive public charging infrastructure is not necessary. The acceptance of electromobility among those users that were included in the surveys of the social scientific research is positive. Most of them showed a lot of enthusiasm and enjoyed participating in the survey. The users see a great potential especially for their daily commute and are eager to use "their" electric vehicle on an almost daily basis. However, these results are based on a group of users that have a rather similar background and are open-minded towards new technology as well as the use of regenerative energy.

The results of the social scientific research show that electromobility is an appropriate alternative for daily transport. The eight Model Regions Electromobility in Germany helped to make electromobility visible on the streets in daily life and showed that the acceptance among users is positive. A first step to a sustainable introduction of electromobility could be accomplished through the Model Region projects. However, electromobility is more than just a new form of mobility. Users have to change their attitude towards mobility from owning their private vehicle to a more complex intermodal and multimodal system. Thus, more research is necessary in order to figure out best ways of how to make people change towards this new form of mobility. It is going to be a process that cannot be completed over a short period of time.

"Elektrolöwe 2010" already suggested that most peoples' mobility behaviour is suitable for electromobility. Yet, the range problem is still on people's mind. People have to be made aware of the fact that the cruising range is sufficient for most purposes. However, for some journeys electric cars are not the ideal vehicle. In order to make people change to electric vehicles, a concept is necessary to make these journeys possible without any extra costs or efforts. Future research should deal with those problems.

References

ADFC—Allgemeiner Deutscher Fahrrad-Club (2010) ADFC-Information zu Pedelecs und E-Bikes. Bremen. http://www.adfcbw.de. Accessed 23 Aug 2011
Federal Government of Germany (2009) Federal Government of Germany 2009, Berlin, http://www.bmvbs.de/cae/servlet/contentblob/27978/publicationFile/104/national-electromobility-development-plan.pdf. Accessed 11 Sept2011
Hessisches Statistisches Landesamt (2011a) Statistische Berichte—Haushalte und Familien in Hessen 2010—Ergebnisse der 1 %—Mikrozensus—Stichprobe. Wiesbaden
Hessisches Statistische Landesamt (2011b) Altersaufbau der Bevölkerung in Hessen. Wiesbaden. http://www.statistik-hessen.de. Accessed 25 Aug 2011
Infas—Institut für angewandte Sozialwissenschaft GmbH, DLR—Deutsches Zentrum für Luft- und Raumfahrt e.V (2008) Mobilität in Deutschland 2008, Bonn und Berlin
Regionalmanagement Nordhessen GmbH (2011) eMobilität in Hessen
Schaefer PK, Knese D (2012) Elektrolöwe 2010—Der hessische Elektroautofahrer. Straßenverkehrstechnik 4(2012):238–243
Technische Universität Dresden (2008) Mobilität in Städten—System repräsentativer Verkehrsbefragungen 2008. Lehrstuhl für Verkehrs- und Infrastrukturplanung, Dresden

Author Biographies

Petra K. Schaefer completed her doctor's degree at the Technical University Darmstadt in 2004. She then worked as a project manager at the ZIV (Centre for Integrated Transport Systems). Since 2007 she is professor for transportation planning and public transport at Frankfurt University of Applied Sciences. Under her guidance the Department New Mobility has been conducting various projects on electromobility as well as traffic management for public events.

Kathrin Schmidt graduated from the Technical University Darmstadt in 2009 with a Diploma's degree in Civil Engineering. She has been a member of the Department New Mobility until February 2012 where she was responsible for the social-scientific accompanying research in the model region Rhine-Main.

Dennis Knese graduated in 2007 from the University of Bremen with a bachelor's degree in Geography and in 2009 from the University of Applied Sciences Wiesbaden with a master's degree in environmental management and infrastructure planning. After a year in Canada where he worked as a research assistant at the Vancouver Economic Development Commission, he started in 2010 as a research assistant for transport planning, urban planning and electromobility at the Frankfurt University of Applied Sciences. He was responsible for the project "Elektrolöwe 2010". Since 2013 he is PhD candidate, working on the integration of electromobility into urban planning and street space design.

Printed by Publishers' Graphics LLC
DBT130929.20.07.58